THE MEASUREMENT OF AIRBORNE PARTICLES
Richard D. Cadle

ANALYSIS OF AIR POLLUTANTS
Peter O. Warner

ENVIRONMENTAL INDICES
Herbert Inhaber

URBAN COSTS OF CLIMATE MODIFICATION
Terry A. Ferrar, Editor

CHEMICAL CONTROL OF INSECT BEHAVIOR: THEORY AND APPLICATION
H. H. Shorey and John J. McKelvey, Jr.

MERCURY CONTAMINATION: A HUMAN TRAGEDY
Patricia A. D'Itri and Frank M. D'Itri

SULFUR IN THE ENVIRONMENT, Parts I and II
Jerome O. Nriagu

ENERGY UTILIZATION AND ENVIRONMENTAL HEALTH
Richard A. Wadden, Editor

METHODOLOGICAL APPROACHES TO DERIVING ENVIRONMENTAL AND OCCUPATIONAL HEALTH STANDARDS
Edward J. Calabrese

FOOD, CLIMATE AND MAN
Margaret R. Biswas and Asit K. Biswas, Editors

CHEMICAL CONCEPTS IN POLLUTANT BEHAVIOR
Ian J. Tinsley

RESOURCE RECOVERY AND RECYCLING
A. F. M. Barton

QUANTITATIVE TOXICOLOGY
V. A. Filov, A. A. Golubev, E. I. Liublina, and N. A. Tolokontsev

ATMOSPHERIC MOTION AND AIR POLLUTION
Richard A. Dobbins

ATMOSPHERIC MOTION AND AIR POLLUTION

An elevated inversion over Hartford, Connecticut, at 0645 hr on October 2, 1972. Photographed from 12 km west of Hartford by Mr. G. C. Atamian, Talcott Mountain Science Center. (Reproduced with permission of G. C. Atamian.)

ATMOSPHERIC MOTION AND AIR POLLUTION

An Introduction for Students of Engineering and Science

RICHARD A. DOBBINS

Professor of Engineering
Brown University
Providence, Rhode Island

A WILEY-INTERSCIENCE PUBLICATION

JOHN WILEY & SONS

New York • Chichester • Brisbane • Toronto

Library of Congress Cataloging in Publication Data:

Dobbins, Richard A.
Atmospheric motion and air pollution.

(Environmental science and technology)
Includes bibliographies and index.
1. Air—Pollution. 2. Atmospheric diffusion.
3. Atmospheric circulation. I. Title. II. Series.
TD883.1.D6 628.5′3′015515 79-952
ISBN 0-471-21675-5

Printed in the United States of America

10 9 8 7 6 5 4 3 2 1

To Iona Blake Dobbins

SERIES PREFACE

Environmental Science and Technology

The Environmental Science and Technology Series of Monographs, Textbooks, and Advances is devoted to the study of the quality of the environment and to the technology of its conservation. Environmental science therefore relates to the chemical, physical, and biological changes in the environment through contamination or modification, to the physical nature and biological behavior of air, water, soil, food, and waste as they are affected by man's agricultural, industrial, and social activities, and to the application of science and technology to the control and improvement of environmental quality.

The deterioration of environmental quality, which began when man first collected into villages and utilized fire, has existed as a serious problem under the ever-increasing impacts of exponentially increasing population and of industrializing society. Environmental contamination of air, water, soil, and food has become a threat to the continued existence of many plant and animal communities of the ecosystem and may ultimately threaten the very survival of the human race.

It seems clear that if we are to preserve for future generations some semblance of the biological order of the world of the past and hope to improve on the deteriorating standards of urban public health, environmental science and technology must quickly come to play a dominant role in designing our social and industrial structure for tomorrow. Scientifically rigorous criteria of environmental quality must be developed. Based in part of these criteria, realistic standards must be established and our technological progress must be tailored to meet them. It is obvious that civilization will continue to require increasing amounts of fuel, transportation, industrial chemicals, fertilizers, pesticides, and countless other products; and that it will continue to produce waste products of all descriptions. What is urgently needed is a total systems approach to modern civilization through which the pooled talents of scientists and engineers, in cooperation with social scientists and the medical profession, can be focused on the development of order and equilibrium in the presently disparate segments of the human environment. Most of the skills and tools that are needed are already in existence. We surely have a right to hope a technology that has created such manifold environmental problems is also capable of solving them. It is our hope that this Series in Environmental Sciences and Technology will not only serve to make this challenge more explicit to the established professionals, but that it also will help to stimulate the student toward the career opportunities in this vital area.

Robert L. Metcalf
Werner Stumm

PREFACE

This textbook contains the material upon which a course of one semester's duration offered at Brown University in recent years has been based. The course has been primarily of interest to senior students in the biomedical, civil, environmental, and mechanical engineering options, as well as to students from various pure science concentration programs. The goal of this text is to provide these students with a fundamental level of technical knowledge and the background that is necessary to comprehend many of the articles that appear in the various specialized scientific journals that are devoted to this subject.

The preparation of these students usually has been the core curriculum that is the central pedagogical structure for students in the Division of Engineering at Brown University. Included in this group of courses are four semesters of college-level calculus and one semester each of thermodynamics, fluid mechanics, and chemistry. This block of courses constitutes a nominal prerequisite for the material presented here. Students majoring in the pure sciences have on occasion expressed interest in the course, and their preparation is often comparable to the above except that the preparation in fluid mechanics is inadequate. The specific requirements from this area of study are a familiarity with the law of hydrostatics, the continuity equation, Euler's equations, and elementary viscous flow concepts. Students lacking this background have been encouraged to perform independent study of these topics provided their preparation in mathematics includes multivariable calculus and an introduction to partial differential equations. In the first half of the text there is a moderate reliance on vector notation, and the derivative vector forms, with which some students may not be totally comfortable, are usually introduced without presumption of prior exposure. The assimilation of both subject matter and mathematical tools is, of course, mutually reinforcing.

Some students have been interested in this course as an introduction to the elements of geophysical fluid mechanics which is the subject of the first six chapters. Others have desired a knowledge of the formation and transport of pollutants in the lower atmosphere which is covered in the last half of the book. The course on air pollution is specifically served by the material in Chapter 1, the topics leading to and including the concept of geostrophic flow in Chapter 2, Chapter 4, an abbreviated discussion of Chapter 5, the Reynolds equations and the neutral and convective boundary layers from Chapter 6, and a detailed coverage of the material that is presented in Chapters 7 through 12. The varying interests of the different groups within the same class have been accommodated through the use of term projects appropriate to their specific motivations, and a written report has been required in lieu of a final examination.

Widely available references, usually either specialized texts or journal articles, have been cited where deemed suitable throughout this work. In some cases, it has been advisable to refer to less accessible reports issued by government agencies. A deliberate effort has been made to display comparisons between experiments and theory whenever they pertain to the discussion. Certain didactic films produced by the National Committee for Fluid Mechanic Films serve as valuable study aids to the material presented in the first half of this text. These films also have been listed as references and are readily available for rental or purchase. Other uncited films in this excellent series are also of potential value.

The development of this text owes a primary debt of gratitude to many students over the past several years. A number of former students of the course have been associated, either through summer or permanent employment, with the Division of Air Pollution Control of the State of Rhode Island. Through contact with these former students, I have been able to keep informed of current developments in legislative matters, on legal implementation and enforcement problems, and, more important, to reflect on the relationship between what is taught in the classroom and what is observed in practice. The informal association with this office, which coincidently resulted in one specific study of certain features of the regional air pollution, has been invaluable.

I also acknowledge the valuable comments and criticisms of portions of this text from several colleagues. Professor Merwin Sibulkin of the Division of Engineering, Brown University, and Professor Gary R. Newman of Southeastern Massachusetts University have made important suggestions on the material in the earlier chapters. Professor James C. Baird, Jr., of the Departments of Chemistry and Physics, Brown University, has provided valuable comments on Chapter 8. Mr. George C. Holzworth, Chief of the Geophysical Research Branch of the Environment Sciences Research Laboratory, U.S. Environmental Protection Agency, whose research is frequently referred to in Chapter 12, has made important suggestions on the presentation of that material.

The principal burden of typing was capably discharged by Mrs. Judith Bardsley Follansbee. The final editing was completed with the diligent assistance of Ms. Rosemary J. Golini.

All of these persons have contributed significantly, but I alone am responsible for the selection of material, its presentation, and any shortcomings of this work.

RICHARD A. DOBBINS

Providence, Rhode Island
August 1979

CONTENTS

ATMOSPHERIC MOTION AND AIR POLLUTION

1

AEROSTATICS AND ATMOSPHERIC STABILITY

The motion of the atmosphere is responsible for the timeless variability and extremes of our local weather. Yet for certain purposes it is possible to overlook this motion and to consider the earth's mantle of air to be a fluid subject only to the law of hydrostatics. This viewpoint, which is justified in detail in Chapter 2, results from the fact that the vertical equation of motion is significantly truncated if, as is often the case, the vertical acceleration can be neglected. The hypsometric equation that is of great practical value is a consequence of this simplification. A more important product of an examination of the hydrostatic behavior of the atmosphere is the development of concepts relating to the atmospheric stability which play a central role throughout the subsequent discussion.

1.1. THE HYPSOMETRIC EQUATION

A fluid at rest is subject to the law of hydrostatics,

$$\frac{dp}{dz} = -\rho(z)g(z, \phi) \tag{1.1}$$

where p = pressure, z = geometric altitude or the height above a reference plane which is normally mean sea level, $\rho(z)$ = the density that is altitude dependent, and $g(z, \phi)$ is the acceleration due to gravity which depends on both z and latitude ϕ. The quantity $g(z, \phi)$ is defined as the acceleration of a body freely falling in a vacuum, and it is the vector sum of the accelerations associated with the Newtonian gravitational force and the centrifugal force that results from the rotation of the earth about its axis.

The direction associated with $g(z, \phi)$ is that of a plumb line, and it is everywhere normal to the ocean's surface which responds freely to the

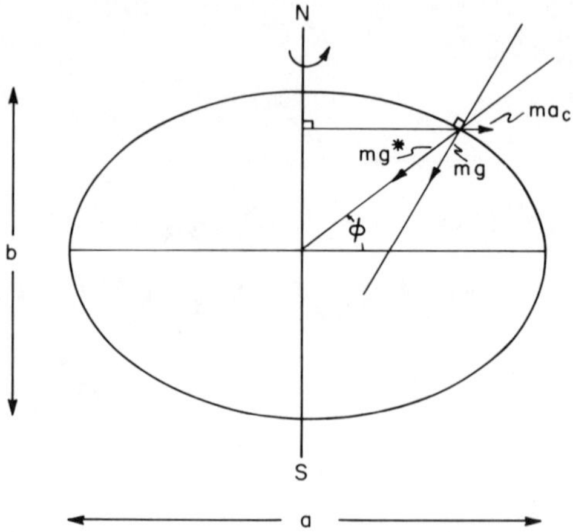

Figure 1.1. The forces acting on an object at rest with respect to the surface of the rotating earth. The departure of the earth from spherical shape is shown greatly exaggerated. The actual value of (a − b)/a is 1/297.

forces acting on it (see Fig. 1.1). The Newtonian gravitational force, mg^*, produces an acceleration always directed toward the center of mass of the earth while the centrifugal force, ma_c, produces an acceleration that is normal to the polar axis and proportional to the distance from it. The latter therefore vanishes at the poles where only Newtonian gravity acts. The two forces are colinear only at the poles and the equator; at other latitudes they are not colinear and their resultant is not pointed toward the center of the earth. The departure from sphericity of the earth results from the action of the centrifugal force. The magnitude of this effect, as measured by the decrement between the equatorial and the polar diameters divided by the latter, is one part in 297.

Both the gravitational and centrifugal forces are conservative, which means that the work done in moving a body in their fields is independent of the path followed between two specified endpoints. Such forces can be expressed as the gradient of a scalar potential function. The geopotential altitude, defined by

$$H(z, \phi) = \frac{1}{G} \int_0^z g(z, \phi) dz$$

where $G = g(0, 45°) = 9.806 \text{ m/s}^2$, is the potential function which includes the combined influence of the Newtonian gravitation and the centrifugal force resulting from the earth's rotation. For our purposes it will be adequate to consider $g = 9.806 \text{ m/s}^2$ for the altitudes and latitude bands of interest.

This approximation would imply that geopotential altitude equals the physical height, $H = z$, for all latitudes. More detailed discussions of the magnitude of $g(z, \phi)$ and its precise latitude and altitude dependence are given in Refs. 1 and 2.

The hydrostatic equation may be regarded as a relationship between two dependent variables and the independent variable z, and further information is required before this relationship can be used. We know that dry air is a thermally perfect gas at the temperatures and pressures of interest. Thus the equation of state

$$p = \rho R T \tag{1.2}$$

is obeyed where T is the absolute temperature and R is the gas constant for air. Standard meteorological units will be used in which pressure is expressed in kilopascals (kPa) [1 pascal (Pa) = 1 newton per square meter (N/m^2)] or in millibars (mb) [1 bar = 10^{+6} dynes per square centimeter (dyn/cm^2)], ρ is in kilograms per cubic meter (kg/m^3) and temperature is in kelvins (°K). The gas constant depends on the composition of air which is variable. The main constituents of dry air are given in Table 1.1; this composition results in a molecular weight of 28.96. The value of the gas constant is then 287.0 m^2/s^2 °K. Thus if the pressure is one standard atmosphere which is equal to 101.325 kPa (or 1013.25 mb) and the temperature is 288°K or 15°C, we find the density of dry air is 1.225 kg/m^3. Since the molecular weight of water is 18.01, the density of humid air is slightly less than the density of dry air.

We eliminate the density of air from Eqs. 1.1 and 1.2 and recast the result as

$$dz = -\frac{RT(z)}{gp(z)}\,dp$$

This equation has considerable practical utility because rawinsondes, that is, a set of balloon-borne meteorological instruments, can measure with

Table 1.1. The Major Constituents of Dry Air

Constituent	Mass (%)	Moles (%)
N_2	75.49	78.98
O_2	23.15	20.95
Ar	1.29	0.934
CO_2	0.047	0.031

moderate accuracy and transmit back to earth the values of pressure, temperature, and dew point at frequency time intervals. Ground-based radio equipment is used to track the instantaneous position of the rawinsonde so that the winds aloft can also be determined. The data secured include p as a known tabular function of T and we may integrate the above relationship to obtain the hypsometric equation,

$$z - z_0 = +\frac{R}{g}\int_p^{p_0} T(p)\,\frac{dp}{p} \tag{1.3a}$$

which can be used to determine the heights at which the observations are made. In practice this equation will be used in incremental form. For fine increments

$$z_i - z_{i-1} = +\frac{R}{g}\,\frac{T_i + T_{i-1}}{2}\,\frac{p_{i-1} - p_i}{\frac{1}{2}(p_{i-1} + p_i)} \tag{1.3b}$$

is suitable. For somewhat coarser increments in Δz and Δp the form

$$z_i - z_{i-1} = +\frac{R}{g}\,\frac{T_i + T_{i-1}}{2}\ln\left(\frac{p_{i-1}}{p_i}\right) \tag{1.3c}$$

is preferred if temperature does not differ greatly from its mean value.

A more detailed version of the hypsometric equation that includes the influence of the presence of water vapor on the value of R and the variation of g with altitude and latitude is given in Ref. 3.

1.2. THE LAPSE RATE

Observations show that in certain portions of the atmosphere the temperature decreases linearly with increasing altitude. This result is expressed by

$$\frac{dT}{dz} = -\lambda \tag{1.4a}$$

or, if λ = constant for all z, in the integrated form

$$T = T_0 - \lambda(z - z_0) \tag{1.4b}$$

where λ, called the lapse rate, is the rate of *decrease* of temperature with altitude. We note that

$$\frac{dp}{dz} = \frac{dp}{dT}\frac{dT}{dz} = -\lambda\frac{dp}{dT} = -g\rho$$

and eliminating ρ in the last equation by means of the equation of state, we find

$$\frac{dp}{p} = \frac{g}{\lambda R}\frac{dT}{T}$$

hence

$$\frac{T}{T_0} = \left(\frac{p}{p_0}\right)^{\lambda R/g} \tag{1.5}$$

where the subscript 0 refers to the values at mean sea level (MSL). Equation 1.5 provides the relationship between pressure and temperature that applies if the lapse rate is constant. Substituting Eq. 1.5 into Eq. 1.3a and integrating, we find

$$z - z_0 = \frac{T_0}{\lambda}\left[1 - \left(\frac{p}{p_0}\right)^{\lambda R/g}\right] \tag{1.6}$$

as the integrated version of the hypsometric equation when the lapse rate is constant. If the lapse rate is zero above a certain altitude so that T is constant for all z, we find directly from the hypsometric equation

$$z - z_1 = \frac{RT}{g}\ln\frac{p_1}{p} \tag{1.7}$$

where p_1 = the pressure at z_1.

For mainly engineering purposes a "standard atmosphere" was defined as one in which $\lambda = +6.5°\text{K/km}$ for $z \lesssim 11$ km, $T_0 = 288.15°\text{K}$, and $p_0 =$ 101.325 kPa (1013.25 mb). Above this lowest region, called the troposphere, is the stratosphere where the lapse rate is zero and the temperature is a constant value of 216.65°K. These layers will be described in more detail in Chapter 5. This particular standard atmosphere has more recently been designated as the midlatitude Spring-Fall supplementary atmosphere, Ref. 2, and other "supplementary" atmospheres are proposed for certain seasons and latitude bands. Of course we recognize that day-to-day departures of the atmosphere from its mean value constitute the highly variable weather phenomena to which man is accustomed. The utility of the standard atmospheres is therefore highly limited. Extracts of Ref. 2 are included in Appendix II and illustrate the mean seasonal variations and thermal structure at high altitudes.

1.3. THE STATIC STABILITY OF THE ATMOSPHERE

The flotation characteristics of a submerged solid object can be described by the application of Archimedes' principle which states that the (upward) buoyancy force is equal to the weight of the fluid displaced. This principle is established by integrating the pressure given by the law of hydrostatics over the surface of the submerged object. In the language of Archimedes' principle a submerged object will float or sink according to whether the upward buoyancy force is greater or less than the downward gravitational force. We will use this principle to determine the stability of a fluid possessing a vertical temperature gradient. Initially the fluid is considered to be incompressible which means that its density is independent of pressure. The density of the incompressible fluid may vary because of its finite thermal coefficient of expansion when the temperature field is nonuniform. (The density of an incompressible fluid may also vary as a result of varying concentration gradients of a solute.)

Let us assume that water is at rest in a vessel and possesses a vertical temperature gradient with the higher temperature at the higher elevations. A density gradient will also be present owing to the finite thermal coefficient of expansion and the lighter fluid will be at the higher elevations ($d\rho/dz < 0$). A fluid possessing a predominantly vertical gradient of density is said to be a stratified fluid. We wish to examine whether the fluid is in a stable configuration with the above described stratification. To determine the static stability of the stratified fluid we imagine a parcel, or finite aggregate, of fluid to be displaced upward from its initial position without transferring heat to the surrounding cooler fluid. Heat exchange from the parcel may be totally suppressed by enclosing it in a hypothetical weightless, perfectly insulating membrane. The application of Archimedes' principle leads to the conclusion that the upwardly displaced parcel will experience a net downward force because its density is greater than the warmer surroundings. Since the upward displacement results in the development of a force tending to restore the parcel to its original position, we describe the stratification as statically stable.

If the cooler water is at the top of the vessel, hence $d\rho/dz > 0$, then the experiment will yield a different result. The upward displacement of a parcel of warm water from a lower layer will result in a net upward force because the insulated parcel would find itself surrounded by water at a lower temperature and greater density. The small upward displacement would result in a force tending to cause further upward motion. The stratification of the fluid in this case corresponds to a condition that is described as statically unstable.

The terms "statically stable" and "statically unstable" are used because the fluid is assumed to be at rest and because we wish to emphasize that the argument will not apply to a fluid in which there is horizontal motion. This case will be discussed in Chapter 6.

We may summarize the above results by saying that the presence of a thermal gradient to produce $d\rho/dz < 0$ would imply static stability under the assumption that the density of the test parcel does not change when its height is altered. If the water were at constant temperature, $d\rho/dz = 0$, no net force develops when the parcel is raised and the situation is described as one of neutral static stability. If the cool water were on top, $d\rho/dz > 0$, then a warm parcel displaced adiabatically upward would find itself surrounded by heavier water and it would then continue to rise upward in this statically unstable case.

We now examine the static stability of the atmosphere, but in this case the argument is complicated by the compressibility of air which causes its density to be dependent on pressure as well as temperature. Thus when an air parcel is displaced upward, its volume and density will change so that its pressure can adjust to that of the surrounding atmosphere. To investigate this case, we retain several features of the argument we used in considering the incompressible fluid. We will assume the test air parcel is again raised to the new ambient pressure adiabatically, as if enclosed by a weightless, nonconducting, and tensionless membrane. Our criterion for stable equilibrium will be that the density of the elevated parcel must be higher than the ambient density. Since both parcel and the ambient atmosphere initially possess the same density, the criterion for static stability will be

$$\left(\frac{d\rho}{dz}\right)_{\text{par}} > \left(\frac{d\rho}{dz}\right)_{\text{atm}} \tag{1.8}$$

When this criterion is applied to water, the quantity on the left vanishes and we recover the result stated above. In calculating $(d\rho/dz)_{\text{par}}$ we will assume not only is the parcel insulated, but that other sources of dissipation are absent and therefore the process is isentropic. In this case results from thermodynamics provide the relationships

$$\frac{p}{p_1} = \left(\frac{\rho}{\rho_1}\right)^{\gamma} \tag{1.9}$$

$$\frac{T}{T_1} = \left(\frac{\rho}{\rho_1}\right)^{\gamma - 1} \tag{1.10}$$

and

$$\frac{T}{T_1} = \left(\frac{p}{p_1}\right)^{\gamma - 1/\gamma} \tag{1.11}$$

where γ = the ratio of specific heats. The last of these relations is known as the dry adiabatic relationship in the meteorological context. The quantity $(\gamma - 1)/\gamma$ is equal to R/c_p, where c_p is the specific heat at constant pressure and for air $\gamma = 1.400$.

Using the equation of state, the definition of lapse rate, and the hydrostatic equation, we find the vertical gradient of the ambient density is given by

$$\left(\frac{d\rho}{dz}\right)_{\text{atm}} = +\frac{\rho}{T}\left(\lambda - \frac{g}{R}\right)$$

To find $(d\rho/dz)_{\text{par}}$ we note that

$$\left(\frac{d\rho}{dz}\right)_{\text{par}} = \left(\frac{d\rho}{dp}\right)_{\text{par}} \left(\frac{dp}{dz}\right)_{\text{atm}}$$

because dp/dz is imposed by the atmosphere and is therefore equal to $-\rho g$. The quantity $(d\rho/dp)_{\text{par}}$ is found by differentiating Eq. 1.9, enabling us to evaluate

$$\left(\frac{d\rho}{dz}\right)_{\text{par}} = -\frac{\rho}{T}\frac{g}{R\gamma}$$

Substituting the two vertical gradients into Eq. 1.8, we find

$$\lambda < \frac{g}{R}\frac{\gamma - 1}{\gamma}$$

as the range of values of the lapse rate which corresponds to the case of static stability. We define the dry adiabatic lapse rate, Γ, by the relationship

$$\Gamma \equiv \frac{\gamma - 1}{\gamma}\frac{g}{R} = +9.8°\text{K/km} \qquad (1.12)$$

Therefore,

1. $\lambda < \Gamma$ corresponds to stable equilibrium when the lapse rate is said to be subadiabatic.
2. $\lambda = \Gamma$ corresponds to neutral stability when the lapse rate is equal to the dry adiabatic value.
3. $\lambda > \Gamma$ corresponds to the statically unstable case when the lapse rate is said to be superadiabatic.

The standard atmosphere displays a condition of weak static stability because its $\lambda = +6.5°\text{K/km}$ is less than $+9.8°\text{K/km}$. In the stratosphere of the standard atmosphere the lapse rate is zero and it corresponds to a case of moderately high static stability. In the lowest layer of the troposphere, conditions may prevail in which the lapse rate is negative and the thermal profile is "inverted" in the sense that temperature increases with altitude rather than decreases with altitude. This condition will correspond to high static stability and is described as a thermal inversion. When a thermal inversion occurs there is a strong tendency of air pollutants to be confined near the ground level where they are most often generated.

The presence of condensable moisture in the air does revise the above description of static stability. It is clear that the upward motion of a humid parcel may cause condensation. We omit discussion of this topic, which is described in Ref. 4.

1.4. POTENTIAL TEMPERATURE

The potential temperature, θ, is defined as the temperature that a parcel of air, originally at an arbitrary geometric altitude, would assume if it were compressed or expanded adiabatically to the pressure of 1000 mb. Thus specializing $p_1 = 1000$ and $T_1 = \theta$ in the dry adiabatic equation, we obtain

$$\theta(z) = T(z)\left(\frac{1000}{p(z)}\right)^{(\gamma-1)/\gamma} \tag{1.13}$$

where $p(z)$ is units of millibars. The adiabatic lapse rate corresponds to an atmospheric temperature profile for which potential temperature is independent of height. This statement can be confirmed by differentiating Eq. 1.13 with respect to z and using the equation of hydrostatics, the gas law, and the definitions of the lapse rate and the dry adiabatic lapse rate. The result is

$$\frac{1}{\theta}\frac{d\theta}{dz} = +\frac{(\Gamma - \lambda)}{T} \tag{1.14}$$

and we find that potential temperature is independent of altitude when $\lambda = \Gamma$.

Often the gradient of potential temperature is of interest and it is to be calculated from the above Eq. 1.13 in general. In the special case of altitudes near sea level where $p = p_0 - \Delta p$, with $\Delta p \ll p_0$, the above equation reduces by the binomial expansion to the following expression:

$$\frac{d\theta}{dz} = \frac{(\Gamma - \lambda)}{\left(1 - \dfrac{\Delta p}{p_0}\right)^{(\gamma-1)/\gamma}} \approx \left(1 + \frac{\gamma - 1}{\gamma}\frac{\Delta p}{p_0}\right)(\Gamma - \lambda) \tag{1.15}$$

For small height deviations from sea level corresponding to

$$\left[\frac{\gamma - 1}{\gamma}\right]\frac{\Delta P}{p_0} \ll 1,$$

Eq. 1.15 reduces to the form often used,

$$\frac{d\theta}{dz} \approx \Gamma - \lambda \tag{1.16}$$

The criteria for static stability in terms of potential temperature are summarized as follows:

1. $\lambda < \Gamma, \dfrac{d\theta}{dz} > 0$ stable
2. $\lambda = \Gamma, \dfrac{d\theta}{dz} = 0$ neutral stability
3. $\lambda > \Gamma, \dfrac{d\theta}{dz} < 0$ unstable

The dependence of potential temperature on altitude for an atmosphere of constant lapse rate can be determined if we combine Eqs. 1.4b and 1.14 to obtain

$$\frac{1}{\theta}\frac{d\theta}{dz} = \frac{\Gamma - \lambda}{T_0 - \lambda z} \tag{1.17}$$

If the lapse rate is constant, the variation of potential temperature with altitude,

$$\theta(z) = T_0\left(1 - \frac{\lambda z}{T_0}\right)^{-(\Gamma - \lambda)/\lambda} \tag{1.18}$$

is found by integration.

1.5. POTENTIAL DENSITY

The potential temperature is a property that is widely used in traditional meteorological work. To examine the properties of buoyant effluents it is useful to define potential density in an analogous fashion. That is, the potential density of a parcel is the density it would assume if it were compressed or expanded adiabatically to a pressure of 1000 mb. Specializing p_1 to 1000 and ρ_1 to ρ^* in Eq. 1.9, we find the expression for the potential density is given by

$$\rho^*(z) = \rho(z)\left(\frac{1000}{p(z)}\right)^{1/\gamma} \tag{1.19}$$

where again p is in units of millibars. We evaluate the vertical derivative, and again using the equation of hydrostatics, the gas law, and the definitions of λ and Γ, we find

$$\frac{1}{\rho^*}\frac{d\rho^*}{dz} = -\frac{\Gamma - \lambda}{T} \tag{1.20}$$

Again the neutrally stable case corresponds to a constant potential density profile. However, a *negative* value of $d\rho^*/dz$ corresponds to a statically stable atmosphere. (This is contrary to the case of dT/dz and $d\theta/dz$, which are both positive when the atmosphere is statically stable.) Thus in terms of potential density the static stability criteria become

1. $\Gamma > \lambda, \dfrac{d\rho^*}{dz} < 0$ stable
2. $\Gamma = \lambda, \dfrac{d\rho^*}{dz} = 0$ neutrally stable
3. $\Gamma < \lambda, \dfrac{d\rho^*}{dz} > 0$ unstable

We summarize these stability criteria graphically in the results depicted in Fig. 1.2. Note that potential density of the compressible fluid plays an analogous role to the density of an incompressible fluid with respect to the criteria for static stability.

The dependence of potential density on altitude for an atmosphere of constant lapse rate can be found by a procedure identical to the one used in the case of the potential temperature. For the vertical dependence of potential density we find, if λ = constant,

$$\rho^*(z) = \rho_0\left(1 - \frac{\lambda z}{T_0}\right)^{+(\Gamma - \lambda)/\lambda} \tag{1.21}$$

Now suppose a large mass of buoyant effluent is released at sea level. To analyze its subsequent motion we will idealize the problem by saying that the effluent is enclosed by a weightless, tensionless, perfectly insulating membrane that prevents both mixing and heat transfer between the buoyant effluent and the surrounding atmospheric gas. The results of this hypothetical experiment can be visualized readily on the plot of potential density versus altitude (Fig. 1.3). The densities of both the effluent and the atmosphere at sea level, where $p \approx 1000$ mb, are equal to their respective potential

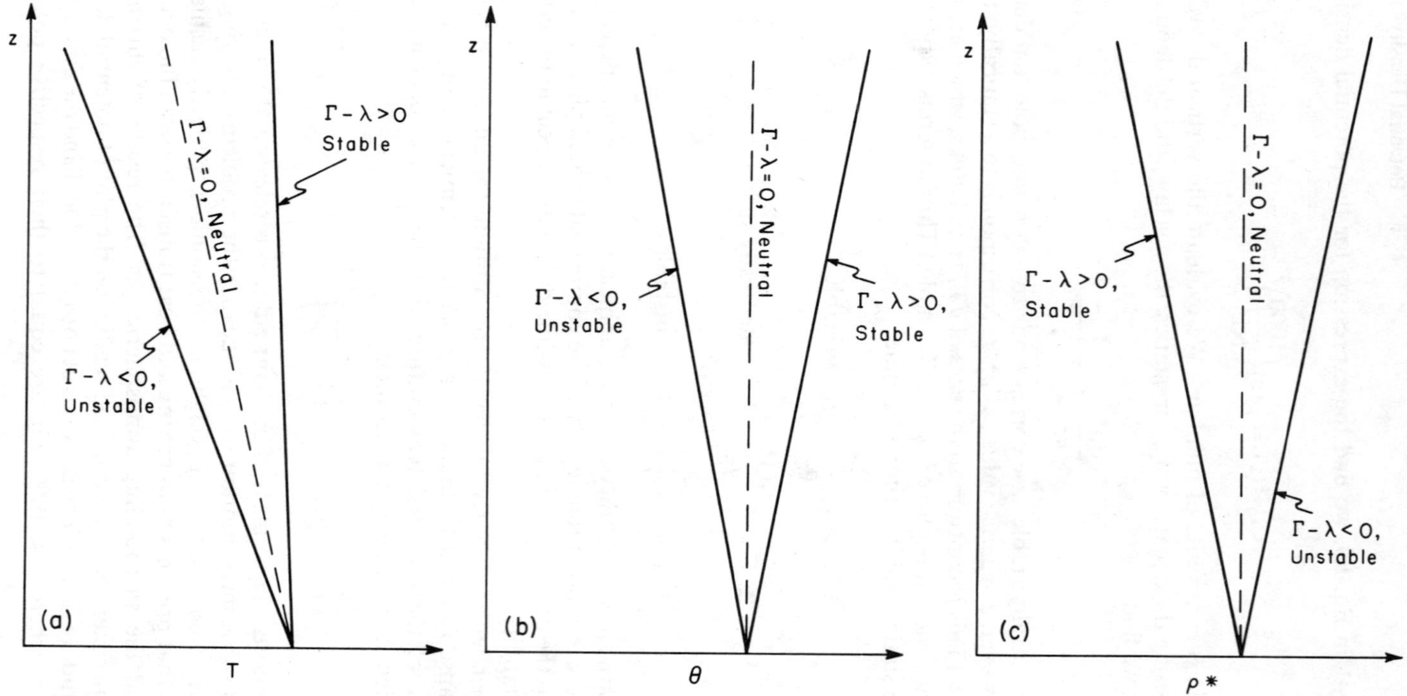

Figure 1.2. The criterion for static stability in terms of (*a*) temperature, (*b*) potential temperature, and (*c*) potential density.

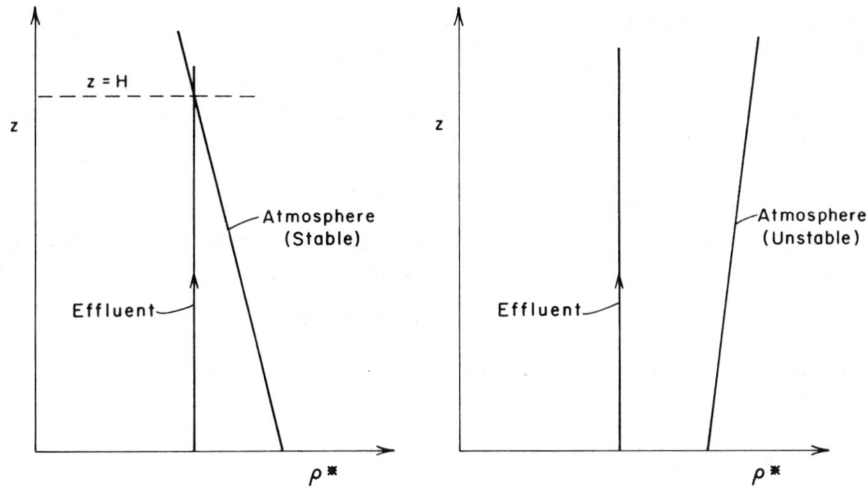

Figure 1.3. An idealization of the upward motion of a buoyant effluent. The effluent has a lower density than air at sea level, and its potential density is unchanged during its upward motion if mixing is ignored. An equilibrium height H is reached only if the atmosphere possesses a statically stable stratification (subadiabatic lapse rate).

densities by definition. The effluent is buoyant because the density at sea level is lower than the ambient density. As the effluent moves upward, the expansion of the gas parcel is adiabatic by virtue of our assumptions and therefore its potential density, ρ^*, is unchanged.

Thus if the atmosphere is neutrally stable or unstable, from Fig. 1.3 we see that the density difference is not reduced as the parcel rises through the atmosphere. When the atmosphere is stable and ρ^* decreases with altitude, the potential densities of effluent and atmosphere are the same at $z = H$, and the actual densities will then also be identical. The parcel thus achieves neutral buoyancy at this elevation. Thus in the absence of mixing, a buoyant effluent is confined by a stable inversion to the altitude where its sea level (potential) density is equal to the potential density of the atmosphere; its vertical motion is unimpeded in the neutral or unstable atmosphere. (The mixing that would actually occur causes the potential density of the effluent to increase as the parcel rises.)

1.6. THE BRUNT–VÄISÄLLÄ FREQUENCY

The buoyant parcel of gas enclosed by the massless, tensionless membrane possesses a natural frequency of oscillation about its equilibrium position.

To calculate this frequency of oscillation we recall that in a linear spring-mass system the square of the circular frequency of oscillation is the quotient of the spring constant, or force per unit displacement, divided by the mass. The mass of the buoyant effluent is equal to the density, ρ_{par}, times its volume, V, and both of these quantities are altitude dependent,

$$m = \rho_{\text{par}}(z)V(z)$$

The net force (downward) is, by Archimedes principle, equal to the difference between the buoyancy force and the gravitational force

$$F = +mg - \rho_{\text{atm}}(z)V(z)g$$

The vertical derivative of the net force is

$$\left(\frac{dF}{dz}\right)_{\text{par}} = -g\frac{d}{dz}[\rho_{\text{atm}}(z)V(z)] = -g\frac{d}{dz}\left[\rho_{\text{atm}}\frac{m}{\rho(z)}\right]$$

With the aid of Eq. 1.19 we find

$$\left(\frac{1}{m}\frac{dF}{dz}\right)_{\text{par}} = -g\frac{d}{dz}\left[\frac{\rho_{\text{at}}}{\rho_{\text{par}}}\right] = -g\frac{d}{dz}\left[\frac{\rho^*_{\text{atm}}}{\rho^*_{\text{par}}}\right]$$

The potential density of the parcel is independent of height, may be factored, and is set equal to the potential density of the atmosphere at the equilibrium height $z = H$. Therefore,

$$\left(\frac{1}{m}\frac{dF}{dz}\right)_{\text{par}} = -\frac{g}{\rho^*}\frac{d\rho^*}{dz}$$

where ρ^* refers to the atmospheric potential density.

The natural frequency of oscillation of the buoyant effluent that has reached its equilibrium height is thus dependent only on the properties of the air and is given by

$$N \equiv \sqrt{-\frac{g}{\rho^*}\frac{d\rho^*}{dz}}, \qquad \frac{d\rho^*}{dz} < 0 \tag{1.22}$$

This quantity is called the Brunt–Väisällä frequency, and from Eqs. 1.14 and 1.20 it may also be expressed as

$$N = \sqrt{+\frac{g}{\theta}\frac{d\theta}{dz}}, \qquad \frac{d\theta}{dz} > 0 \tag{1.23}$$

and in the useful form

$$N = \sqrt{\frac{g}{T}(\Gamma - \lambda)}, \qquad \Gamma - \lambda > 0 \tag{1.24}$$

The frequency N has units of radians per second, and therefore the natural frequency of oscillation in cycles per second is $N/2\pi$ and the corresponding period of oscillation T_N is $2\pi/N$. The concept of the natural frequency of oscillation of a neutrally buoyant plume about its equilibrium position will be useful in subsequent discussion of the motion of rising plumes.

1.7. THE GRAPHICAL PRESENTATION OF ATMOSPHERIC DATA

In meteorological practice a graphical presentation of the information obtained in a low level sounding is often presented in the "Adiabatic Chart D-31" of the U.S. National Weather Service. A reduced-size version of this chart is enclosed loosely. This chart is made up of two graphs which are shown separately in Fig. 1.4. To understand the basis of these charts we note that Eq. 1.6 can be used to show that atmosphere properties can be related by

$$\ln \frac{p_0}{p} = + \frac{zg}{RT_0}\left(1 + \frac{T_0 - T}{2T_0} + \cdots\right)$$

where p_0 and T_0 are the actual ground-level pressures and temperatures and T is the temperature at height z where the pressure is p. For the low

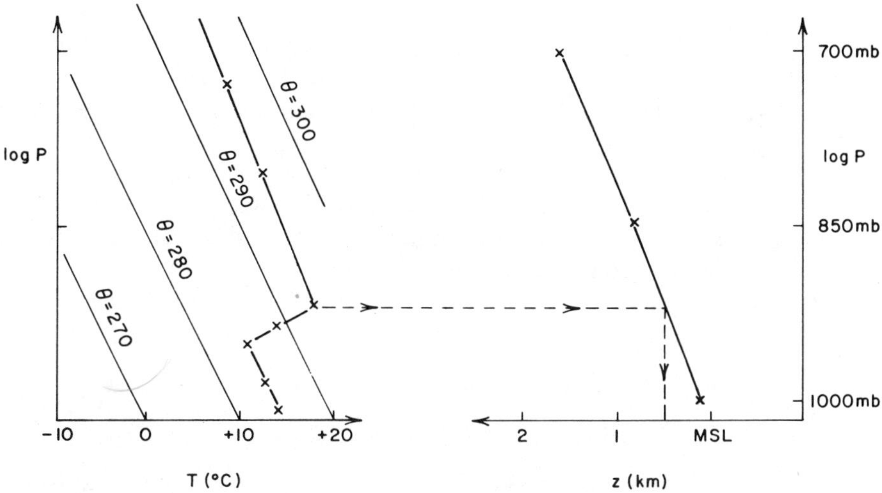

Figure 1.4. The use of "Adiabatic Chart D-31" of the U.S. National Weather Service for the presentation of rawinsonde data. The chart consists of a graph of log p versus T with lines of constant θ and an overlapping graph of log p versus z.

levels of the atmosphere, $T_0 - T \ll T_0$ and

$$\ln \frac{p_0}{p} \approx \frac{g}{RT_0} z$$

which is the equation for the isothermal atmosphere. The right-hand portion of Chart D-31 is a graph of ln p versus z measured positive to the left. This portion of the graph is normally drawn as a two-segment, straight line graph. The heights corrected to height above mean sea level of the 700 mb, 850 mb, and 1000 mb pressure readings, as obtained from the reduction of the rawinsonde data using the hypsometric equation (Eqs. 1.3a, b, or c), are used typically to obtain the three points necessary to draw the two-segment line. The left-hand graph, a plot of ln p versus T, is used to display the temperatures at the various pressure levels as reported by rawinsonde observations. Lines of constant potential temperature can be found from Eq. 1.13 for specified values of p and T and possess a slight curvature because of the logarithmic-linear nature of the graph. The lines of constant θ are labeled in kelvins (°K). A segment of a temperature profile that is parallel to a line $\theta =$ constant corresponds to neutral stability, whereas more steeply inclined profiles represent the statically stable, subadiabatic lapse rate. When the information of the atmospheric sounding is presented on Chart D-31 it is possible to visualize readily the temperature profile and the implied static stability. In addition, the heights of prominent features of the temperature profile, such as inversions, can be obtained as illustrated in Fig. 1.4.

REFERENCES

1. R. G. Fleagle and J. A. Businger, *An Introduction to Atmospheric Physics* (Academic Press, New York, 1963).
2. E.S.S.A., N.A.S.A., U.S.A.F., *U.S. Standard Atmospheric Supplements, 1966* (U.S. Government Printing Office, Washington, D.C., 1966).
3. V. A. Belinskii, *Dynamic Meteorology*, translated by M. Artman et al., 1961 (State Publishing House of Technical-Theoretical Literature, Moscow, 1948).
4. S. L. Hess, *Introduction to Theoretical Meteorology* (Holt & Company, New York, 1959).

PROBLEMS

1.1. The enclosed data of pressure and temperature were telemetered from a rawinsonde launched by the Environmental Meteorological Support Unit (EMSU) at Los Angeles Airport. Use the appropriate version of the hypsometric equation to calculate to slide rule accuracy the altitude

Partial Extract of EMSU Data Los Angeles, Calif. at 11:34 PST, March 16, 1972

P (kPa)	z(ASL) (m)	T (°C)	Dew Point (°C)
101.2	33.5	15.8	11.1
100.0		13.5	8.0
98.0		11.5	6.8
97.4		12.4	6.7
96.8		20.5	3.5
95.2		23.9	−1.4
93.5		24.5	−2.1
85.0		17.1	−8.2
74.0		7.2	−15.8
70.0		5.2	−17.8

at which the 85.0 kPa (850 mb) observations were made. Assume $g =$ constant and ignore the influence of water vapor on the molecular weight of air.

1.2. Plot the above data on the Adiabatic Chart D-31.

1. Using the results of Problem 1.1 obtain a straight line approximation to ln p versus z.
2. Use the raw data to obtain a curve of ln p versus T (°C).
3. Comment on the stability of various layers of the atmosphere and identify the altitudes of these layers using the graph.

1.3. Plot the potential temperature and potential density profiles for the data provided. Comment on the static stability of the atmosphere described by this data.

1.4. What are the values of $d\theta/dz$ at the lower boundaries of the troposphere and the stratosphere of the standard atmosphere?

1.5. Derive a relationship between the entropy and the potential temperature. Formulate a criterion for the static stability of the atmosphere in terms of the vertical entropy gradient.

1.6. Show that the criterion for static stability,

$$\left(\frac{d\rho}{dp}\right)_{\text{par}} < \left(\frac{d\rho}{dp}\right)_{\text{atm}}$$

leads to the same value of the adiabatic lapse rate that we have already formulated.

1.7. Calculate the Brunt–Väisällä frequency and the corresponding time period at $z = 400$ m for the following cases:
   1. The standard atmosphere.
   2. The atmospheric conditions measured over Los Angeles on March 16, 1972. Calculate the local lapse rate based on the nearest two tabular entries.

1.8. A proposed data gathering system would employ a balloon-borne telemetry unit that would transmit temperature at frequent fixed time intervals. No other data would be acquired by the instrumentation package that is carried aloft. The balloon would be tracked by visual observations of two persons using theodolites separated by a precisely known distance. The observers are to provide the elevation and azimuth angles at each instant that the temperature is transmitted. From these observations the position and altitude of the balloon are known functions of time. Derive the finite difference equation necessary to determine pressure (and therefore the other scalar quantities—density, potential temperature, and potential density) versus altitude.

2

SOME CONCEPTS OF ATMOSPHERIC FLUID DYNAMICS

The motion of an object is readily determined with respect to a frame of reference which is at rest or travels at uniform velocity with respect to the distant stars. In such a coordinate system the motion of an object that is not subject to external forces is controlled by its inertia and it will remain either at rest or in motion at constant velocity. The frame of reference is then said to be an inertial coordinate system. The surface of the earth, which is the frame of reference that is used to observe the motion of the atmosphere, is not an inertial coordinate system because of the rotation of the earth about its axis. The equations expressing the conservation of momentum of the atmosphere are significantly altered by the enforced use of a noninertial coordinate system. In this chapter we first examine the influence of the motion of the coordinate system on the measurement of acceleration and on the equations of motion. A series of approximations are then presented which, when applicable, simplify the equations that describe the large scale flow in the troposphere. It will be demonstrated that the simplified equations continue to display explicitly the important role played by the rotation of the earth-based coordinate system.

2.1. THE MEASUREMENT OF ACCELERATION IN A ROTATING COORDINATE SYSTEM

Consider a particle moving at constant angular velocity, Ω, on a circle whose axis is colinear with the z axis (see Fig. 2.1). We wish to find the velocity vector, $\mathbf{v}$, of the particle in terms of the instantaneous position vector, $\mathbf{r}(t)$. The magnitude of the velocity will be given by the product $\Omega\rho_s$ where ρ_s is the distance of the particle from the z axis and which in turn is equal to the

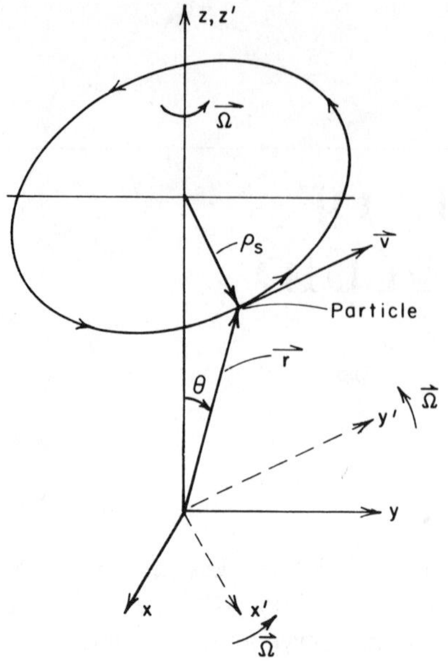

Figure 2.1. A particle in uniform circular motion about the z axis.

magnitude of $\mathbf{r}$ times $\sin\theta$. We define $\mathbf{\Omega}$ as a vector parallel to the axis of rotation, positive in the sense of the right-hand screw convention (x moving toward y corresponding to a positive rotation), and of magnitude equal to the angular speed (radians/s). Therefore

$$|\mathbf{v}| = |\mathbf{\Omega}|\rho_s = |\mathbf{\Omega}||\mathbf{r}|\sin\theta$$

and the direction of $\mathbf{v}$ will always be perpendicular to $\mathbf{r}$ and $\mathbf{\Omega}$. In view of the definition of the vector cross product, these facts are succinctly expressed as

$$\mathbf{v} = \mathbf{\Omega} \times \mathbf{r} \tag{2.1}$$

The rotation about an axis can be decomposed into the simultaneous angular rotations about two separate axes as a result of the distributive property of the vector cross product. Thus,

$$\mathbf{v} = \mathbf{\Omega}_3 \times \mathbf{r} = (\mathbf{\Omega}_1 \times \mathbf{r}) + (\mathbf{\Omega}_2 \times \mathbf{r})$$

where

$$\mathbf{\Omega}_1 + \mathbf{\Omega}_2 = \mathbf{\Omega}_3$$

Let us add to the picture in Fig. 2.1 a rotating coordinate system (x', y', z') that moves at a constant rate of rotation with the vector $\mathbf{r}$. The velocity with respect to the prime coordinates, $\mathbf{v}_{\text{rel}}$, is zero since the position vector is at rest in this system. The velocity of particle in the fixed or inertial coordinate system, $\mathbf{v}_{abs}$, is $\mathbf{\Omega} \times \mathbf{r}$ from the above discussion.

At this point we allow the position vector to move in an arbitrary fashion with respect to the rotating (primed) coordinates. The velocity with respect to the primed coordinates is by definition the time derivative of the position vector.

With respect to the fixed coordinates system, the velocity, $\mathbf{v}_{abs}$, is the sum of the velocity resulting from the rotation of the primed coordinate system, $\mathbf{\Omega} \times \mathbf{r}$, plus the velocity of the particle with respect to this system. Thus,

$$\mathbf{v}_{abs} = \mathbf{v}_{rel} + \mathbf{\Omega} \times \mathbf{r}$$

or

$$\mathbf{v}_{abs} = \left[\left(\frac{d}{dt}\right)_{rel} + \mathbf{\Omega} x\right]\mathbf{r} \tag{2.2}$$

where the notation $(d/dt)_{rel}$ designates the time derivative with respect to the rotating coordinate system.

The acceleration in the absolute (inertial) coordinate system can be evaluated by applying the operator indicated by the square bracket in Eq. 2.2 to the absolute velocity vector. Thus,

$$\mathbf{a}_{abs} = \frac{d}{dt}(\mathbf{v}_{abs}) = \left[\left(\frac{d}{dt}\right)_{rel} + \mathbf{\Omega} x\right](\mathbf{v}_{rel} + \mathbf{\Omega} \times \mathbf{r})$$

Performing the operations indicated above, we find absolute acceleration,

$$\mathbf{a}_{abs} = \mathbf{a}_{rel} + \mathbf{\Omega} \times (\mathbf{\Omega} \times \mathbf{r}) + 2\mathbf{\Omega} \times \mathbf{v}_{rel} \tag{2.3}$$

The first term on the right of Eq. 2.3 is the acceleration of the particle in the rotating coordinate system, the second is easily shown to be $-\Omega^2\rho_s$, the familiar centripetal acceleration, and the final term is called the Coriolis acceleration. A contrast between the nature of the last two acceleration terms is apparent; the centripetal acceleration is independent of $\mathbf{v}_{rel}$ and proportional to ρ_s, while the Coriolis acceleration is independent of $\mathbf{r}$ or ρ_s and proportional to $\mathbf{v}_{rel}$.

The physical significance of the Coriolis acceleration can be understood by considering the problem of throwing a ring from a moving carousel to a fixed target. At time $t = 0$ the ring is hurled toward the target and, after leaving the hand of the person, it travels at constant speed and zero absolute acceleration in the horizontal plane, if we agree to neglect the drag resulting from air resistance. As the ring moves toward the target, the carousel rotates beneath the projectile which traces out a curved trajectory on the moving

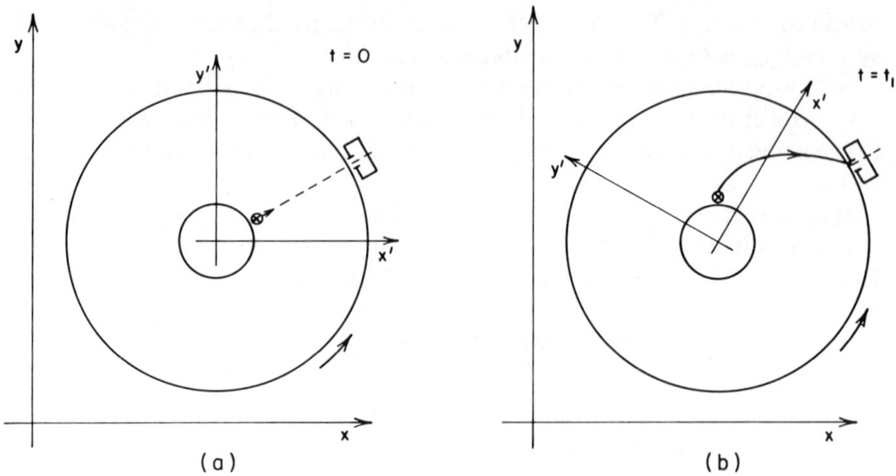

Figure 2.2. An object is hurled at a fixed target from a rotating carousel at $t = 0$. (*a*) In inertial coordinates (x, y) the particle undergoes rectilinear motion. (*b*) The locus of the points on the carousel over which the object passes forms a curved trajectory in the rotating coordinates (x', y') that move with the carousel.

platform as indicated in Fig. 2.2*b*. The horizontal acceleration relative to the merry-go-round is, from Eq. 2.3,

$$\mathbf{a}_{\text{rel}} = -\mathbf{\Omega} \times (\mathbf{\Omega} \times \mathbf{r}) - 2\mathbf{\Omega} \times \mathbf{v}_{\text{rel}}$$

With respect to the carousel the ring is subjected to two kinds of acceleration. This first is the radial acceleration, which also acts on the person who is throwing the object. The second term is the Coriolis acceleration, which, for the indicated directions of $\mathbf{v}_{\text{rel}}$ and $\mathbf{\Omega}$, would be directed to the right of the path.

During the flight of the projectile, the absolute acceleration in the horizontal plane is zero. An observer on the ground sees the ring traveling in rectilinear motion in the horizontal plane. The observer on the carousel viewing the object against the rotating frame of reference sees the curved trajectory described above and associates it with the operation of "forces" that result from the rotation of the coordinate system. To amplify this viewpoint we may write Newton's law in the moving coordinate system as

$$m[\mathbf{a}_{\text{rel}} + \mathbf{\Omega} \times (\mathbf{\Omega} \times \mathbf{r}) + 2\mathbf{\Omega} \times \mathbf{v}_{\text{rel}}] = \sum \mathbf{F}_i$$

where m is the mass of the particle and $\mathbf{F}_i$ are the forces acting on it. We transpose the two acceleration terms to recast the law in the form more reminiscent of Newton's law in the inertial coordinate system:

$$m\mathbf{a}_{\text{rel}} = \sum \mathbf{F}_i + m\mathbf{a}_{\text{cf}} + m\mathbf{a}_{\text{cor}}$$

where $\mathbf{a}_{cf} = -\boldsymbol{\Omega} \times (\boldsymbol{\Omega} \times \mathbf{r})$ and $\mathbf{a}_{cor} = -2\boldsymbol{\Omega} \times \mathbf{v}_{rel}$. The term $m\mathbf{a}_{cf}$ is called the "centrifugal force" and $m\mathbf{a}_{cor}$ is the "Coriolis force," both of which are not real forces and are therefore called pseudoforces. The convention of treating the acceleration terms that result from the rotation of the coordinate system as "forces" is widespread and will be observed here.

2.2. EQUATIONS OF FLUID MOTION IN A ROTATING COORDINATE SYSTEM

The vector form of Euler's equations of motion that apply to an inviscid fluid in an inertial coordinate system is

$$\frac{D\mathbf{q}}{Dt} = -\frac{\nabla p}{\rho} + \mathbf{f}_m \tag{2.4}$$

where ∇p is the gradient of the pressure, ρ is the fluid density which may vary with position and time, and $\mathbf{f}_m$ is any body force expressed in terms of its magnitude per unit mass of fluid. (Throughout our discussion we will neglect the viscosity of the fluid for reasons that will be discussed in Chapter 6.) The quantity $D\mathbf{q}/Dt$ is the acceleration of a fluid particle and is called by various names, one of these being the particle derivative. The above equation applies to the Eulerian description of the flow in which its properties, for example, the velocity vector or one of its components, are expressed as functions of the position coordinates and time. In the Lagrangian description the properties of a particular fluid particle or aggregate are specified as it assumes various positions in space. Both the Eulerian and Lagrangian descriptions are useful for our purposes, but we will rely most heavily on the former at present. However, the Lagrangian description proves useful in calculating the spreading of a pollutant in a turbulent flow.

We digress briefly to review the concept of the particle derivative in more detail. If any quantity, say u, the x component of velocity, is a function of x, y, z, and t, the chain rule of differentiation states that

$$\frac{du}{dt} = \frac{\partial u}{\partial x}\frac{dx}{dt} + \frac{\partial u}{\partial y}\frac{dy}{dt} + \frac{\partial u}{\partial z}\frac{dz}{dt} + \frac{\partial u}{\partial t}$$

This expression is to be associated with the following picture: the scalar quantity u depends on time and the position vector whose coordinates are x, y, and z. The above expression indicates that the position vector is moving with components of speed dx/dt, dy/dt, and dz/dt. The quantity du/dt is the derivative that results when the velocity components of the position vector are arbitrary; this type of derivative has little value in its present form.

The above derivative can be converted into a component of the acceleration of a fluid particle at a particular point merely by requiring the position vector to follow the fluid particle at the point in question. This is accomplished by setting the velocity components of the position vector equal to the local values of the velocity of the fluid,

$$\frac{dx}{dt} = u, \qquad \frac{dy}{dt} = v, \qquad \text{and} \quad \frac{dz}{dt} = w$$

We then express the derivative given above as

$$\frac{Du}{Dt} = u\frac{\partial u}{\partial x} + v\frac{\partial u}{\partial y} + w\frac{\partial u}{\partial z} + \frac{\partial u}{dt}$$

The special notation employing the upper case D, signifying the derivative following the fluid particle, is a widely used convention that is not usually observed in the meteorological literature. Applying this argument to the other two components of velocity and adding these terms vectorily, we obtain the full expression for the time derivative of the velocity of the fluid particle located at the point $\mathbf{r}$ at time t as

$$\frac{D\mathbf{q}}{Dt} = \frac{\partial \mathbf{q}}{\partial t} + (\mathbf{q} \cdot \nabla)\mathbf{q}$$

where

$$(\mathbf{q} \cdot \nabla)(\;\;) \equiv u\frac{\partial}{\partial x}(\;\;) + v\frac{\partial}{\partial y}(\;\;) + w\frac{\partial}{\partial z}(\;\;)$$

The instantaneous time derivative of the velocity of the fluid particle is obviously the instantaneous acceleration of the fluid particle.

The particle derivative of any scalar quantity η, such as temperature of air or the concentration of a pollutant or other minor chemical species, is again found by applying the chain rule of differentiation and then requiring the position vector to move at the local fluid velocity,

$$\frac{D\eta}{Dt} = \frac{\partial \eta}{\partial t} + \mathbf{q} \cdot \nabla\eta$$

The term $-\mathbf{q} \cdot \nabla\eta$, which is equal to $-q|\nabla\eta|\cos\theta$ where θ is the angle between the vectors $\mathbf{q}$ and $\nabla\eta$, is associated with the transport of the property η by the motion of the fluid and is called the advection of η. The alignment of $\nabla\eta$ and the velocity vector in a moving fluid determine the magnitude of the advection. Thus, referring to Fig. 2.3, the $\nabla\eta$ vector is always normal to

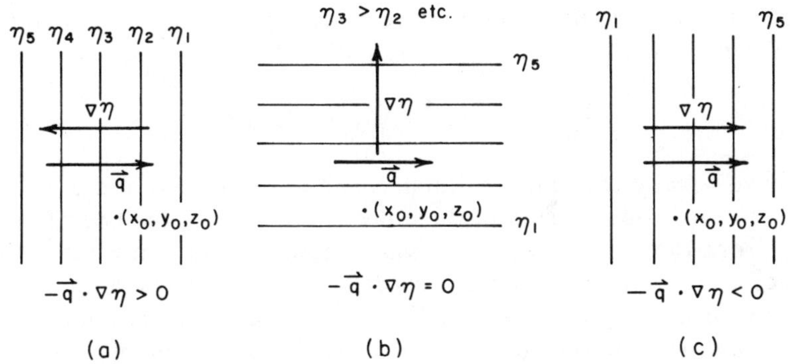

Figure 2.3. The advection of the scaler property of a fluid. The local rate of change of η at (x_0, y_0, z_0) is influenced by the orientation of the $\mathbf{q}$ and $\nabla\eta$ vectors.

the lines of constant η (isopleths) and is directed toward the higher values. If $\mathbf{q}$ and $\nabla\eta$ are antiparallel (Fig. 2.3*a*) then $-\mathbf{q}\cdot\nabla\eta$ is a maximum positive value. The advection term vanishes if $\mathbf{q}$ and $\nabla\eta$ are normal to one another (Fig. 2.3*b*) and is a maximum negative value when these two vectors are parallel (Fig. 2.3*c*).

The rate of change of η at any fixed point in space is called the local rate of change, and from the above discussion we see that it is the sum of the particle derivative plus the advection

$$\frac{\partial\eta}{\partial t} = \frac{D\eta}{Dt} - \mathbf{q}\cdot\nabla\eta$$

If the local rate of change of η is equal to zero then either $\eta =$ constant for all x, y, z, and t or the rate of change of η for each fluid parcel is equal to the advective term itself. If the value of η is constant for each fluid parcel then $D\eta/Dt = 0$ and the local rate of change of η is equal to the advection. In this case, for example, a local value of η will increase most rapidly when $\mathbf{q}$ and $\nabla\eta$ are antiparallel, as depicted in Fig. 2.3*a*, and will decrease most rapidly when $\mathbf{q}$ and $\nabla\eta$ are parallel as shown in Fig. 2.3*c*.

We proceed to specialize Eq. 2.4 to apply to a parcel of air in the earth's atmosphere using a coordinate system fixed in the rotating earth. First we replace the acceleration of the fluid particle with respect to an inertial coordinate system by the three acceleration terms, Eq. 2.3, that apply when the rotating coordinate system is used. In our discussion of the hydrostatic equation in Chapter 1 it was noted that the acceleration of gravity, $g(z, \phi)$, at the surface of the earth is a combination of the acceleration resulting from the Newtonian gravity and the centripetal acceleration. These two terms can be combined as the single conservative body force term which is equal to $-G\nabla H$, where H is the geopotential. Equation 2.4 now becomes

$$\frac{D\mathbf{q}}{Dt} = -\frac{\nabla p}{\rho} - G\nabla H - 2\mathbf{\Omega} \times \mathbf{q} \tag{2.5}$$

The above equation is simply a statement that the acceleration of the fluid particle results from the sum of three body force terms; the body force resulting from the existence of a pressure gradient, the conservative body force resulting from the combined effect of Newtonian gravitational force and the centrifugal pseudoforce, and finally the Coriolis pseudoforce that results from the finite velocity of the fluid particle in the moving coordinate system which is fixed in the surface of the rotating earth. We have followed the convention of renaming $\mathbf{q}_{\text{rel}}$ as $\mathbf{q}$ for convenience; the presence of the Coriolis term will signify that velocity is measured with respect to a moving coordinate system.

The tangent plane coordinate system is one of several systems frequently used in meteorological work. In this system Cartesian coordinates are established with the origin at the surface of the earth, the x and y axes tangent to the surface directed to the east and to the north, respectively (see Fig. 2.4). Using this coordinate system we may specialize Eq. 2.5 into forms that are particularly useful. We first resolve the angular velocity vector into its components in the x, y, and z directions. From Fig. 2.4b we find that the angular velocity vector can be resolved into components about the y and z axes,

$$\mathbf{\Omega} = 0\mathbf{i} + \Omega \cos \phi\mathbf{j} + \Omega \sin \phi\mathbf{k}$$

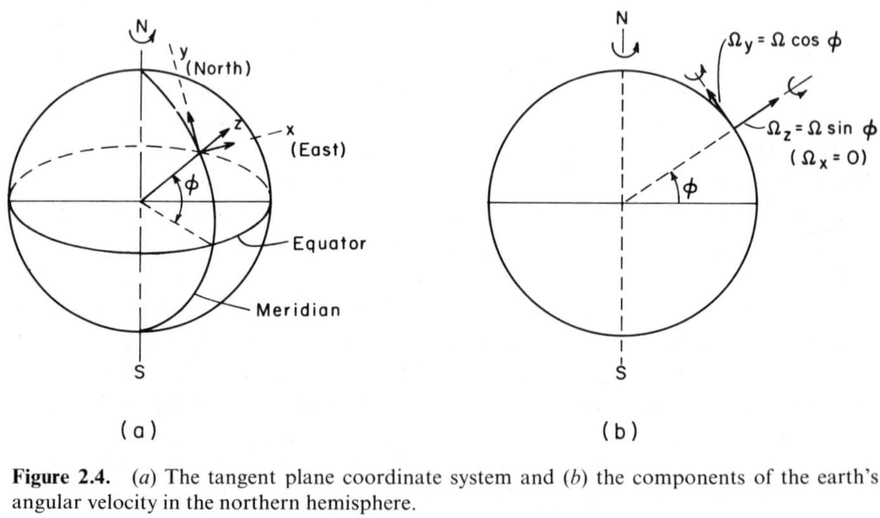

Figure 2.4. (a) The tangent plane coordinate system and (b) the components of the earth's angular velocity in the northern hemisphere.

The Coriolis acceleration term is evaluated in the tangent plane coordinates as

$$\begin{aligned} 2\mathbf{\Omega} \times \mathbf{q} &= \begin{vmatrix} \mathbf{i} & \mathbf{j} & \mathbf{k} \\ 0 & 2\Omega \cos \phi & 2\Omega \sin \phi \\ u & v & w \end{vmatrix} \\ &= (2w\Omega \cos \phi - 2v\Omega \sin \phi)\mathbf{i} \\ &\quad + (2u\Omega \sin \phi)\mathbf{j} \\ &\quad - (2u\Omega \cos \phi)\mathbf{k} \end{aligned}$$

The quantity $2\Omega \sin \phi$ will prove to be a recurring group which is called the Coriolis parameter and is represented by the letter f. The gradient of the geopotential altitude, $G\nabla H$, is the acceleration of gravity times the perpendicular unit vector, $-g\mathbf{k}$. Equations 2.5 in component form in the tangent plane coordinates become

$$\frac{Du}{Dt} = -\frac{1}{\rho}\frac{\partial p}{\partial x} + vf - wf \cot \phi \tag{2.6a}$$

$$\frac{Dv}{Dt} = -\frac{1}{\rho}\frac{\partial p}{\partial y} - uf \tag{2.6b}$$

and

$$\frac{Dw}{Dt} = -\frac{1}{\rho}\frac{\partial p}{\partial z} - g + uf \cot \phi \tag{2.6c}$$

The angular velocity of the earth, to be used in calculating f, is the angle it rotates in one second. If the earth were not revolving about the sun, its angular velocity would be found by dividing the number of radians in a full circle (2π radians) by the number of seconds in one day ($24 \times 3600 = 86{,}400$ s, the duration of the solar day). However, a correction is necessary to account for the revolution of the earth about the sun once every 365.25 days. This is apparent if we imagine the earth to be rotating about its axis at such an angular velocity as to always present the same side to the sun. Its angular velocity would then be found by dividing 2π radians by the number of seconds in one year. Because we measure our rotational velocity with respect to the sun about which we revolve, the actual period to rotate 2π radians is $86{,}400 \times 365.25/366.25$ or 86,164 s, the length of the sidereal day. The angular velocity of the earth is therefore $2\pi/86{,}164$ or $7.2921 \times 10^{-5}\ \text{s}^{-1}$. For the midlatitude regions of the earth the value of $2 \sin \phi \approx 1.4$ and therefore $f \approx 1.0 \times 10^{-4}\ \text{s}^{-1}$.

Atmospheric phenomena of large scale clearly cannot be described by the tangent plane coordinate system, which fails to include the curvature of the earth; a spherical coordinate system could be used to express Eq. 2.5 in a form more appropriate for such events. Some phenomena of more moderate scale may be described in tangent plane coordinates if the variation of the Coriolis parameter, f, with y is considered. In the beta plane approximation $f \approx f_0 + \beta y$ where $\beta = \partial f/\partial y$. Finally, the small scale atmospheric effects of interest to our present purposes can usually be described by considering f to be constant, and this approximation will be used throughout these notes.

2.3. THE CONTINUITY EQUATION

The continuity equation for a compressible fluid with no sources or sinks is

$$\frac{\partial \rho}{\partial t} + \nabla \cdot (\rho \mathbf{q}) = 0 \tag{2.7a}$$

Expanding the divergence term, we may express this equation as

$$\frac{1}{\rho}\frac{D\rho}{Dt} + \nabla \cdot \mathbf{q} = 0 \tag{2.7b}$$

We now distinguish between several classes of fluids with respect to the manner in which the density varies:

1. The density is constant throughout the flow field, hence from Eq. 2.7 we find

$$\nabla \cdot \mathbf{q} = \frac{\partial u}{\partial x} + \frac{\partial v}{\partial y} + \frac{\partial w}{\partial z} = 0 \tag{2.8}$$

An illustration of this case is the flow of water with no variations in its temperature or the concentration of a solute. In this case the fluid is described as possessing a constant density or is said to be homogeneous.

2. The density of any one fluid particle is constant but there are fluid particles of varying density. This situation is illustrated by the flow of two immiscible fluids such as oil and water. A second illustration is the flow in the ocean of water whose temperature and salinity content is variable and where length scales are large enough so that the diffusion of heat and mass are negligible. The particle derivative of the density, $D\rho/Dt$, is zero since the density of a specified aggregate does not change from one moment to the next. In this case the continuity equation again reduces to Eq. 2.8. The fluid particle density is constant even though the density is not uniform and the term homogeneous is not applicable.

In the atmosphere it is clear that the density varies over a wide range in the vertical direction. In addition, we know that the pressure and temperature of air vary from one position to another on the surface of the earth. The equation of state, Eq. 1.2, therefore tells us that density will generally vary in the horizontal direction, although we expect this horizontal variation of density would be much smaller than the vertical variation. We also know that sound waves propagate freely through air, and these waves are a manifestation of the variation of density with pressure. Thus, it is clear that Eq. 2.7 is the form of the continuity equation that will be generally applicable. Under some circumstances, which are discussed below, the simpler form of the continuity equation may be applied in the atmosphere.

2.4. COMMON APPROXIMATIONS OF ATMOSPHERIC FLUID DYNAMICS

The above Eq. 2.6 together with the continuity equation constitute a set of nonlinear, simultaneous, partial differential equations that possess limited value in the form presented thus far. At the moment there are more unknowns than equations and no solution is possible. Their present form is also disadvantageous because certain terms are present that are either invariably small or frequently small. Thus, simpler forms of the equations of motion are obtainable that possess practical utility and that provide important physical insight into the dominant mechanisms that influence the motion of the atmosphere. More detailed discussions of these topics are given in Refs. 1 and 2, p. 256.

2.4.1. The Barotropic Fluid

The equations of motion, Eq. 2.6, and the continuity equation represent four equations in the five unknowns u, v, w, p, and ρ and they are not a closed set of equations without one additional independent relationship between two or more of these quantities. A simple closure relationship is afforded if there exists an explicit functional relationship between p and ρ of any unspecified form. A fluid for which $p = p(\rho)$ or $\rho = \rho(p)$ is said to be a barotropic fluid. A specific example would be a fluid that undergoes only isentropic processes and for which the relationship $p = c\rho^{\gamma}$ applies.

If $p = p(\rho)$ then it follows that

$$\frac{\partial p}{\partial x} = \frac{dp}{d\rho}\frac{\partial \rho}{\partial x}, \qquad \frac{\partial p}{\partial y} = \frac{dp}{d\rho}\frac{\partial \rho}{\partial y}, \qquad \frac{\partial p}{\partial z} = \frac{dp}{d\rho}\frac{\partial \rho}{\partial z}$$

hence

$$\nabla p = \frac{dp}{d\rho}\nabla \rho \tag{2.9}$$

Thus, in a barotropic flow field, the surfaces of constant pressure (isobaric surfaces) are parallel to the surfaces of constant density (isoteric surfaces), and the lines of constant pressure (isobars) are parallel to the lines of constant density (isoteres). From the equation of state,

$$p = \rho RT$$

we find

$$\frac{dp}{p} = \frac{d\rho}{\rho} + \frac{dT}{T}$$

and, therefore,

$$\frac{1}{p}\nabla p = \frac{1}{\rho}\nabla\rho + \frac{1}{T}\nabla T \tag{2.10}$$

Now if the isobaric surfaces (and lines) are parallel to the isoteric surfaces (and lines), it follows from the above equation that they are both parallel to the isothermal surfaces (and lines). An important property of the barotropic flow field is illustrated in Fig. 2.5*a* where a circular mass of fluid is depicted with ∇p and $\nabla\rho$ parallel; hence the isobars are parallel to the isoteres. The center of action of the net pressure force acting on the circular parcel passes through the center of mass and no torque is exerted on the fluid mass by the pressure field. Examination of the nonbarotropic case depicted in Fig. 2.5*b* indicates the contrary, that is, the net pressure force

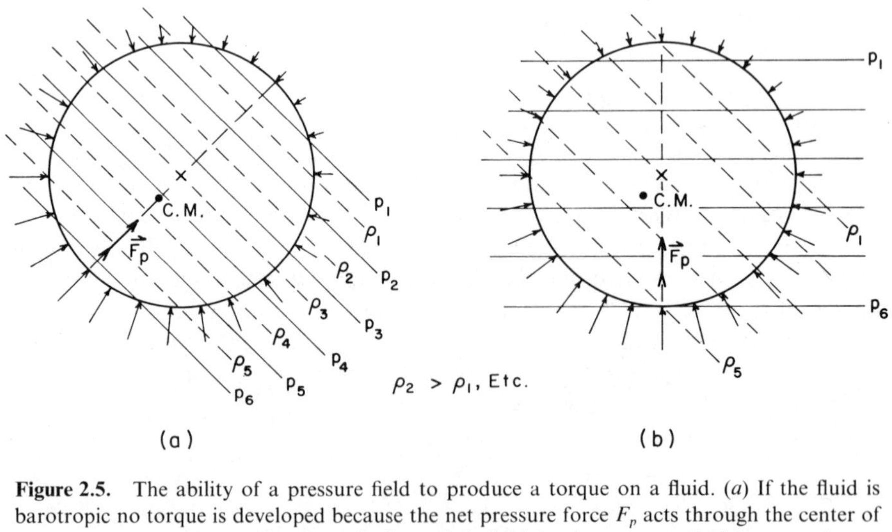

Figure 2.5. The ability of a pressure field to produce a torque on a fluid. (*a*) If the fluid is barotropic no torque is developed because the net pressure force F_p acts through the center of mass. (*b*) A torque is developed on the baroclinic fluid by the pressure field because the net pressure force does not act through the center of mass.

does not act through the center of mass and therefore a torque is developed. The latter type of flow field, that is, one in which the isobaric surfaces intersect the isoteric surfaces, is called a baroclinic fluid. The torque developing property of the baroclinic field will be examined in more detail in the next chapter.

In general, the atmosphere must be considered to be a baroclinic fluid. However, in some regions of the atmosphere and when the effects of solar heating and heat exchanges with cloud and rain droplets may be ignored, the atmosphere may be considered to be barotropic. This approximation was one of the important simplifications which was exploited in the first successful attempts to predict the motion of large scale weather phenomena on purely theoretical grounds.

2.4.2. The Hydrostatic Approximation

At this point the magnitude of various terms in the equations of motion in tangent plane coordinates, Eq. 2.6, will be examined. We start by asserting that the typical wind speed, U, at moderate altitudes, is 10 m/s (as opposed to 1 m/s or 100 m/s). The horizontal length scale, L, is representative of the distance from the center of a high pressure region to the center of a low pressure region—the scale size of a large scale wind-pressure system—and its magnitude is 1000 km. The vertical length scale, H_v, is the height of the troposphere—about 10 km.

In order to make an estimate of the magnitude of the vertical velocity component, we write the continuity equation in the following form

$$\frac{\partial \rho}{\partial t} + \underbrace{u\frac{\partial \rho}{\partial x} + v\frac{\partial \rho}{\partial y}}_{} + \underbrace{\rho\frac{\partial u}{\partial x} + \rho\frac{\partial v}{\partial y}}_{} = -w\frac{\partial \rho}{\partial z} - \rho\frac{\partial w}{\partial z}$$

$$\left(\frac{.05\rho_0 U}{L}\right) \quad \left(\frac{U.05\rho_0}{L}\right) \quad \left(\rho_0\frac{U}{L}\right) \quad \left(W\frac{\rho_0}{H_v}\right) \quad \left(\rho_0\frac{W}{H_v}\right)$$

The first term is the change in the magnitude of density at a fixed point owing to pressure and temperature changes. For typical surface level atmospheric conditions the change in atmospheric density is about 5%. The time period over which this occurs is the horizontal length scale divided by the wind speed. Thus, we conclude that

$$\frac{\partial \rho}{\partial t} \approx \left(\frac{\Delta \rho}{\Delta t}\right)_{x,\,y,\,z} \approx \frac{0.05\rho_0}{L/U} = 0.05\,\frac{\rho_0 U}{L}$$

The terms $u(\partial \rho/\partial x)$ and $v(\partial \rho/\partial y) \approx U(0.05\rho_0/L)$ on the basis that the horizontal density gradient is about 5% of the sea level mean density divided by the horizontal length scale. The terms $\rho(\partial u/\partial x)$ and $\rho(\partial v/\partial y)$ will be $\rho_0(U/L)$ in magnitude and are notably larger than the other terms on the left side. The term $w(\partial \rho/\partial z)$ will be of magnitude W times the vertical density gradient,

$$\frac{\partial \rho}{\partial z} \approx \frac{\Delta \rho}{\Delta z} \approx \frac{\rho_0 - 0}{0 - H_v} \approx -\frac{\rho_0}{H_v}$$

The term $\rho(\partial w/\partial z)$ is of magnitude $\rho_0(W/H_v)$.

The terms on the right side may be, at most, the magnitude of the largest terms on the left, although they may also be smaller. From this consideration we deduce that

$$W \lesssim \frac{H_v}{L} U$$

and since $H_v/L \approx 10^{-2}$ it is apparent that the vertical wind is, at most, of the order 0.1 m/s. These conclusions are, of course, not intended to be applied in regions of the atmosphere where violent atmospheric effects, such as thunderstorms, are occurring nor in the vicinity of mountains.

We use this result to examine Eq. 2.6c,

$$\frac{Dw}{Dt} = -\frac{1}{\rho}\frac{\partial p}{\partial z} - g + uf \cot \phi$$

The particle derivative of w will be the maximum value of W divided by the time required for a parcel of the atmosphere to travel a distance equal to the height of the atmosphere. Hence

$$\frac{Dw}{Dt} \gtrsim \frac{W}{H_v/W} = \frac{W^2}{H_v} = 10^{-6} \text{ m/s}^2$$

The term $uf \cot \phi \approx Uf \approx 10^{-3}$ m/s^2. Since $g \approx 10$ m/s^2 it is clear that both Dw/Dt and $uf \cot \phi$ can be ignored and that the vertical pressure gradient balances the gravitational acceleration. The equation for the vertical component of momentum thus reduces to the hydrostatic equation. We rewrite Eqs. 2.6 noting that the first equation can be simplified because $w \ll v$,

$$\frac{Du}{Dt} = -\frac{1}{\rho}\frac{\partial p}{\partial x} + vf \tag{2.11a}$$

$$\frac{Dv}{Dt} = -\frac{1}{\rho}\frac{\partial p}{\partial y} - uf \tag{2.11b}$$

$$g = -\frac{1}{\rho}\frac{\partial p}{\partial z} \tag{2.11c}$$

The hydrostatic approximation is important especially because it allows the use of the hydrostatic equation in reducing the radiosonde observations of the earth's atmosphere in its usual state of unrest.

2.4.3. The Horizontal Flow Approximation

On occasions the vertical component of velocity is small compared with our upper bound order of magnitude estimate of 0.1 m/s. In this case a term like $w(\partial u/\partial z)$ is small compared with $u(\partial u/\partial x)$ and $v(\partial u/\partial y)$ and therefore

$$\frac{Du}{Dt} = \frac{\partial u}{\partial t} + u\frac{\partial u}{\partial x} + v\frac{\partial u}{\partial y} + w\frac{\partial u}{\partial z} \approx \frac{\partial u}{\partial t} + u\frac{\partial u}{\partial x} + v\frac{\partial u}{\partial y}$$

The latter group, in which the term $w(\partial u/\partial z)$ is missing, is designated as $D_h u/Dt$. Again, if $w \ll 0.1$ m/s, then $w(\partial v/\partial z) \ll u(\partial z/\partial z)$ and

$$\frac{Dv}{Dt} \approx \frac{D_h v}{Dt}$$

Hence Eqs. 2.11 reduce to the horizontal flow equations

$$\frac{D_h u}{Dt} = -\frac{1}{\rho}\frac{\partial p}{\partial x} + fv$$

$$\frac{D_h v}{Dt} = -\frac{1}{\rho}\frac{\partial p}{\partial y} - fu \tag{2.12}$$

and

$$g = -\frac{1}{\rho}\frac{\partial p}{\partial z}$$

The first two of the above equations can be conveniently expressed by vector notation as

$$\frac{D_h \mathbf{q}_h}{Dt} = -\frac{1}{\rho}\nabla_h p - \mathbf{f} \times \mathbf{q}_h \tag{2.13}$$

where

$$\mathbf{q}_h = u\mathbf{i} + v\mathbf{j}, \qquad \text{and} \quad \mathbf{f} = f\mathbf{k}$$

It is important to note that the above equations do not imply that $w = 0$, but rather that the terms involving w are small compared with other convective terms and they need not be included in the horizontal momentum equations. The flow is sometimes said to be quasihorizontal to emphasize the fact that the vertical component of the wind is small but nonzero. We would expect the horizontal flow approximation to be valid near flat regions of the earth's surface at nighttime when convective plumes induced by solar heating of the surface are absent.

2.4.4. The Geostrophic Approximation

We rewrite Eq. 2.13 and examine the magnitude of the terms on the left side

$$\frac{\partial \mathbf{q}_h}{\partial t} + u\frac{\partial \mathbf{q}_h}{\partial x} + v\frac{\partial \mathbf{q}_h}{\partial y} = -\frac{1}{\rho}\nabla_h p - \mathbf{f} \times \mathbf{q}_h$$

$$\left(\frac{U^2}{L}\right) \quad \left(\frac{U^2}{L}\right) \quad \left(\frac{U^2}{L}\right) \qquad\qquad (fU)$$

The magnitude of terms such as $\partial \mathbf{q}_h/\partial t$ and $u(\partial \mathbf{q}_h/\partial x)$ will be U^2/L and the magnitude of $\mathbf{f} \times \mathbf{q}$ is fU. We define the Rossby number as the ratio of the convective acceleration terms to the Coriolis acceleration and its numerical value is U/fL. According to the magnitudes assigned above, the Rossby number will be 10^{-1} for mid- or high latitudinal regions and the relative acceleration of the fluid parcel may be ignored. The pressure gradient then balances the Coriolis force, a condition described as geostrophic balance, and the equation of motion reduces to

$$\frac{1}{\rho}\nabla_h p = -\mathbf{f} \times \mathbf{q}_G \tag{2.14}$$

The wind vector under these circumstances is designated as the geostrophic wind, $\mathbf{q}_G$. Applying the operator $\mathbf{k} \times$ to both sides of the above, we obtain the geostrophic wind as (note that $\mathbf{k} \times \mathbf{k} \times \mathbf{q}_G = -\mathbf{q}_G$)

$$\mathbf{q}_G = \frac{1}{f\rho}\mathbf{k} \times \nabla_h p \tag{2.15}$$

We find a seeming paradox—the "geodynamic paradox"—that the wind direction is at right angles to the pressure gradient. This is in contrast to the flow of a fluid in a pipe where the pressure gradient and velocity vectors are parallel or nearly so. The paradoxical character results from the confusion of velocity with acceleration. If it is recognized that $\mathbf{f} \times \mathbf{q}_G$ is actually an acceleration term and that the velocity and acceleration are mutually perpendicular, the paradoxical character is removed. (Uniform circular motion provides another common example of a motion where the acceleration and velocity are mutually perpendicular.)

If the condition of geostrophic balance were to be interpreted to imply $D\mathbf{q}_h/Dt = 0$, then the consequence would be that $\mathbf{q}_h$ is constant in magnitude and direction and the streamlines would be great circles, that is, along geostrophic lines. This condition is unrealistic and it is more descriptive of the atmosphere to say that the horizontal wind is the sum of two components,

$$\mathbf{q}_h = \mathbf{q}_G + \mathbf{q}_A$$

where $\mathbf{q}_A$ is the ageostrophic wind that results from the nonzero relative acceleration. If $\mathbf{q}_A \ll \mathbf{q}_G$ then we say the condition of geostrophic balance is approximately fulfilled and the flow may be described as quasigeostrophic, often a good approximation in regions of the atmosphere not too close to the surface of the earth.

To understand the condition of geostrophic balance, we imagine a hypothetical situation in which a large body of fluid is at rest in a rotating coordinate system at $t = 0$ when a constant pressure gradient is instantaneously applied. We analyze the subsequent motion of a parcel of the fluid in Fig. 2.6 with the aid of the equation of horizontal flow. Eq. 2.13. Since the fluid is initially at rest, no Coriolis force is present and the acceleration is antiparallel to the pressure gradient. The velocity subsequently increases to the point where the Coriolis force becomes appreciable. This force is directed to the right of the instantaneous velocity vector. Because the pressure gradient is imposed and constant, the acceleration term must *decrease* as $\mathbf{f} \times \mathbf{q}_h$ increases with the increasing magnitude of $\mathbf{q}_h$. Thus, at $t = 1$ unit the acceleration is slightly to the right of the direction of the pressure force, F_p. At a later time, say $t = 2$ units, the Coriolis force is moderate and still at right angles to $\mathbf{q}_h$, and the acceleration has reduced markedly. After a long time has elapsed the Coriolis term has developed in magnitude to the point that it nearly balances the pressure gradient. The acceleration is reduced almost to zero, the motion is nearly along the isobar, and the condition of geostrophic balance is reached. Atmospheric time scales are sufficiently long to allow the development of the geostrophic flow condition. Of course, small scale vigorous meteorological events will not conform to this description.

Equation 2.15 indicates that the direction of the velocity vector is such that, in the northern hemisphere, the *low* pressure region is on the *left*-hand side when the observer has his back to the wind. This statement is a familiar guide to mariners and others and is known as the Buys–Ballot rule.

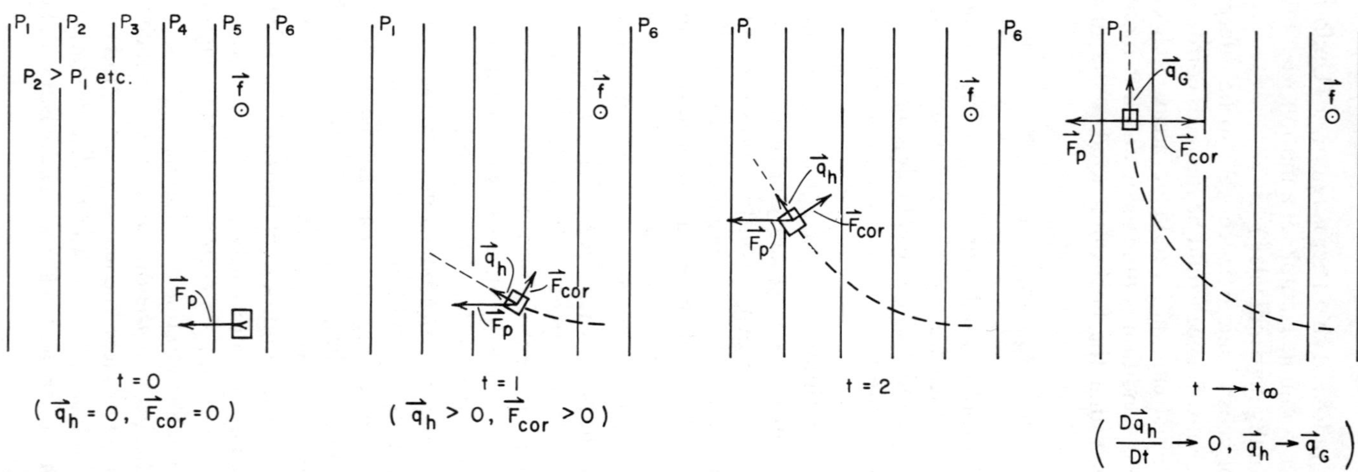

Figure 2.6. The development of the condition of geostrophic balance (northern hemisphere). If the pressure field does not change in time, or is slowly changing, a near equilibrium is achieved in which the pressure force is approximately opposed by the Coriolis force.

2.5. THE PROUDMAN–TAYLOR THEOREM

As we have seen, the hydrostatic approximation is usually applicable to the atmosphere while the barotropic and geostrophic approximations are applicable on some occasions. The simultaneous application of these three approximations leads to an interesting result known, in a slightly different context, as the Proudman–Taylor theorem. We start by differentiating Eq. 2.15, the expression for the geostrophic wind, with respect to z,

$$\frac{\partial \mathbf{q}_G}{\partial z} = -\frac{1}{f\rho^2}\frac{\partial \rho}{\partial z}\mathbf{k} \times \nabla_h p + \frac{1}{f\rho}\mathbf{k} \times \nabla_h \frac{\partial p}{\partial z} \tag{2.16}$$

We note that for the barotropic fluid in hydrostatic equilibrium,

$$\frac{\partial \rho}{\partial z} = \frac{d\rho}{dp}\frac{\partial p}{\partial z} = -\rho g \frac{d\rho}{dp}$$

and

$$\nabla_h\left(\frac{\partial p}{\partial z}\right) = -g\nabla_h \rho = -g\frac{d\rho}{dp}\nabla_h p$$

Substituting these last two relationships into Eq. 2.16, we find $\partial \mathbf{q}_G/\partial z = 0$ and therefore $\mathbf{q}_G$ = constant, that is, possesses the same magnitude and direction for all values of z. This result is essentially the Proudman–Taylor theorem which, in slightly more general terms, states that the (relative) velocity does not vary in the direction of the axis of rotation and the flow is two dimensional in planes perpendicular to the axis of rotation (see Fultz, Ref. 3). The Proudman–Taylor theorem, first enunciated by Proudman and then demonstrated experimentally by Taylor, is normally formulated for geostrophic motions of a homogeneous fluid. The barotropic approximation, if applicable, allows the theorem to be used to describe atmospheric motions which are also geostrophic. Examination of radiosonde data shows that the atmosphere occasionally displays this type of behavior over limited altitude zones not too close to the earth's surface where its roughness and convective effects may influence the flow. Carefully executed laboratory demonstrations of the Coriolis force, low Rossby number flows, and the Proudman–Taylor theorem are displayed in the didactic film by Fultz, Ref. 4.

The flow that results when $\partial \mathbf{q}/\partial z = 0$ is, of course, quite different from that described by the horizontal flow approximation (Section 2.4.3). In the latter case the horizontal wind components, u and v, will depend on altitude and the flow is three dimensional (even if $w = 0$, a condition not required by the horizontal flow approximation). Obviously the horizontal flow approximation may be expected to describe atmospheric flows more frequently than the Proudman–Taylor theorem which requires that the fluid is barotropic and that the Rossby number is low so that the geostrophic approximation is applicable.

REFERENCES

1. J. G. Charney, "On the Scale of Atmospheric Motions," *Geofys. Publ.*, **17**, 3–17 (1948).
2. V. A. Belinskii, *Dynamic Meteorology*, translated by M. Artman et al., 1961 (State Publishing House of Technical-Theoretical Literature, Moscow, 1948).
3. D. Fultz, "Rotating Flows," in *Illustrated Experiments in Fluid Mechanics*, A. H. Shapiro, Editor (M.I.T. Press, Cambridge, Mass., 1972), pp. 143–150.
4. D. Fultz, "Rotating Flows," Film No. 21624, Encyclopedia Britannica Education Corporation, Chicago, Ill. (1973).

PROBLEMS

2.1. A puck travels a horizontal distance of 200 m on a heading due east in a time interval of 10 s at constant velocity. We wish to calculate the lateral deflection of the puck owing to the Coriolis acceleration at New York where the latitude is 40° 45′ north. Air resistance and surface friction are to be neglected.

1. Make a sketch of the Coriolis force components acting on the puck.
2. What reasonable approximation is necessary to obtain a simple solution to this problem?
3. Find the lateral deflection of the puck.
4. How does the initial compass heading enter into this problem?

2.2. On a piece of graph paper (10 × 10 to the inch) establish an $x - y$ coordinate system and connect the points (25, 20), (25, 25), (30, 25), and (30, 20) to form a rectangle. Consider the rectangle to be an object that will move at constant velocity in this inertial coordinate system. This initial position corresponds to $t = 0$.

1. If the particle velocity is $-2.0\,\mathbf{j}$ in./s, plot the position of the particle after the end of one second and at the end of a two second interval.
2. On a separate piece of tracing paper draw a rotating coordinate system, $x' - y'$, whose angular velocity Ω_0 is $+\pi/4\ \text{s}^{-1}$. At $t = 0$ the $x - y$ and $x' - y'$ coordinates coincide. Determine graphically the position of the object in the $x' - y'$ coordinates at $t = 1$ and $t = 2$ s by superimposing the rotating system over the inertial

coordinate system. Sketch the particle trajectory. Does the lateral deflection agree with the sign predicted by the Coriolis force?

3. What is the angular velocity Ω_{rel} and its sign about an axis in the object in the *rotating* coordinate system?
4. What is the relationship between Ω_0 and Ω_{rel}?
5. If the particle rotated with angular velocity Ω_{abs} in the absolute coordinate system, how do you expect Ω_0, Ω_{rel}, and Ω_{abs} would be related?

2.3. A racing car weighing 2000 lb travels at 200 mph in a straight line at Sydney, Australia, where the latitude is about 34° south. What is the magnitude and direction of the lateral Coriolis force?

2.4. Find the horizontal deflection owing to the Coriolis force when a rifle bullet travels 100 m to a target in the northerly direction. The speed of the projectile is to be considered constant equal to its average speed of 500 m/s relative to the surface of the earth. The latitude is 35° north.

2.5. An object dropped from a stationary balloon falls to earth at a terminal relative velocity of 50 m/s from an altitude of 1000 m. The latitude is 30° north. What is the magnitude and the direction of the displacement that is due solely to the Coriolis effect when the object strikes the surface of the earth?

2.6. An object undergoes approximately drag-free projectile motion in the northern hemisphere.

1. Write the equations of motion of the particle including *all* terms that result from the Coriolis and gravity forces. Use Cartesian coordinates with x in the direction of an initial easterly motion, and z in vertical direction.
2. Make a free body diagram for the projectile during the downward portion of its flight showing all force components. Use two planes if necessary to show these forces.
3. Compare the equations with Eqs. 2.6 that apply to the atmosphere and comment on the similarity and the differences between them.

2.7. Starting from the appropriate equation, formulate the Buys–Ballot rule for the southern hemisphere where the vertical component of the earth's angular velocity vector is pointed downward.

2.8. Express Eq. 2.15 in component form and by appropriate manipulation show, if f = constant, that

$$\frac{\partial u_G}{\partial x} + \frac{\partial v_G}{\partial y} = 0 \quad \text{or} \quad \nabla_h \cdot \mathbf{q}_G = 0$$

Such a flow field is said to be horizontally divergence free or horizontally nondivergent.

3

THE DYNAMICS OF VORTICITY

The presence of a rotational component in the movement of a fluid could in principle be made apparent by a small neutrally buoyant cross. If the arms of the cross always point in the same direction as it moves with the fluid then the flow is said to be irrotational. An angular movement of the cross would reveal the presence of angular motion, or vorticity, in the fluid. A curvature of the streamlines of a flow does not directly reveal the presence of vorticity. As we will see below, a fluid motion in which the streamlines are concentric circles may, under certain circumstances, be an irrotational flow. Fluid motions which are irrotational are of importance in certain branches of engineering, but atmospheric flows are usually always highly rotational with length scales, that is, the average distance to the center of rotation, that vary over a wide range. The turbulent flow near the surface of the earth has rotationality whose length scale or eddy size is of the order of 10 mm or smaller. The rotationality of large scale weather patterns has a length scale in the range of 100 to 1000 km. The motion of rotational flows is often more easily understood by keeping quantitative account of the vorticity and the mechanisms of its production, destruction, and amplification. In this chapter we introduce a mathematical formalism to describe rotational flows, and examine the dynamics of vorticity in an inertial, and, finally, in a rotating coordinate system.

3.1. THE DEFINITION OF VORTICITY

We consider, with reference to Fig. 3.1, the definition

$$\Delta\theta_z = \frac{\Delta\alpha - \Delta\beta}{2}$$

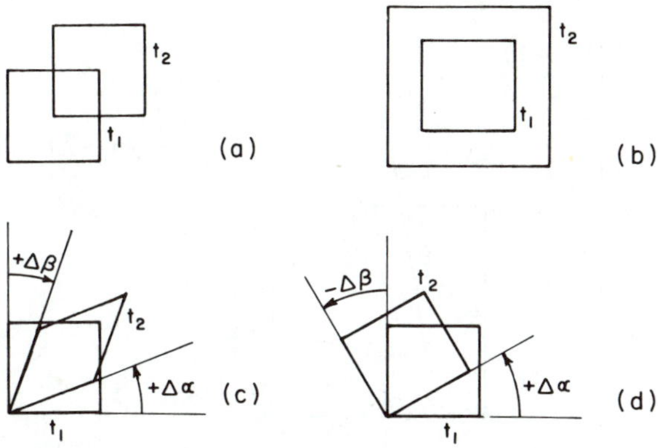

Figure 3.1. The basic displacements of a fluid element: (*a*) translation, (*b*) dilatation, (*c*) angular deformation, and (*d*) rotation. The incremental rotation of the fluid element, $(\Delta\alpha - \Delta\beta)/2$, is zero except for the last case when it is equal to $\Delta\alpha$.

for the rotation of a fluid element about an axis parallel to the z axis where the angles $\Delta\alpha$ and $\Delta\beta$ are defined positively in the directions shown in Fig. 3.1*c*. For the basic displacements of pure translation and dilation $\Delta\alpha$ and $\Delta\beta$ are both zero; for pure angular deformation $\Delta\alpha$ and $\Delta\beta$ are equal in magnitude but of opposite sign. In all three of these cases the incremental rotation of the fluid element is zero. When the fluid element undergoes a pure rotation the value of $\Delta\beta = -\Delta\alpha$ and the above definition gives $\Delta\theta_z = \Delta\alpha$. Any arbitrary displacement of a fluid element can be expressed as the sum of these four distinct basic displacements.

The angular velocity of the fluid element about an axis parallel to the z axis is defined as

$$\left(\frac{d\theta}{dt}\right)_z = \lim_{\Delta t \to 0} \frac{1}{2}\left(\frac{\Delta\alpha}{\Delta t} - \frac{\Delta\beta}{\Delta t}\right) = \frac{1}{2}\left(\frac{d\alpha}{dt} - \frac{d\beta}{dt}\right)$$

The quantity $d\alpha/dt$ is evaluated in terms of the velocity components and their derivatives by noting, with reference to Fig. 3.2,

$$\Delta\alpha = \frac{jf}{ej}$$

where

$$jf = \left[v_0 + \left(\frac{\partial v}{\partial x}\right)_0 \Delta x\right]\Delta t - v_0 \Delta t$$

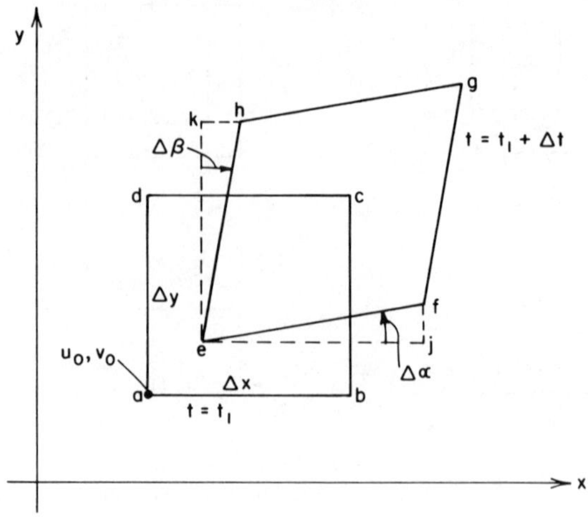

Figure 3.2. The distortion of a fluid element undergoing an arbitrary displacement.

and

$$ej = \Delta x + \left[u_0 + \left(\frac{\partial u}{\partial x}\right)_0 \Delta x\right]\Delta t - u_0 \Delta t$$

Thus we find

$$\Delta\alpha = \frac{\left(\frac{\partial v}{\partial x}\right)_0 \Delta t}{1 + \left(\frac{\partial u}{\partial x}\right)_0 \Delta t}$$

and

$$\frac{\Delta\alpha}{\Delta t} = \frac{\left(\frac{\partial v}{\partial x}\right)_0}{1 + \left(\frac{\partial u}{\partial x}\right)_0 \Delta t}$$

Therefore

$$\frac{d\alpha}{dt} = \lim_{\Delta t \to 0} \frac{\Delta\alpha}{\Delta t} = \frac{\partial v}{\partial x}$$

The quantity $d\beta/dt$ is found by repeating the argument or by noting that we may obtain this result by interchanging x and y, and u and v, respectively.

Thus,

$$\frac{d\beta}{dt} = \frac{\partial u}{\partial y}$$

and the z component of the angular velocity is found to be

$$\left(\frac{d\theta}{dt}\right)_z = \frac{1}{2}\left(\frac{\partial v}{\partial x} - \frac{\partial u}{\partial y}\right)$$

The argument can be repeated for rotations in the $y - z$ plane about the x axis and the $x - z$ plane about the y axis. The corresponding results, which apply to a fluid element undergoing arbitrary motion, are

$$\left(\frac{d\theta}{dt}\right)_x = \frac{1}{2}\left(\frac{\partial w}{\partial y} - \frac{\partial v}{\partial z}\right)$$

$$\left(\frac{d\theta}{dt}\right)_y = \frac{1}{2}\left(\frac{\partial u}{\partial z} - \frac{\partial w}{\partial x}\right)$$

Thus, the angular velocity vector of the fluid parcel at the point x, y, z is defined by

$$\left(\frac{d\boldsymbol{\theta}}{dt}\right) = \left(\frac{d\theta}{dt}\right)_x \mathbf{i} + \left(\frac{d\theta}{dt}\right)_y \mathbf{j} + \left(\frac{d\theta}{dt}\right)_z \mathbf{k}$$

where the components are given by the above relations.

In fluid dynamics it is customary to deal with the quantity that is twice the angular velocity vector. This quantity is the vorticity vector, $\boldsymbol{\omega}$, and it is defined by

$$\boldsymbol{\omega} = \nabla \times \mathbf{q} \tag{3.1}$$

Since

$$\nabla \times \mathbf{q} = \left(\frac{\partial w}{\partial y} - \frac{\partial v}{\partial z}\right)\mathbf{i} + \left(\frac{\partial u}{\partial z} - \frac{\partial w}{\partial x}\right)\mathbf{j} + \left(\frac{\partial v}{\partial x} - \frac{\partial u}{\partial y}\right)\mathbf{k} \tag{3.2}$$

it is apparent that

$$\boldsymbol{\omega} = 2\left(\frac{d\boldsymbol{\theta}}{dt}\right) \tag{3.3}$$

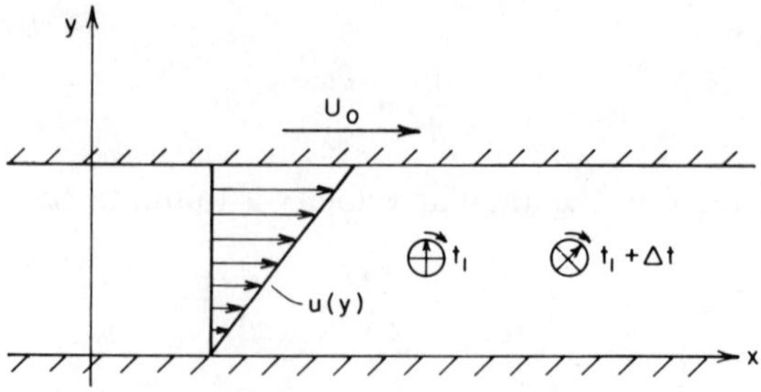

Figure 3.3. The motion of a fluid in simple couette flow. All fluid elements possess negative vorticity which is uniform at all points in the flow field.

The vorticity of a fluid at a point is then twice the local angular velocity of the fluid element. Since this definition is based on purely kinematic considerations, it applies to fluids of any type—incompressible or compressible, in steady motion or unsteady motion—and to portions of the flow field where viscosity may be neglected or must be included. A fluid for which $\boldsymbol{\omega} = 0$ is said to be irrotational and this concept is useful for certain classes of problems in engineering. The atmosphere, however, displays strong rotational effects.

The simplest example of a rotational flow is afforded by the Couette flow, see Fig. 3.3, for which

$$u = U_0 \frac{y}{h}$$

From the definition of vorticity, we find in this case

$$\boldsymbol{\omega} = -\frac{U_0}{h}\mathbf{k}$$

and hence the magnitude of the vorticity vector is a constant for all values of x and y. The negative sign indicates that the direction of rotation is negative with respect to the right-hand convention. An object floating on the surface of such a flow would undergo angular motion at the speed of $|(\mathbf{d\theta/dt})| = \frac{1}{2}|\boldsymbol{\omega}| = -U_0/2h$ as a result of the velocity gradient. Such an object can be termed "a vorticity meter" in the sense that a zero angular velocity will indicate an irrotational flow and the magnitude of the vorticity is numerically equal to twice its spin rate. Since the object floats with a fixed fluid parcel, it measures the rate of change of vorticity of a particular fluid parcel, namely the quantity $D\boldsymbol{\omega}/Dt$. We will later derive the general relationship for this quantity. Shapiro (Ref. 1) has exploited the use of the vorticity meter to provide striking demonstrations of certain rotational flows.

In the case of Couette flow the magnitude of the vorticity vector is independent of position and the physical dimensions of the vorticity meter are not important. This type of flow field is exceptional; in most cases the magnitude of the vorticity is a function of position. The vorticity meter will then indicate an average value of the vorticity field which will vary across the width of the meter. For example, in a laminar boundary layer the vorticity is highest near the wall and decreases monatomically to the outer edge of the boundary layer. To measure an accurate value of vorticity at any point it is necessary that the meter be small in comparison to the thickness of the boundary layer. (The problem is totally analogous with the measurement of a temperature field with a thermometer which is moderate in size compared with the curvature of the temperature distribution.) For a vorticity meter to read the correct value in the presence of a vorticity gradient, its dimensions must be small enough to achieve the necessary spatial resolution.

3.2. THE CONSERVATION PROPERTIES OF VORTICITY

For any vector **A**,

$$\nabla \cdot \nabla \times \mathbf{A} = 0$$

Substituting the velocity vector, **q**, for **A** we find the equation for the conservation of vorticity,

$$\nabla \cdot \boldsymbol{\omega} = 0 \tag{3.4}$$

The implications of this relationship can be explained by observing the analogies between the conservation of vorticity in a fluid (of any type) and the conservation of mass in a fluid of constant density with no mass sources or sinks. For such a fluid the conservation equation is

$$\nabla \cdot \mathbf{q} = 0 \tag{3.5}$$

We recall that a streamline is a line that is everywhere tangent to the local velocity vector. A streamtube is the volume swept out by the streamlines passing through the perimeter of a differential area normal to the direction of the flow. Analogously we define the vortex line as a line that is everywhere tangent to the local vorticity vector. A vortex tube is the volume swept out by the vortex lines passing through the perimeter of a differential area that is normal to the vorticity vector (see Fig. 3.4).

We recall the divergence theorem that relates the surface and volume integrals of a vector field by

$$\iint_S \mathbf{A} \cdot d\mathbf{S} = \iiint_V (\nabla \cdot \mathbf{A}) dV$$

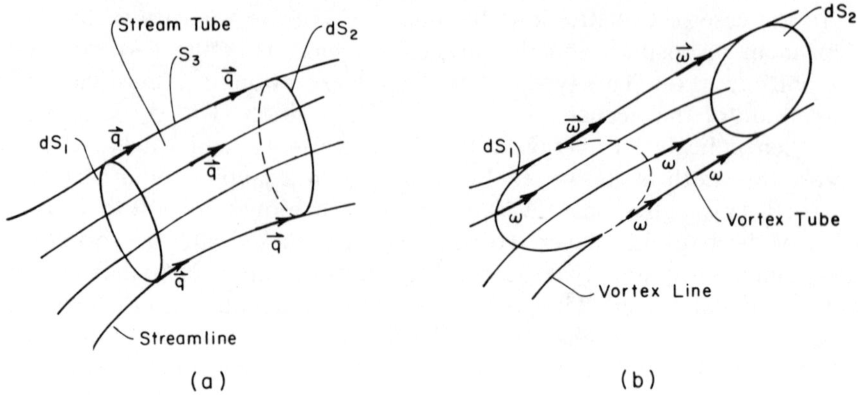

Figure 3.4. (*a*) The velocity vector, the streamline, and the stream tube, and (*b*) the vorticity vector, the vortex line, and the vortex tube.

We let $\mathbf{A} = \mathbf{q}$ and note that $\nabla \cdot \mathbf{q} = 0$ and therefore an integral form of the continuity equation is

$$\iint_S \mathbf{q} \cdot d\mathbf{S} = 0 \tag{3.6}$$

We apply this result to a portion of the streamtube indicated in Fig. 3.4 that is made up of end areas dS_1 and dS_2 and the lateral area S_3. The contribution to the integral, Eq. 3.6, is zero over the lateral area, S_3, because $\mathbf{q} \cdot d\mathbf{S}$ is zero here by definition of the streamtube. We conclude that

$$\mathbf{q}_1 \cdot d\mathbf{S}_1 = \mathbf{q}_2 \cdot d\mathbf{S}_2 = q dS = \text{const} \tag{3.7}$$

Since ρ is constant, we may state the result that the mass flow, $\rho q dS$, is a constant along the streamtube. (If we integrate this expression over the cross-sectional area of a pipe, e.g., we obtain the familiar integrated form of the continuity equation for a one-dimensional flow.) Furthermore, for a given streamtube as dS decreases, then $\mathbf{q}$ increases even to the hypothetical limiting case of $dS \to 0$ and $\mathbf{q} \to \infty$ such that their product remains constant. Thus, in the absence of sources or sinks, the streamline cannot end within the fluid but must extend to infinity in both directions, or close upon itself.

These concepts can be applied to the vorticity of a flowfield. We use the divergence theorem and the statement that $\nabla \cdot \boldsymbol{\omega} = 0$ to find

$$\iint_S \boldsymbol{\omega} \cdot d\mathbf{S} = 0 \tag{3.8}$$

The application of Eq. 3.8 to the vortex tube gives the result

$$\boldsymbol{\omega}_1 \cdot d\mathbf{S}_1 = \boldsymbol{\omega}_2 \cdot d\mathbf{S}_2 = \omega dS = \text{const} \tag{3.9}$$

and we conclude that the strength of the vortex tube ωdS is constant. According to Eq. 3.9, the vorticity vector, $\boldsymbol{\omega}$, must increase if the cross section, dS, of the vortex tube decreases, again to the limiting case of $dS \to 0$ and $\boldsymbol{\omega} \to \infty$ such that their product is constant. Consequently, a vortex tube cannot end in the fluid but must extend to infinity in both directions, or close upon itself, or extend to the boundaries of the fluid where an interaction with the surface through the influence of viscosity will occur. The familiar smoke ring is an example of a region of concentrated vorticity that closes on itself.

For the incompressible fluid the mass flow in the streamtube, $\rho \mathbf{q} dS$, is often the quantity of significance. For a finite bundle of streamtubes, the mass flux is given by an integration over all streamtubes;

$$\dot{m} = \rho \iint \mathbf{q} \cdot d\mathbf{S}$$

The mass flux is the integrated strength of a finite bundle of streamtubes.

The corresponding quantity for the vorticity field is the circulation defined by

$$\Gamma_c = \iint \boldsymbol{\omega} \cdot d\mathbf{S} \tag{3.10}$$

Because of the definition of $\boldsymbol{\omega}$ we make use of Stokes' theorem

$$\Gamma_c = \iint (\nabla \times \mathbf{q}) \cdot d\mathbf{S} = \oint \mathbf{q} \cdot d\mathbf{l} \tag{3.11}$$

This useful result states that the circulation, which may be thought of as the integrated strength of a finite bundle of vortex tubes, can be evaluated by means of the line integral $\oint \mathbf{q} \cdot d\mathbf{l}$ even without a knowledge of the distribution of vorticity over the cross sectional area S that is enclosed by the line integral. It is important to note that $\boldsymbol{\omega}$ must be defined everywhere within the closed path along which the line integral is evaluated and which surrounds the area that is pierced by the vortex lines.

To illustrate the above concepts we investigate the free vortex which has velocity components, in cylindrical coordinates

$$v_\theta = +\frac{\Gamma_0}{2\pi r} \qquad \text{and } u_r = w_z = 0 \tag{3.12}$$

The curl of the velocity in cylindrical coordinates (see Appendix I) gives the vorticity vector

$$\boldsymbol{\omega} = \nabla \times \mathbf{q} = \frac{1}{r}\frac{\partial}{\partial r}\left(\frac{\Gamma_0}{2\pi}\right)\mathbf{e}_z$$

Thus, the vorticity is found to have only a z component which is zero for $r > 0$ and indeterminant when $r = 0$,

$$\begin{aligned} \boldsymbol{\omega} = \omega_z \mathbf{e}_z &= 0 \qquad r > 0 \\ &= \frac{0}{0} \qquad r = 0 \end{aligned}$$

The flow field is therefore irrotational at all points in the field except at $r = 0$, where velocity and vorticity are undefined.

We can gain further insight into this case if we evaluate the circulation for a path that encloses the origin—path q indicated in Fig. 3.5*a*. A minor complication arises here because the vorticity is not defined at the origin and therefore Eq. 3.11 does not apply. We can easily circumvent this difficulty by using the path made of the four portions q, r, t, and s shown in Fig. 3.5*b*. Because the region enclosed by this path excludes the origin where the vorticity is undefined, Stokes' theorem is now applicable. The contribution to the line integral over the two straight segments r and s will vanish because these segments can be separated by an arbitrarily small distance. The value of $\mathbf{q} \cdot d\mathbf{l}$ for any adjacent line segments on paths r and s is numerically equal but of opposite sign and the integrated contributions over r and s vanish. Since the vorticity is everywhere zero inside the new path

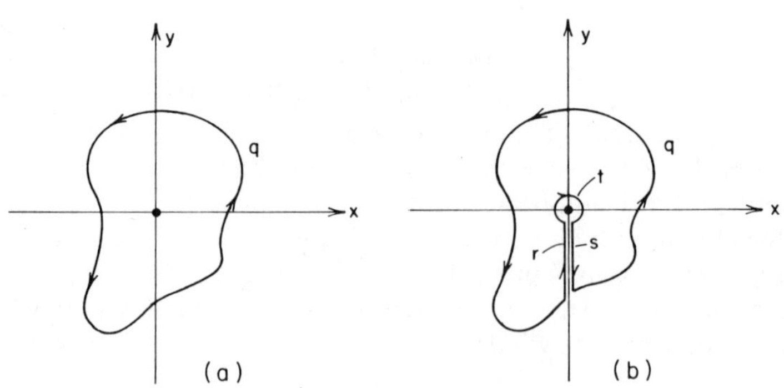

Figure 3.5. The calculation of the circulation on a path enclosing a free vortex at the origin. (*a*) Path q cannot be used because the vorticity is not defined everywhere within it. (*b*) The vorticity is defined within the area enclosed by the path q-r-t-s.

$q - r - t - s$ the circulation for this path vanishes. Hence

$$\Gamma_{cq} + \Gamma_{ct} = 0$$

where

$$\Gamma_{ct} = - \oint_t \mathbf{q} \cdot d\mathbf{l}$$

The latter integral is easily evaluated,

$$\Gamma_{ct} = - \int_0^{2\pi} \left(\frac{\Gamma_0}{2\pi r} \right) r d\theta = -\Gamma_0$$

hence

$$\Gamma_{cq} = -\Gamma_{ct} = +\Gamma_0$$

This result will apply to any path that encloses the origin including one that possesses an arbitrarily small area. The circulation does not change as the area enclosed by the path is reduced to a value approaching zero. The situation represented by the free vortex can therefore be visualized by considering it to be a vortex filament of strength $\Gamma_0 = \omega dS$ in which the cross section dS of the vortex tube has been reduced to zero while the magnitude of the vorticity increases to infinity such that their product is a constant. In mathematical language the distribution of vorticity in a free vortex is represented as the product of two Dirac functions

$$\omega_z = \Gamma_0 \delta(x)\delta(y) \tag{3.13}$$

The Dirac function $\delta(x - x_0)$ has the property that its value is zero everywhere except at $x = x_0$, but its integral over any finite value x is unity,

$$\int_0^{x_1} \delta(x - x_0) dx = 1$$

where

$$0 \lesseqgtr x_0 < x_1$$

We may evaluate the circulation represented by Eq. 3.13 by using Eq. 3.10 in the form

$$\Gamma_c = \int_{-y_1}^{y_1} \int_{-x_1}^{x_1} \omega_z dx dy \tag{3.14}$$

The area of integration is a rectangle of dimensions $2x_1$, $2y_1$ with its center at the origin. Inserting Eq. 3.13 into Eq. 3.14 and using the integral property of the Dirac function we find the circulation,

$$\Gamma_c = \Gamma_0 \int_{-y_1}^{y_1} \int_{-x_1}^{x_1} \delta(x)\delta(y) dx dy = \Gamma_0$$

which is in agreement with the result obtained above. The distribution of vorticity represented by a Dirac function is of little practical utility unless the circulation resulting from the vortex tube is the only quantity of interest.

The central portions of atmospheric distributions of vorticity are roughly approximated by a solid body rotation for which

$$v_\theta = \begin{cases} \Omega_0 r & r \le r_0 \\ 0 & r > r_0 \end{cases}$$

$$u_r = 0$$

and

$$w_z = 0 \tag{3.15}$$

Here Ω_0 is the angular speed of the solid body rotation. The vorticity is then found to be

$$\begin{aligned} \boldsymbol{\omega} = \frac{1}{r}\frac{\partial}{\partial r}(\Omega_0 r^2)\mathbf{e}_z &= 2\Omega_0 \mathbf{e}_z && r \le r_0 \\ &= 0 && r > r_0 \end{aligned}$$

Thus, the vorticity has only a vertical component which is numerically twice the angular spin rate in agreement with Eq. 3.3. The circulation along the path $r = r_0$ of the bundle of vortex tubes represented by Eq. 3.15 is equal to $2\Omega_0 \pi r_0^2$.

The solid body rotation over a finite circular region as described above provides a discontinuous change in vorticity at $r = r_0$ which is not representative of vorticity distributions that appear in nature. A useful function that represents a central solid body rotation with a smooth decay to zero vorticity is given by

$$v_\theta = \frac{\Gamma_0}{2\pi r}\left[1 - \exp\left(-\frac{r^2}{r_0^2}\right)\right] \tag{3.16}$$

$$u_r = 0$$

and

$$w_z = 0$$

This function describes a vorticity distribution that in the central region is a solid body rotation and which decays smoothly with radius (see Problem 3.3). The vorticity distributions represented by Eqs. 3.12, 3.15, and 3.16 are shown in Fig. 3.6.

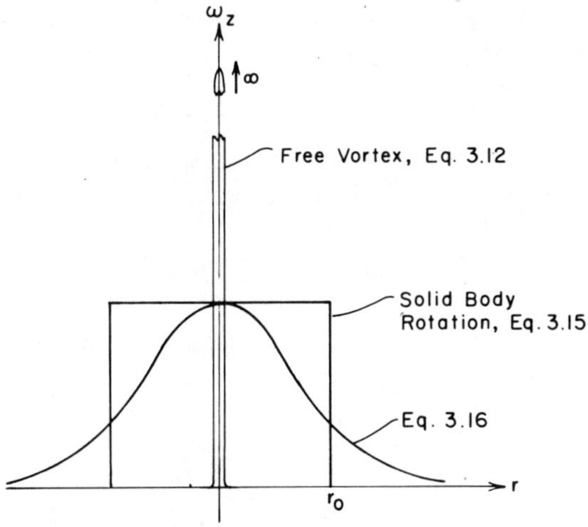

Figure 3.6. Three types of vorticity distribution. The equation numbers refer to the velocity fields given in the text.

3.3. THE DYNAMICS OF VORTICITY

A useful relationship for the rate of change of vorticity of a particular parcel of fluid as it is carried through the flow field, that is $D\boldsymbol{\omega}/Dt$, is obtained by applying the curl operator to the Euler equation. We will consider the case of a flow in an inertial coordinate system and therefore we use Eq. 2.4 in the form

$$\frac{\partial \mathbf{q}}{\partial t} + (\mathbf{q} \cdot \nabla)\mathbf{q} = -\frac{1}{\rho} \nabla p - g\nabla\phi \tag{3.17}$$

Here $g\nabla\phi$ represents any conservative body force such as gravity or a centrifugal force; any nonconservative force such as the Coriolis force is not yet considered. Applying the curl operation to the first term, we obtain

$$\nabla \times \frac{\partial \mathbf{q}}{\partial t} = \frac{\partial}{\partial t}(\nabla \times \mathbf{q}) = \frac{\partial \boldsymbol{\omega}}{\partial t}$$

The second term is expanded as

$$(\mathbf{q} \cdot \nabla)\mathbf{q} = \nabla\left(\frac{q^2}{2}\right) - \mathbf{q} \times (\nabla \times \mathbf{q})$$

and when the curl of both sides is obtained we find

$$\nabla \times [(\mathbf{q} \cdot \nabla)\mathbf{q}] = -\nabla \times (\mathbf{q} \times \boldsymbol{\omega})$$

because the curl of the gradient of any scalar vanishes. The curl of the cross product is conveniently expressed as

$$\nabla \times (\mathbf{q} \times \boldsymbol{\omega}) = (\boldsymbol{\omega} \cdot \nabla)\mathbf{q} + \boldsymbol{\omega}(\nabla \cdot \mathbf{q}) - (\mathbf{q} \cdot \nabla)\boldsymbol{\omega}$$

where we have used the vorticity conservation expression $\nabla \cdot \boldsymbol{\omega} = 0$ to eliminate one term. The curl of the third term in Eq. 3.16 gives

$$\nabla \times \frac{1}{\rho} \nabla p = -\frac{1}{\rho^2} \nabla \rho \times \nabla p$$

plus a term involving $\nabla \times \nabla \rho$ which vanishes. Finally the curl of the last term also vanishes and recognizing

$$\frac{\partial \boldsymbol{\omega}}{\partial t} + (\mathbf{q} \cdot \nabla)\boldsymbol{\omega} = \frac{D\boldsymbol{\omega}}{Dt}$$

we obtain

$$\frac{D\boldsymbol{\omega}}{Dt} = (\boldsymbol{\omega} \cdot \nabla)\mathbf{q} - \boldsymbol{\omega}(\nabla \cdot \mathbf{q}) + \frac{1}{\rho^2} \nabla \rho \times \nabla p \qquad (3.18)$$

This is the Helmholtz equation which gives the rate of change of vorticity of the fluid particle as it moves through the flow field.

We also obtain a relationship for the rate of change of circulation along a path that moves with the fluid particles with which it was initially associated. This quantity is $D\Gamma_c/Dt$ and it can be found from Eq. 3.11 as

$$\frac{D\Gamma_c}{Dt} = \frac{D}{Dt} \oint \mathbf{q} \cdot d\mathbf{l}$$

The particle derivative of the line integral along a path moving with the fluid particles is the line integral of the particle derivative

$$\frac{D\Gamma_c}{Dt} = \oint \frac{D}{Dt}(\mathbf{q} \cdot d\mathbf{l}) = \oint \frac{D\mathbf{q}}{Dt} \cdot d\mathbf{l}$$

because

$$\oint \mathbf{q} \cdot \frac{D(d\mathbf{l})}{Dt} = \oint \mathbf{q} \cdot d\mathbf{q} = \oint d\left(\frac{q^2}{2}\right) = 0$$

Using Euler's equation we find

$$\frac{D\Gamma_c}{Dt} = -\oint \frac{\nabla p}{\rho} \cdot d\mathbf{l}$$

where we have used the property that the line integral around the closed path of the gradient of a scalar is zero. By Stokes' theorem, Eq. 3.11, we express the above as

$$\frac{D\Gamma_c}{Dt} = + \oiint \left[\frac{1}{\rho^2} \nabla\rho \times \nabla p\right] \cdot d\mathbf{S} \tag{3.19}$$

The rotational fluid motions in which viscosity* may be neglected can be described by the above two important Eqs. 3.18 and 3.19, which we now proceed to interpret in detail.

The term $\nabla\rho \times \nabla p$ in Eqs. 3.18 and 3.19 clearly represents the turning moment that results in a baroclinic fluid. For example, if at an instant $t = t_1$, $\boldsymbol{\omega} = 0$ everywhere in a flowfield and thus $\Gamma = 0$ for any closed path within the flowfield, then the baroclinic fluid ($p \neq p(\rho)$, hence $\nabla\rho \times \nabla p \neq 0$) will experience a turning moment because of the nonalignment of the isoteric and isobaric surfaces as discussed previously (Section 2.4.1). Both $D\boldsymbol{\omega}/Dt$ at each point and $D\Gamma/Dt$ for any closed path moving with the fluid will be nonzero at the instant $t = t_1$ even though $\boldsymbol{\omega}$ and Γ vanish at that instant. Thus, the term $\nabla\rho \times \nabla p$ accounts for the production of both vorticity and circulation.

If the fluid is barotropic then Eqs. 3.18 and 3.19 reduce to

$$\frac{D\boldsymbol{\omega}}{Dt} = (\boldsymbol{\omega} \cdot \nabla)\mathbf{q} - \boldsymbol{\omega}(\nabla \cdot \mathbf{q}) \tag{3.20}$$

and

$$\frac{D\Gamma_c}{Dt} = 0 \tag{3.21}$$

and we find that the circulation of any closed path is constant while the magnitude of the vorticity vector may change according to the terms on the right side of Eq. 3.20. We may interpret the changes in $\boldsymbol{\omega}$ as resulting from changes in the geometry of the fluid elements making up the vortex tubes. For example, since $-\nabla \cdot \mathbf{q} = +1/\rho(D\rho/Dt)$, the term $-\boldsymbol{\omega}(\nabla \cdot \mathbf{q})$ indicates that the $D\boldsymbol{\omega}/Dt > 0$ if the fluid parcel is compressed. Thus, the $-\boldsymbol{\omega}(\nabla \cdot \mathbf{q})$ term may be interpreted as an amplification owing to the compression of the fluid elements.

* If the viscosity of the fluid is considered, additional terms would appear in Eqs. 3.18 and 3.19 that would allow for the diffusion of vorticity and the decay of circulation. These terms become important in atmospheric flows mainly for the smallest scale vorticity associated with a turbulent flow field.

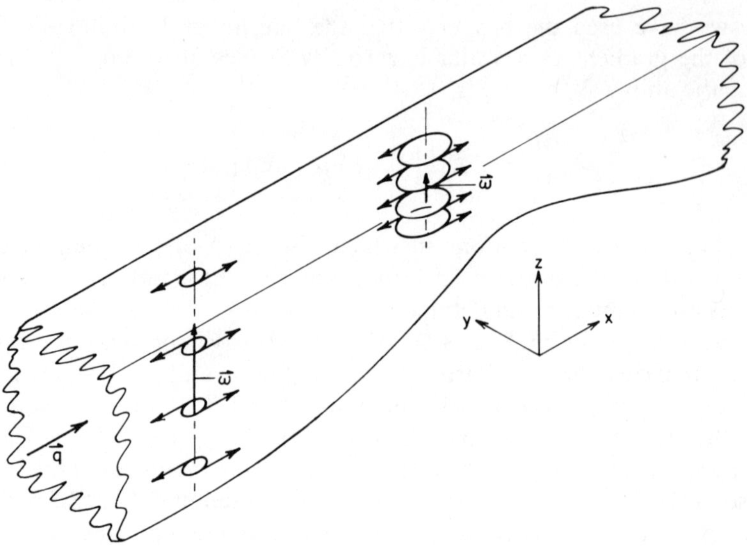

Figure 3.7. A channel flow experiment. Vertical vortex lines that are compressed as a result of the contour of the channel bottom cause weakened vorticity at the point of minimum depth.

If the fluid is incompressible, the Helmholtz equation reduces to

$$\frac{D\boldsymbol{\omega}}{Dt} = (\boldsymbol{\omega} \cdot \nabla)\mathbf{q} \tag{3.22}$$

We are then left with the remaining term $(\boldsymbol{\omega} \cdot \nabla)\mathbf{q}$, which may be interpreted by considering the experiment depicted in Fig. 3.7. A rectangular trough has a hump in the bottom over which water flows at a subcritical velocity.* Vorticity introduced by a paddle so that the axis of rotation, hence the direction of the vortex filaments and vortex tubes, is along the vertical z axis. No vorticity component is added with orientation of the x and y axes. Thus, we wish to specialize Eq. 3.22 to the case where $\boldsymbol{\omega} = \omega_z \mathbf{k}$ and we find

$$\frac{D\omega_z}{Dt} = \omega_z \frac{\partial w}{\partial z}$$

Along any portion of the duct where its lower surface is flat, $w = 0$ and $\partial w/\partial z = 0$, hence $D\omega_z/Dz = 0$ and $\omega_z =$ a constant for any given fluid parcel. A fluid which has no vorticity will gain none; a fluid element with

* That is, the flow velocity is less than the shallow water wave speed $= \sqrt{gh}$ where h is the depth of the water.

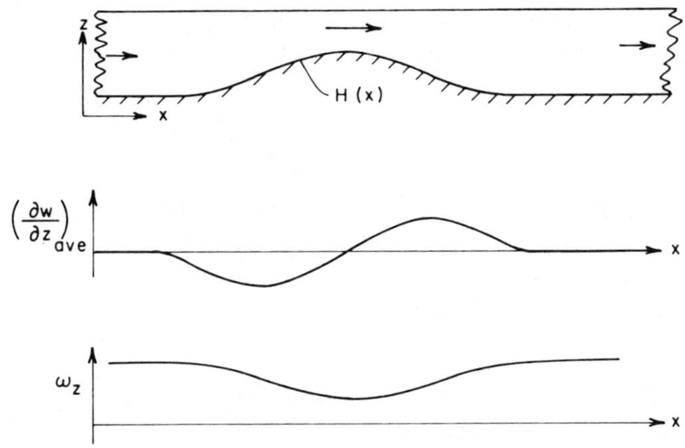

Figure 3.8. The variation of $(\partial w/\partial z)_{av}$ and ω_z in the channel flow experiment depicted in Fig. 3.7 (positive initial vorticity). The averaging of the vertical derivative is performed over depth.

vorticity will neither gain nor lose. As the ramp in the bottom of the channel is encountered, a finite value of vertical velocity, w, is developed but its magnitude decreases as z increases since the level of the free surface changes only slightly. Thus $\partial w/\partial z$, if averaged over height at any one cross section, is negative during the approach to the shallow portion of the trough. If ω_z is initially positive, as indicated in Fig. 3.8, then $D\omega_z/Dt$ is negative and ω_z is reduced as the vortex lines and vortex tubes are reduced in length by the passage of the fluid into the more shallow section of the channel. If ω_z is initially negative then $D\omega_z/Dt$ is positive and thus ω_z is reduced regardless of its initial sign. As the fluid continues its passage along the duct from the shallow portion to the original depth, the value of w is a negative quantity at the bottom of the channel and $\partial w/\partial z$ is generally positive. Vorticity of either sign is intensified by the stretching of the vortex tubes and lines that is imposed by the flow field.

Thus, the term $(\boldsymbol{\omega} \cdot \nabla)\mathbf{q}$ in the Helmholtz equation represents an intensification that results when the vortex tubes are stretched. In the experiment described above, the rate of change of circulation along a path surrounding the vorticity concentration is zero by Eq. 3.21. Therefore, the cross section, S_c, of the vorticity concentration clearly changes in the opposite manner to ω_z such that $\omega_z S_c = \Gamma_c =$ constant throughout the entire flow.

This experiment and other interesting attributes of vorticity are skilfully presented in the film by Shapiro (Ref. 1) who uses a vorticity meter of the type described above to demonstrate the variation of $\boldsymbol{\omega}$ and $D\boldsymbol{\omega}/Dt$ of a vertical vorticity distribution in a channel of variable depth.

Finally, we note that in the case of a barotropic fluid which is initially irrotational, $\boldsymbol{\omega} = 0$ at $t = 0$, Eqs. 3.18 and 3.19 reduce to

$$\frac{D\boldsymbol{\omega}}{Dt} = 0 \quad \text{and} \quad \frac{D\Gamma_c}{Dt} = 0$$

In this case the flow remains irrotational, $\boldsymbol{\omega} = 0$ at all points in the flow field, and $\Gamma = 0$ and remains zero for any closed path moving with the fluid particles. Some of the properties of vorticity and circulation that have been described above make up a set of theorems known as Helmholtz vortex theorems:

1. The strength of a vortex tube is constant along its length.
2. The vortex tube cannot end in a fluid but must extend to the boundaries or close upon itself.
3. If only conservative body forces are present, a barotropic fluid that is initially irrotational will remain irrotational.

The first two of these theorems follow from the conservation properties of vorticity. The third theorem follows from Eq. 3.18 when $p = p(\rho)$ or $\rho =$ constant.

3.4. THE DYNAMICS OF VORTICITY IN A ROTATING COORDINATE SYSTEM

The discussion of vorticity up to this point has presumed that the velocity is measured in an inertial coordinate system. In the atmosphere it is the vorticity relative to the rotating earth that is of interest. From Chapter 2 we recall that

$$\mathbf{q}_{\text{abs}} = \mathbf{q}_{\text{rel}} + \boldsymbol{\Omega} \times \mathbf{r}$$

and to obtain the relationship between absolute and relative vorticity we apply the curl operator, $\nabla \times$, to the above. By the appropriate vector identity, we find

$$\nabla \times \boldsymbol{\Omega} \times \mathbf{r} = 2\boldsymbol{\Omega}$$

and therefore

$$\nabla \times \mathbf{q}_{\text{abs}} = \nabla \times \mathbf{q}_{\text{rel}} + 2\boldsymbol{\Omega}$$

or

$$\boldsymbol{\omega}_{\text{abs}} = \boldsymbol{\omega}_{\text{rel}} + 2\boldsymbol{\Omega}$$

as the simple relationship between relative and absolute vorticity. Since $\nabla \cdot \boldsymbol{\omega}_{\text{abs}} = 0$, then $\nabla \cdot \boldsymbol{\omega}_{\text{rel}} = 0$, and the conservation properties enunciated above for absolute vorticity will also apply to relative vorticity.

We seek the analogous relation to the Helmholtz equation, Eq. 3.18, for the dynamics of the relative vorticity as it moves with the fluid. Again we restrict our attention to the cases where the viscosity of the fluid may be neglected such as is the case for moderate or large scale atmospheric events. Substituting the above expression for absolute vorticity into Eq. 3.18, we find

$$\frac{D(\boldsymbol{\omega}_{\text{rel}} + 2\boldsymbol{\Omega})}{Dt} = [(\boldsymbol{\omega}_{\text{rel}} + 2\boldsymbol{\Omega})\cdot \nabla]\mathbf{q} - (\boldsymbol{\omega}_{\text{rel}} + 2\boldsymbol{\Omega})(\nabla \cdot \mathbf{q}) + \frac{1}{\rho^2}\nabla\rho \times \nabla p \tag{3.23}$$

An alternate procedure would be to use Euler's equations in the form applicable to the rotating coordinate system in which the nonconservative Coriolis force is included and to proceed as in deriving Eq. 3.18 from the beginning. This method, which will yield the same result as Eq. 3.23 after considerable manipulation, may be applied to obtain the particle derivative of the circulation on a closed contour. The result in this case is

$$\frac{D\Gamma_c}{Dt} = +\iint \left(\frac{1}{\rho^2}\nabla\rho \times \nabla p\right)\cdot d\mathbf{s} - \iint [\nabla \times (2\boldsymbol{\Omega} \times \mathbf{q})]\cdot d\mathbf{s} \tag{3.24}$$

From this we realize that the presence of the Coriolis pseudoforce can result in the production of vorticity. Nonconservative forces, of which the Coriolis force is an example, always produce vorticity and for this reason are also called rotational forces.

This information is contained in Eq. 3.23 as well. To show the ability of the angular velocity of the coordinate system to produce relative vorticity, we specialize Eq. 3.23 to apply to a homogeneous or constant-density fluid for which $\nabla \cdot \mathbf{q} = 0$ and $\nabla\rho \times \nabla p = 0$. We will also require that the initial relative vorticity is zero and the value of $\boldsymbol{\Omega}$ is constant throughout space so that $D\boldsymbol{\Omega}/Dt = 0$. Equation 3.23 simplifies to

$$\frac{D\boldsymbol{\omega}_{\text{rel}}}{Dt} = (2\boldsymbol{\Omega}\cdot \nabla)\mathbf{q} \qquad \text{at } t = 0$$

The vertical component of relative vorticity, when only a vertical component of angular velocity is present, is

$$\frac{D\omega_{z,\,\text{rel}}}{Dt} = 2\Omega_z \frac{\partial w}{\partial z} \qquad \text{at } t = 0 \tag{3.25}$$

Thus, we find that relative vorticity is produced, even in the absence of an initial distribution of vorticity, by the interaction of the angular velocity and the strain rate terms. In a stretching motion where $\partial w/\partial z > 0$, a positive rotation will result in the generation of positive relative vorticity which, when present, will also interact with a stretching type strain rate term to further amplify itself. Thus, in the nearsighted view of the rotating coordinate system, the rotational velocity of the coordinate system interacts with the strain rate terms to generate relative vorticity.

The above discussion implies that a channel flow experiment of the type described above, Fig. 3.7, would develop relative vorticity as a result of the earth's rotation when no vorticity is initially present. However, the magnitude of $2\Omega_z = f \approx 10^{-4}\ \mathrm{s}^{-1}$ is too small to permit this effect to be observed in this type of laboratory experiment. On the other hand, Shapiro has demonstrated (Refs. 1 and 2) this effect over a period of one-half hour in a different type of experiment. A large diameter, shallow tank is filled with water which is allowed to come to a state of rest. A plug is removed from the bottom center of the tank and a vorticity meter (see discussion, p. 44) is placed on the surface of the water directly over the outlet. The stretching term $\partial w/\partial z$ is finite and positive at this point because $w \approx 0$ at the surface but is finite and negative at the outlet. Equation 3.25 indicates that positive relative vorticity is generated through the interaction of the strain rate term and the angular velocity of the frame of reference.

3.5. VORTICITY IN LARGE SCALE ATMOSPHERIC MOTIONS

In large scale atmospheric events the vertical component of vorticity is of primary interest. Using the customary notation of ζ for the vertical component of relative vorticity and noting that $2\Omega_z = f$, we express the z component of Eq. 3.23 as

$$\frac{D(\zeta + f)}{Dt} = \omega_x \frac{\partial w}{\partial x} + (\omega_y + 2\Omega_y)\frac{\partial w}{\partial y} - (\zeta + f)\left(\frac{\partial u}{\partial x} + \frac{\partial v}{\partial y}\right) + \frac{1}{\rho^2}(\nabla\rho \times \nabla p)$$

In the present context the Coriolis parameter, f, is referred to as the "earth's vorticity." The terms involving the horizontal components of vorticity (ω_x, ω_y, and $2\Omega_y$) in the above equation represent the conversion of horizontal to vertical vorticity and these terms are generally small. If, in addition, we assume that the atmosphere may be treated as a barotropic fluid, the above equation reduces to

$$\frac{D(\zeta + f)}{Dt} = -(\zeta + f)\left(\frac{\partial u}{\partial x} + \frac{\partial v}{\partial y}\right) \tag{3.26}$$

The quantity $\partial u/\partial x + \partial v/\partial y$ is called the horizontal divergence of the fluid. This term is obviously zero if u and v are independent of x and y and the horizontal cross section and shape of the fluid element remain unchanged as it moves with the flow field. The horizontal divergence will also be zero if the rate of change of u in the x direction is equal and of opposite sign to the rate of change of v in the y direction. In this case a fluid element momentarily possessing, for example, a square cross section is distorted into a parallelogram while the horizontal cross section remains unchanged. Finally, if u increases with increasing x and v increases with increasing y, the horizontal cross section of the fluid parcel moving in the direction of increasing x and y will also increase. Thus, the horizontal divergence term (units of s^{-1}) represents the fractional rate of change of the horizontal cross section of the fluid element; it is positive when the cross section of the fluid element increases as it moves through the flow field and negative when it decreases.

Equation 3.26 therefore can be interpreted in terms of the experiment involving the dynamical vertical vorticity in a channel of variable depth (Fig. 3.7). As the fluid is on the approach portion of the ramp, the horizontal divergence is positive and the term of the right side of Eq. 3.26 is negative if the absolute vertical vorticity, $\zeta + f$, is positive. Therefore, $D(\zeta + f)/Dt < 0$ and the magnitude of the absolute vorticity decreases. If in the laboratory experiment the value of $f \ll \zeta$, then the relative vorticity and absolute vorticity are essentially the same. In the atmosphere, the magnitude of the absolute vertical vorticity in the large scale wind systems ranges from about $4 \times 10^{-5}\ s^{-1}$ to $20 \times 10^{-5}\ s^{-1}$. Since $f \approx 10 \times 10^{-5}\ s^{-1}$ in midlatitude regions, the relative vorticity ranges from about -6×10^{-5} to $+10 \times 10^{-5}$ s^{-1}. Of course, much higher magnitudes of vertical vorticity are present in smaller scale intense phenomena such as tornadoes.

Equation 3.26 has considerable importance in the history of numerical weather forecasting because a variant of this equation was used in the calculations reported in 1950 (Ref. 3) which provided great stimulus for a more intensive effort in this area. The object of numerical weather forecasting is to calculate the future weather by using the observations of the present weather as an initial condition. The calculation is to be accomplished by the application of those equations that are necessary to describe the controlling physical events; it takes advantage of the effectiveness of modern large scale digital computers to handle complex systems of equations and prodigious amounts of data. The calculations of Charney, Fjörtoft, and Neumann in 1950 were based on the barotropic assumption applied to an atmospheric model made up of a single level, or vertical increment, which was intended to approximate the actual atmosphere at the 500 mb level. Multilevel barotropic models remain in service as of this writing primarily as a supplement for the more detailed calculations that are widely used to assist the local weather forecasters.

The calculation by Charney and his colleagues was not the first attempt to predict the weather numerically. An earlier calculation was performed by L. F. Richardson in 1916 to 1918 and reported in his book of 1922 (Ref. 4). Although this calculation was a failure from the standpoint of accuracy of the prediction, it is looked back on as a very advanced attempt. The details of this calculation possess some similarity to the procedures that are used in the contemporary multilevel models that use a set of relations known as the primitive equations. The reason for the failure of Richardson's calculation therefore has been a recent preoccupation of atmospheric scientists. The interesting review by Platzman (Ref. 5) on this subject is detailed on both the description of Richardson's remarkable calculation and on the biographical background as well.

REFERENCES

1. A. H. Shapiro, "Vorticity," 16 mm film, National Committee for Fluid Mechanics Films, distributed by Encyclopedia Brittanica Educational Corp., Chicago, Ill.
2. A. H. Shapiro, Editor, *Illustrated Experiments in Fluid Mechanics* (M.I.T. Press, Cambridge, Mass., 1972).
3. J. G. Charney, R. Fjörtoft, and J. von Neumann, "Numerical Integration of the Barotropic Equation," *Tellus*, **2**, 237–254 (1950).
4. L. F. Richardson, *Weather Prediction by Numerical Process* (reprinted by Dover Publications, Inc., New York, 1965).
5. G. W. Platzman, "A Retrospective View of Richardson's Book on Weather Prediction," *Bull. Amer. Meteorol. Soc.*, **48**, 514–550 (1967).

PROBLEMS

3.1. Describe the vorticity vector, direction and the magnitude, in the following flow fields:

1. The flow of a fluid with kinematic viscosity, ν, down an inclined plane of slope, S, for which

$$u = \frac{gS}{\nu} y\left(L - \frac{y}{2}\right)$$

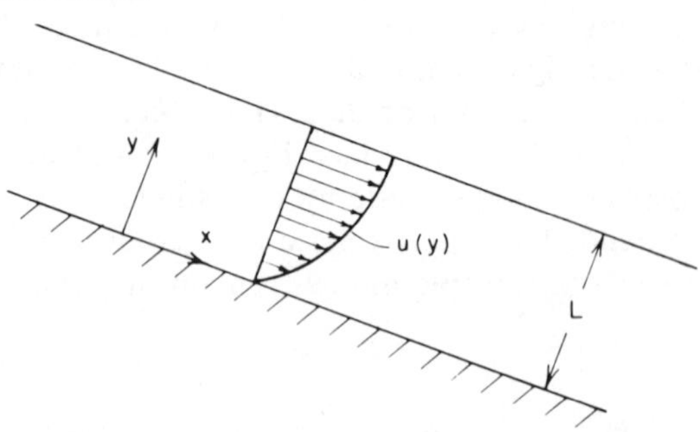

2. The stagnation point flow of a fluid approaching a solid surface located on the $x - y$ plane from the $+z$ direction,

$$u = 2Kx, \qquad v = 2Ky, \qquad w = -4Kz$$

3. An aggregate of fluid undergoing a solid body rotation,

$$v_\theta = \omega_0 r$$
$$u_r = 0$$
$$w_z = 0$$

4. A free vortex in a fluid given by

$$v_\theta = \frac{\Gamma}{2\pi r}$$
$$u_r = 0$$
$$w_z = 0$$

Recall that, in cylindrical coordinates,

$$\nabla \times \mathbf{q} = \left(\frac{1}{r}\frac{\partial w_z}{\partial \theta} - \frac{\partial v_\theta}{\partial z}\right)\mathbf{e}_r + \left(\frac{\partial u_r}{\partial z} - \frac{\partial w_z}{\partial r}\right)\mathbf{e}_\theta + \left(\frac{1}{r}\frac{\partial}{\partial r}(rv_\theta) - \frac{1}{r}\frac{\partial u_r}{\partial \theta}\right)\mathbf{e}_z$$

3.2. A crude model of a tropical hurricane, in coordinates fixed at the eye of the hurricane, consists of the combination of a solid body rotation for the central core and a surrounding free vortex. That is,

$$v_\theta = \omega_c r \qquad \text{for } r < R_c$$

and

$$v_\theta = \frac{\Gamma_0}{2\pi r} \qquad \text{for } r > R_c$$

Recall that for a solid body rotation, $v_\theta^2/r = (1/\rho)(dp/dr)$

1. Sketch the velocity versus distance from the eye of the storm.
2. Sketch the vorticity and circulation versus distance from the eye.
3. Calculate the relationship between pressure at the eye and its value far away ($p = 1$ atm far from the eye where the velocity is zero).

4. If the velocity at the edge of the core is 100 mph, what is the barometric pressure at the center of the storm? Assume the density is constant equal to 1.22 kg/m^3.

3.3. The following algebraic expression represents the velocity distribution in a cyclonic pressure system:

$$v_\theta = +\frac{\Gamma_0}{2\pi r}\left[1 - \exp\left(-\frac{r^2}{r_0^2}\right)\right]$$

1. Show that for $r/r_0 \ll 1$, the motion reduces to a solid body rotation, $v_\theta = \omega_0 r$. Relate ω_0 to the parameters Γ_0 and r_0. (Recall $e^{-x} \simeq 1 - x$ if $x \ll 1$.)
2. Determine the magnitude of the relative vorticity, ζ, as a function of radius.
3. Determine the circulation $\Gamma = \oint \mathbf{q} \cdot d\mathbf{s}$ with a circular path of radius r.
4. Sketch v_θ, ζ, and Γ as a function of radius.

3.4 Solar heating of coastal land surfaces during the summer months can cause a horizontal density gradient. Thus, while isobars tend to remain horizontal, isoteres (lines of constant density) slope upward in the seaward direction.

1. What is the direction of the vorticity developed? (Show sketch.)
2. At sea level will the direction of the wind be onshore or offshore? (Simplify the appropriate equation by assuming there is initially no vorticity. Neglect earth's rotation.)

3.5. Water flows through an initially straight channel into a diverging section with negligible change in height. A concentrated distribution of vertical vorticity is introduced upstream.

1. Specialize the appropriate equation to apply to this flow field.
2. How does the magnitude of the vortex change as it passes into the wider section?

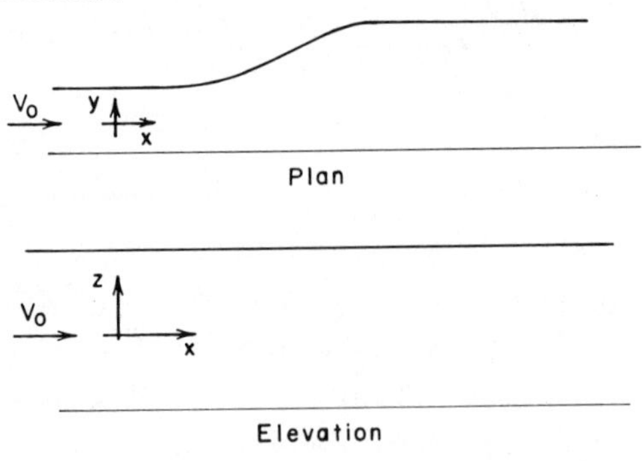

4

LARGE SCALE WIND-PRESSURE SYSTEMS

The large scale atmospheric motion results from the unequal heating of the earth by the sun. The average angle of incidence of the sun's rays is perpendicular to the earth's surface at the equator and nearly parallel to the surface at the polar regions. This differential heating causes a net gain of radiation at the equator that balances the loss of radiation at the poles. The atmosphere transports this energy from the equator to the poles by means of the large scale circulation that determines our local weather patterns. At the surface of the earth the large scale flow patterns are commonly described in terms of high pressure regions, low pressure regions, and frontal zones that are mainly associated with the latter. The large scale wind-pressure systems which move across the earth's surface also possess an individual life cycle during which their strength initially intensifies, then achieves a mature or arrested state of development, and finally decays. The science of meteorology deals with the motion and history of these large scale weather systems in detail. Our purpose here is to compare the basic dynamics of the large scale wind-pressure systems and to consider their influence on the ability of the atmosphere to disperse pollutants.

The geostrophic wind, which is given by

$$\mathbf{q}_G = \frac{1}{f\rho}\mathbf{k} \times \nabla_h p$$

was earlier described as the wind that results when the pressure and Coriolis force act in near equilibrium on a fluid parcel which experiences no acceleration. Actually the isobars are usually curved and a lateral component of acceleration is required to produce a departure from rectilinear motion. In the following discussion we will consider the pressure distribution to possess circular symmetry and the isobars, therefore, to be concentric circles. We will assume that the isobars coincide with the streamlines as in geostrophic flow. Therefore, a component of acceleration toward the center of the pressure system must be present. The equation of motion is then found by equating centripetal acceleration to the sum of the pressure and Coriolis forces. In vector form we have

$$-\frac{q_{gr}^2}{r}\mathbf{e}_r = -\frac{1}{\rho}\frac{\partial p}{\partial r}\mathbf{e}_r - \mathbf{f} \times \mathbf{q}_{gr} \tag{4.1}$$

where $\mathbf{e}_r$ is a unit vector directed radially away from the axis of rotation and q_{gr} designates the gradient wind which results under these circumstances. Since the convective terms are absent in the equation, the flow has a low Rossby number even though the fluid is accelerated centripetally. This equation applies to both the cyclone and the anticyclone according to whether $\mathbf{q}_{gr}$, which possesses only an azimuthal component because the streamlines are parallel to the isobars, is positive or negative. We will deal with these two cases separately.

4.1. THE LOW PRESSURE (CYCLONIC) SYSTEM

Suppose the wind direction is counterclockwise about the center of the pressure system as depicted in Fig. 4.1. Although the atmospheric isobars are never circles as shown in Fig. 4.1, the discussion we give below will

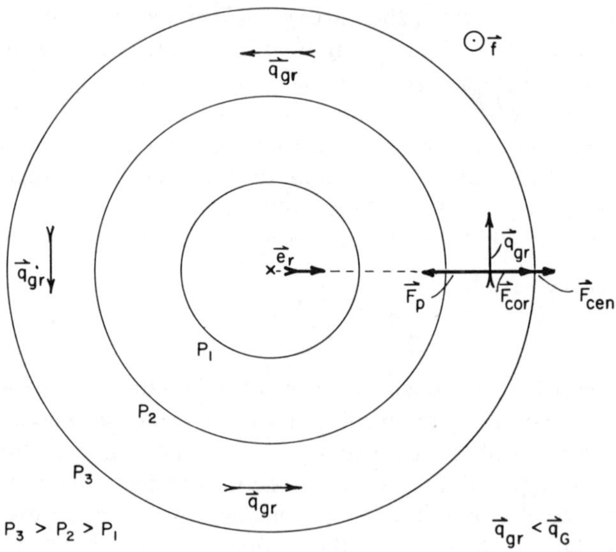

Figure 4.1. The force balance in the cyclonic system (northern hemisphere).

apply in practice if the local radius of curvature of the actual isobar is used. In the northern hemisphere **f** points vertically upward and the Coriolis force will be directed radially outward and opposite to the pressure force. Since the pressure force is directed inward and pressure gradient is directed positively outward, the low pressure region is at the center of the system. The inward centripetal acceleration can be recast as a centrifugal pseudo-force directed outward as depicted in Fig. 4.1. In this language the force balance on the fluid particle in scalar form is

$$\frac{1}{\rho}\frac{\partial p}{\partial r} = f q_{\text{gr}} + \frac{q_{\text{gr}}^2}{r} \tag{4.2}$$

This wind-pressure system is designated as a cyclonic system because the Coriolis and centrifugal forces reinforce each other. The direction of rotation is positive and possesses the same sense as the earth's rotation. The relative vorticity is clearly positive and adds to the vorticity owing to the earth's rotation to yield a relatively large positive absolute vorticity ($\zeta + f \approx 12$ to $20 \times 10^{-5}\ \text{s}^{-1}$).

We solve Eq. 4.2 to obtain the explicit relationship for the gradient wind,

$$q_{\text{gr}} = \sqrt{\left(\frac{fr}{2}\right)^2 + \frac{r}{\rho}\frac{\partial p}{\partial r}} - \frac{fr}{2} \tag{4.3}$$

We define

$$V_{\max} = \frac{fr}{2}$$

and note that the magnitude of the geostrophic wind (as $q_{\text{gr}}^2/r \to 0$ in Eq. 4.2 at large radial distances) is given by

$$q_G = \frac{1}{f\rho}\frac{\partial p}{\partial r}$$

Equation 4.3 can then be recast as

$$q_{\text{gr}} = V_{\max}\left(\sqrt{1 + \frac{2q_G}{V_{\max}}} - 1\right) \tag{4.4}$$

Equation 4.4 may be simplified when $q_G \ll V_{\max}$ either because the pressure gradient is weak or because the radius of curvature of the isobar is large. Using three terms in the binomial expansion of the radical in Eq. 4.4, we find

$$q_{\text{gr}} \approx q_G\left(1 - \frac{1}{2}\frac{q_G}{V_{\max}}\right) \qquad \text{if } q_G \ll V_{\max} \tag{4.5}$$

4.2. THE HIGH PRESSURE (ANTICYCLONIC) SYSTEM

If the wind direction is clockwise the Coriolis force will be pointed radially inward in the northern hemisphere. The pressure force again will oppose and therefore the high pressure region is at the center of the system. The centrifugal force opposes the Coriolis force and the pressure system is designated as an anticyclone. The scalar force balance (Fig. 4.2) then becomes

$$fq_{gr} = \frac{q_{gr}^2}{r} + \frac{1}{\rho}\left|\frac{\partial p}{\partial r}\right| \tag{4.6}$$

The pressure gradient is negative when pressure decreases with increasing radius and the absolute value is used here to make the pressure force positive in the same sense as the centrifugal force. We solve Eq. 4.6 to obtain

$$q_{gr} = \frac{fr}{2} - \sqrt{\left(\frac{fr}{2}\right)^2 - \frac{r}{\rho}\left|\frac{\partial p}{\partial r}\right|} \tag{4.7}$$

The positive sign is discarded in front of the radical sign because $q_{gr} \to 0$ as $dp/\partial r \to 0$. Using the above notation for V_{max} and q_G, we recast Eq. 4.7 as

$$q_{gr} = V_{max}\left(1 - \sqrt{1 - 2\frac{q_G}{V_{max}}}\right) \tag{4.8}$$

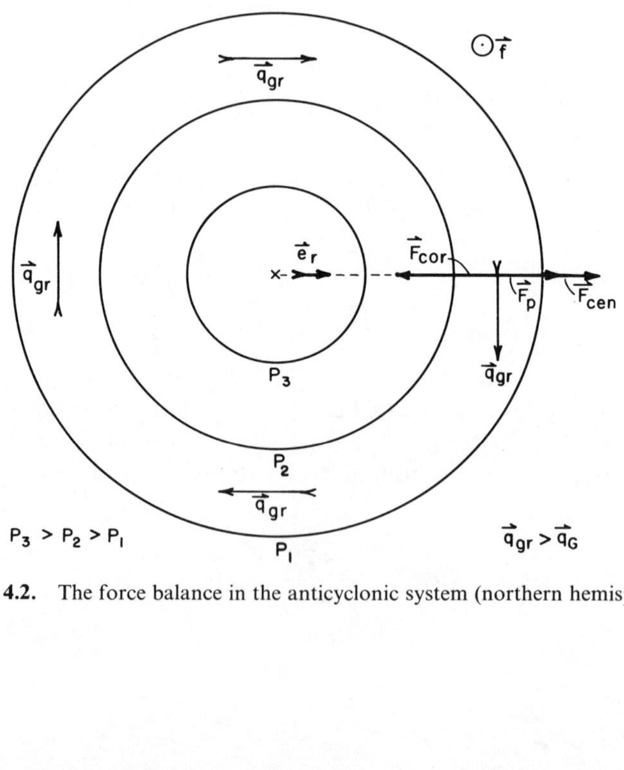

Figure 4.2. The force balance in the anticyclonic system (northern hemisphere).

The direction of rotation of the atmosphere is, in the case of the anticyclone, opposite to the earth's rotation and the relative vorticity is of opposite sign to the vorticity owing to the earth's rotation, f. The absolute vorticity is positive but smaller in magnitude ($\zeta + f \approx 2$ to $6 \times 10^{-5}\ \mathrm{s}^{-1}$) compared with the absolute vorticity of the cyclonic pressure system. When pressure gradients are weak or the radius of curvature of the isobar is large, Eq. 4.8 for the gradient wind simplifies to

$$q_{gr} \approx q_G\left(1 + \frac{1}{2}\frac{q_G}{V_{max}}\right) \qquad \text{for } q_G \ll V_{max} \tag{4.9}$$

This relationship will be more frequently applicable than the corresponding equation for the cyclone because anticyclonic pressure gradients usually are weaker and because the anticyclonic isobars usually display less pronounced curvature.

4.3. FURTHER DIFFERENCES BETWEEN CYCLONIC AND ANTICYCLONIC PRESSURE SYSTEMS

An important difference between the cyclonic and anticyclonic pressure systems is that high wind speeds are always associated with the former. Hurricanes and tornadoes are examples of low pressure, cyclonic systems which are accompanied by destructive wind speeds. On the other hand, the anticyclonic system is usually accompanied by light to moderate winds. The very low geostrophic winds that occur in the United States during the summer months often accompany anticyclonic systems possessing very weak pressure gradients.

A second major difference between the two wind systems is apparent by considering the direction of the flow induced by the pressure gradient itself during the development of the pressure system and during the establishment of a Coriolis force sufficient to balance the pressure gradient. In the developing low pressure system the flow is directed radially inward tending to cause a convergence ($\nabla_h \cdot \mathbf{q}_h < 0$) and an upward motion ($w > 0$) at the eye of the cyclone. The cooling of the atmosphere resulting from upward motion will cause moist air to produce precipitation. On the other hand, the anticyclonic pressure gradient produces a small component of velocity away from the center of the pressure system tending to produce a divergence ($\nabla_h \cdot \mathbf{q}_h > 0$) and downward motion ($w < 0$) at the eye of the anticyclone. The downward motion, called a subsidence, tends to reduce the relative humidity of the air because of the warming produced by compression of air moving vertically downward to a level of higher pressure.

The subsidence associated with the anticyclone has an important effect upon the thermal lapse rate. As an air parcel descends in the anticyclone, it is warmed adiabatically. The parcel originally at height z_2 will be warmer at the lower height z_1 by an amount equal to the dry adiabatic lapse rate, multiplied by the change in height, $\Gamma(z_2 - z_1)$. If the lapse rate is subadiabatic, as is normally the case, then the warming associated with the subsidence will result in a decrease in the lapse rate tending to render it more strongly subadiabatic as depicted in Fig. 4.3*a*. The above effect is readily visualized in terms of the potential temperature profile as depicted in Fig. 4.3*b*. When the lapse rate is subadiabatic, the potential temperature increases with altitude. Lowering of the air parcels by a subsidence will cause no change in the potential temperature because the compression is adiabatic. The slope of the potential temperature, $d\theta/dz$, which is normally weakly positive, will become more strongly positive.

The effect of the increase of dT/dz and $d\theta/dz$ (or decrease of $d\rho^*/dz$) accompanying the subsidence of the anticyclone is to tend to produce a stable stratification. The temperature inversion that may result with the anticyclone is termed a subsidence inversion and experience indicates (Ref. 1) the most severe episodes of air pollution occur when large anticyclones with mild pressure gradients move slowly across heavily populated

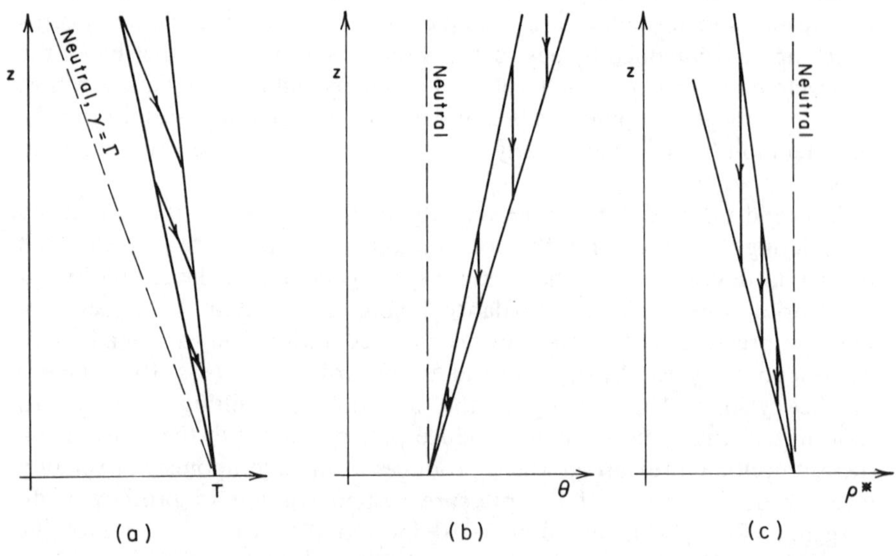

Figure 4.3. The influence of a subsidence on a stable lapse rate: (*a*) temperature, (*b*) potential temperature, and (*c*) potential density versus altitude. Each parcel is compressed adiabatically during the downward motion and the stable lapse rate becomes more strongly stable.

areas. The term "anticyclonic gloom" (Ref. 2) is a highly descriptive phrase that is applied to this condition.

The importance of the large scale weather on the removal of air pollutants, which are normally emitted at a somewhat constant rate, is clear from the above discussion. The low pressure, cyclonic system is frequently accompanied by moderate winds, precipitation, and positive values of the vertical wind. The upward vertical wind will tend to change a subadiabatic lapse rate toward the neutrally stable adiabatic lapse rate. All of these attributes of the cyclonic pressure system will promote the removal of air pollutants released at or above ground level. The high pressure anticyclonic system is frequently accompanied by light winds, an absence of precipitation, and a subsidence that tends to augment the normal stable stratification of the atmosphere. These features of the anticyclone combine to retard the dispersion of air pollutants and to cause their concentrations to increase in the vicinity of their sources. In addition, the sunlight that is present with the fair weather accompanying the anticyclone will promote the formation of those pollutants that are created by photochemical reactions. Further discussion of the role of the large scale weather phenomena on air pollution will be given in Chapter 6.

REFERENCES

1. J. Korshover, "Climatology of Stagnating Anticyclones East of the Rocky Mountains 1936–1965," Environmental Science Services Administration, U.S. Department of Commerce, PHS Pub. No. 999-AP-34 (PB 174709), 1967.
2. R. S. Scorer, *Air Pollution* (Pergamon Press, Oxford, 1968).

PROBLEMS

4.1. Prepare a sketch of the force balance on a fluid parcel located in the southern hemisphere in a low pressure system. Prepare a similar sketch for a fluid parcel in a high pressure system, south of the equator.

4.2. Based on your answer to Problem 4.1, discuss the ability of high and low pressure systems to disperse air pollutants in the southern hemisphere.

4.3. Synoptic weather maps issued by the U.S. Weather Service show isobars for increments of pressure equal to 4 mb. Derive a dimensional relationship for the geostrophic wind of the form

$$q_G = C \frac{1}{2 \sin \phi} \left(\frac{\rho_0}{\rho}\right)\left(\frac{1}{\Delta r}\right)$$

where q_G has units of m/s, and Δr has units of km. Δp is equal to 4.0 mb and ρ_0 is the standard sea level density. Determine the value of c. Recall that, in scalar form,

$$q_G = \frac{1}{f\rho}\frac{\partial p}{\partial r} \approx \frac{1}{2\sin\phi\Omega\rho}\frac{\Delta p}{\Delta r}$$

hence

$$q_G \approx \frac{\Delta p}{\rho_0 \Omega}\left(\frac{1}{2\sin\phi}\right)\left(\frac{\rho_0}{\rho}\right)\frac{1}{\Delta r}$$

4.4. Using the answer to Problem 4.3, calculate the geostrophic wind at Nashville, Tenn., on October 2, 1972 when the separation of the 1020 and 1024 mb isobars associated with a large anticyclone was 700 km. The temperature and pressure there were 8°C and 1021.9 mb. The latitude of Nashville is 36.1°N (see weather map on page 314).

4.5. Calculate the gradient wind at Nashville for the conditions given in Problem 4.4 if the radius of curvature of the isobar is 600 km.

5

THERMAL RADIATION AND THE ATMOSPHERIC STRUCTURE AND CHEMISTRY

The earth as a whole is presumed to be in radiative equilibrium with interstellar space. We mean by this statement that the quantity of radiant energy that the earth receives, which originates almost entirely at the sun, is equal to the radiation emitted from the earth's surface and atmosphere that directly escapes to the near or far reaches of the universe. Of course, the radiation received at a particular location on the earth may be unequal to the radiation that is locally emitted. The equality of the incoming and outgoing radiation applies to long-term averages of both incoming and outgoing components over the entire planet.

For example, at the equator the sun at midday is located at high altitude angles so that the incoming radiation far exceeds the outgoing radiation. Even over a long-term average there is a net influx of radiation in the equatorial region. At the polar regions of the earth where the altitude angle of the sun is always small, there is a long-term net efflux of radiative energy from the surface and the atmosphere; this outward transfer of radiative energy balances the radiative flux that arrives at the equatorial latitudes.

Thus, the sun may be viewed thermodynamically as a heat source and the interstellar space to which the earth radiates its energy as a heat sink. The atmospheric motion is driven by the energy arriving at the equator and this energy is transported by this motion to the polar regions where it is rejected. In this chapter we review some important concepts of radiation and discuss some of the influences of the radiative energy transfer on the properties of the atmosphere and atmospheric motion.

5.1. CONCEPTS OF RADIATION

Visible, ultraviolet, and infrared radiation are all examples of electromagnetic radiation which are distinguished from one another by the frequency or wavelength of the oscillatory electromagnetic fields with which they are associated. Visible radiation, for example, ranges in wavelength, λ, from 0.40 to 0.73 micron (μm) where 1 μm* is defined as 10^{-6} m. The shorter of these wavelengths corresponds to a deep violet color and also represents the longest wavelength limit of the ultraviolet portion of the spectrum. Red light occupies the portion of the visible spectrum whose wavelengths are in the vicinity of 0.73 μm which represents the short wavelength limit of the infrared portion of the spectrum that extends out to 100 μm. The term thermal radiation is used to describe the ultraviolet, visible, and infrared portions of the electromagnetic spectrum because almost all of the energy that is involved in radiant heat exchanges is contained in these spectral regions. Radiation whose spectral content falls outside of these regions also contributes to heat exchange but contains only a negligible fraction of the total heat transfer. Figure 5.1 depicts the electromagnetic spectrum and notes certain spectral regions that are of special interest.

The wavelength of light in a vacuum and its frequency, ν, are related by the familiar equation

$$\nu\lambda = c$$

where c is the vacuum speed of light equal to $3.000 \times 10^{+8}$ m/s. Thus, visible light at a vacuum wavelength of 0.5 μm corresponds to a frequency of $6.00 \times 10^{+14}$ s^{-1} or 600 terahertz (THz). The more common frequency unit is the reciprocal of the wavelength expressed in units of centimeters. Thus, the wavelength 0.5 μm corresponds to the frequency of 20,000 cm^{-1}. The quantum or photon is the fundamental energy unit in the corpuscular viewpoint of radiation. Its magnitude is $h\nu$ where h is Planck's constant which is equal to 6.626×10^{-34} Js. Thus, high frequency, short wavelength radiation of the ultraviolet portions of the spectrum corresponds to photons of high energy while the infrared spectral region corresponds to photons of lower energy.

A surface or a region of space which absorbs all radiant energy incidence upon it is defined to be a perfect absorber or blackbody. A fundamental principle† of radiation known as Kirchhoff's law states that, if thermal equilibrium persists, no object may emit more radiant energy than a blackbody. The latter can be constructed by forming a cavity with walls made up

* One micron is equal to 10^{+4} angstrom units (Å) = 10^{+3} nanometer (nm).

† This abbreviated discussion of thermal radiation may be supplemented by reference to standard texts (Refs. 1 to 3).

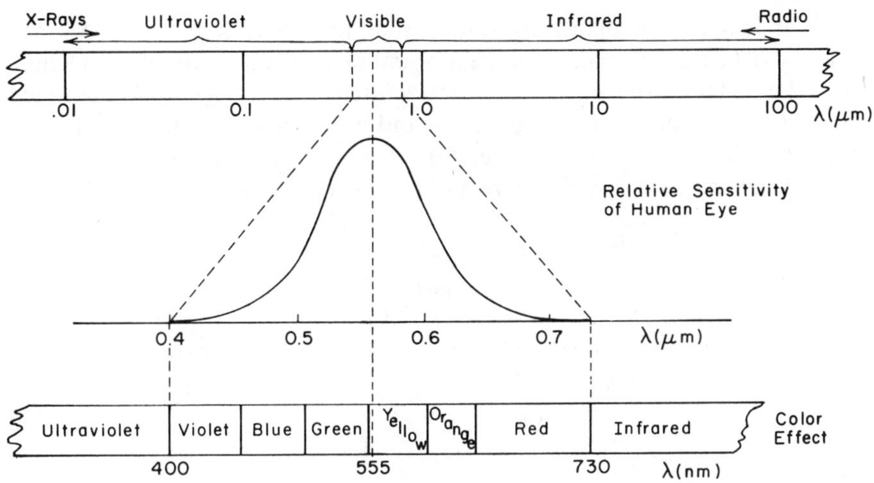

Figure 5.1. The electromagnetic spectrum and visual sensation.

of a diffusely reflecting, at least partially absorbing, material and a small aperture. A photon of radiation that enters the aperture may not be absorbed when it first impinges upon the interior wall, but the smallness of the aperture assures that escape from the cavity is unlikely before many diffuse reflections occur. Because the wall is partially absorbing the probability that a photon is absorbed rather than escaping the chamber approaches a certainty. The aperture-enclosure configuration described above is a good approximation to a perfect absorber and, by Kirchhoff's law, is therefore a perfect emitter or blackbody.

Some materials may approach the blackbody as a perfect absorber-emitter. However, merely because an object is black in color, which implies that visible light of all wavelengths is absorbed with high efficiency, does not necessarily mean that infrared wavelengths are absorbed. Likewise a surface that is highly reflective to visible light may prove to be absorbing to infrared radiation. The term blackbody in the present context therefore has a specific technical meaning that does not necessarily bear a relationship to the color sensation perceived by the eye.

The energy emitted by a blackbody per unit of area normal to the direction of propagation, per unit of solid angle in the wavelength interval $|d\lambda|$ is given by

$$B_\lambda(T)|d\lambda| = \frac{2c^2h\lambda^{-5}}{\exp\left(\dfrac{hc}{\lambda kT}\right) - 1}|d\lambda| \tag{5.1}$$

where k is the Boltzmann constant equal to 1.3805×10^{-23} J/°K. The quantity $B_\lambda(T)$ is the corresponding amount of energy per unit of wavelength and is defined to be the blackbody spectral intensity. The algebraic expression for $B_\lambda(T)$ as a function of temperature and wavelength is called the Planck function. Since $\lambda = c/\nu$ and therefore $|d\lambda| = (c/\nu^2)|d\nu|$, the blackbody spectral intensity expressed in terms of frequency becomes

$$B_\nu(T) = \frac{2c^{-2}h\nu^3}{\exp\left(\dfrac{h\nu}{\mathrm{k}T}\right) - 1}$$

The Planck function in terms of wavelength is sometimes expressed as

$$B_\lambda(T) = \frac{2c_1\lambda^{-5}}{\exp\left(\dfrac{c_2}{\lambda T}\right) - 1} \tag{5.2}$$

where $c_1 = c^2h = 0.5954 \times 10^{-16}$ Wm^2 and $c_2 = hc/\mathrm{k} = 0.01439$ m°K.

If $B_\lambda(T)$ is plotted versus λ for a prescribed T we find the function is zero at small and large λ and possesses a maximum at intermediate values. As temperature increases, the value of $B_\lambda(T)$ is greater at all wavelengths and the position of its maximum shifts toward shorter wavelengths (Fig. 5.2). The position of the maximum in T can be found by the usual means to be

$$T\lambda_{\max} = 2898\text{°K } \mu m \tag{5.3}$$

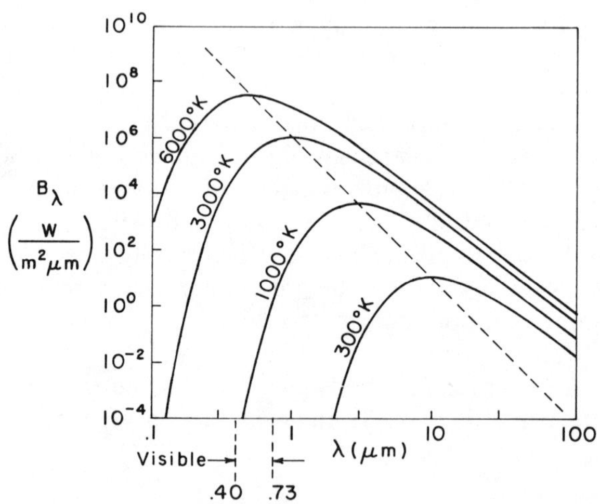

Figure 5.2. The Planck function $\mathbf{B}_\lambda$ versus wavelength for selected temperatures (log-log presentation). The temperatures correspond to the sun (~6000°K), a tungsten lamp (~3000°K), a barely incandescent object (~1000°K), and an object at room temperature (~300°K). The dashed line is the locus of the maxima of $_\lambda$ that is given by Eq. 5.3.

The consequences of this result are confirmed by common experience. The sun whose outer layer, the photosphere, is approximated by a blackbody at a temperature of 5762°K and has its maximum spectral emissive power at about 0.5 μm which is near the center of the visible region of the spectrum. An incandescent lamp operates at about 2900°K and produces its peak spectral intensity at about 1.0 μm in a spectral region where the eye has no sensitivity. The visible radiation of the incandescent lamp is of a yellow cast compared with the white light produced by the sun. A cherry red heating unit at about 920 to 1020°K (1200 to 1380°F) has its strongest emission in the 3 μm spectral region and the visible light it emits will consist predominately of radiation near the long or red wavelength limit of the visible spectrum. Cooler objects, near room temperature, 300°K, also emit thermal radiation with peak spectral intensities in the 10 μm region and visible radiation of such low intensity, as to be totally undetected by the eye and small in magnitude compared with light reflected in typical daytime levels of illumination.

A surface that emits blackbody radiation possesses a directional spectral emissive power which is the product of B_λ times the cosine of the angle between the direction of the ray and the normal to the surface. The energy emitted by a black surface per unit surface area, per unit time, per unit wavelength into a hemispherical solid angle is found by an integration over 2π steradians to be $\pi B_\lambda(T)$ or $\pi B_\nu(T)$. This quantity is defined to be the hemispherical spectral emissive power.

Finally, the energy emitted per unit surface area, per unit time, into a hemispherical half space in all wavelengths is defined to be the blackbody hemispherical total emissive power. This quantity is found by an integration of the hemispherical spectral emissive power over wavelength which yields the Stefan–Boltzmann law,

$$W_b = \int_0^\infty \pi B_\lambda(T) d\lambda = \sigma T^4 \tag{5.4}$$

where

$$\sigma = \frac{2C_1\pi^5}{15C_2^4} = 5.669 \times 10^{-8} \text{ W/m}^2{}^\circ\text{K}^4$$

is the Stefan–Boltzmann constant. The fourth power dependency of emission was first found experimentally (Stefan, 1879) and later established theoretically (Boltzmann, 1884). The radiation from material surfaces is sometimes calculated by approximating their emissive power as a fixed fraction of the emission of a blackbody. This fraction, called the emissivity, is then multiplied by σT^4 to obtain the radiative heat flux from the surface. An object that emits a fixed fraction of the blackbody radiation at all wavelengths is called a greybody. The greybody approximation is suitable for some material surfaces but is totally unsuitable for gases at atmospheric temperatures which have emissivities that are highly dependent on wavelength (or frequency).

The energy emitted by a blackbody at a given temperature in a specified wavelength interval expressed as a fraction of the energy emitted over all wavelengths is often of interest. This quantity is the fraction,

$$\mathscr{F}(\lambda_1, \lambda_2) = \frac{\int_{\lambda_1}^{\lambda_2} \pi B_\lambda(T) d\lambda}{\int_0^{\infty} \pi B_\lambda(T) d\lambda} \tag{5.5}$$

The denominator is the blackbody hemispherical total emissive power given by the Stefan–Boltzmann law. The numerator may be expressed as the difference between two integrals with a common lower limit, hence

$$\mathscr{F}(\lambda_1, \lambda_2) = \frac{\pi}{\sigma T^4}\left[\int_0^{\lambda_2} B_\lambda(T) d\lambda - \int_0^{\lambda_1} B_\lambda(T) d\lambda\right] \tag{5.6}$$

Substituting Eq. 5.1 into the above expression, we find the integration variable can be converted to the product λT. Thus, we define

$$\mathscr{F}(0, \lambda_i T) \equiv 15\left(\frac{C_2}{\pi}\right)^4 \int_0^{\lambda_i T} (\lambda T)^{-5} \frac{d(\lambda T)}{\left[\exp\left(\frac{C_2}{\lambda T}\right) - 1\right]} \tag{5.7}$$

and, therefore,

$$\mathscr{F}(\lambda_1, \lambda_2) = \mathscr{F}(0, \lambda_2 T) - \mathscr{F}(0, \lambda_1 T) \tag{5.8}$$

The quantity $\mathscr{F}(0, \lambda_i T)$ is widely tabulated and is presented graphically in Fig. 5.3.

For example, we can calculate the fraction of the sun's radiation lying below 4.0 μm, if we take the sun's temperature to be approximately 6000°K, by evaluating $\mathscr{F}(0, \lambda T)$ using the value of 24,000 μm°K as the independent variable to enter Fig. 5.3. We find that 99% of the total solar radiation lies below 4.0 μm. Assuming the earth temperature is 300°K, we find 0.2% of its radiation is emitted at wavelengths shorter than 4.0 μm while 99.8% of the total radiation is emitted at longer wavelengths. The 4.0 μm wavelength is often taken as the dividing line between the "short wavelength" radiation emitted by the sun and the "long wavelength" radiation emitted by the earth.

The definitions and units of the various quantities that are used to describe blackbody radiation are summarized in Table 5.1.

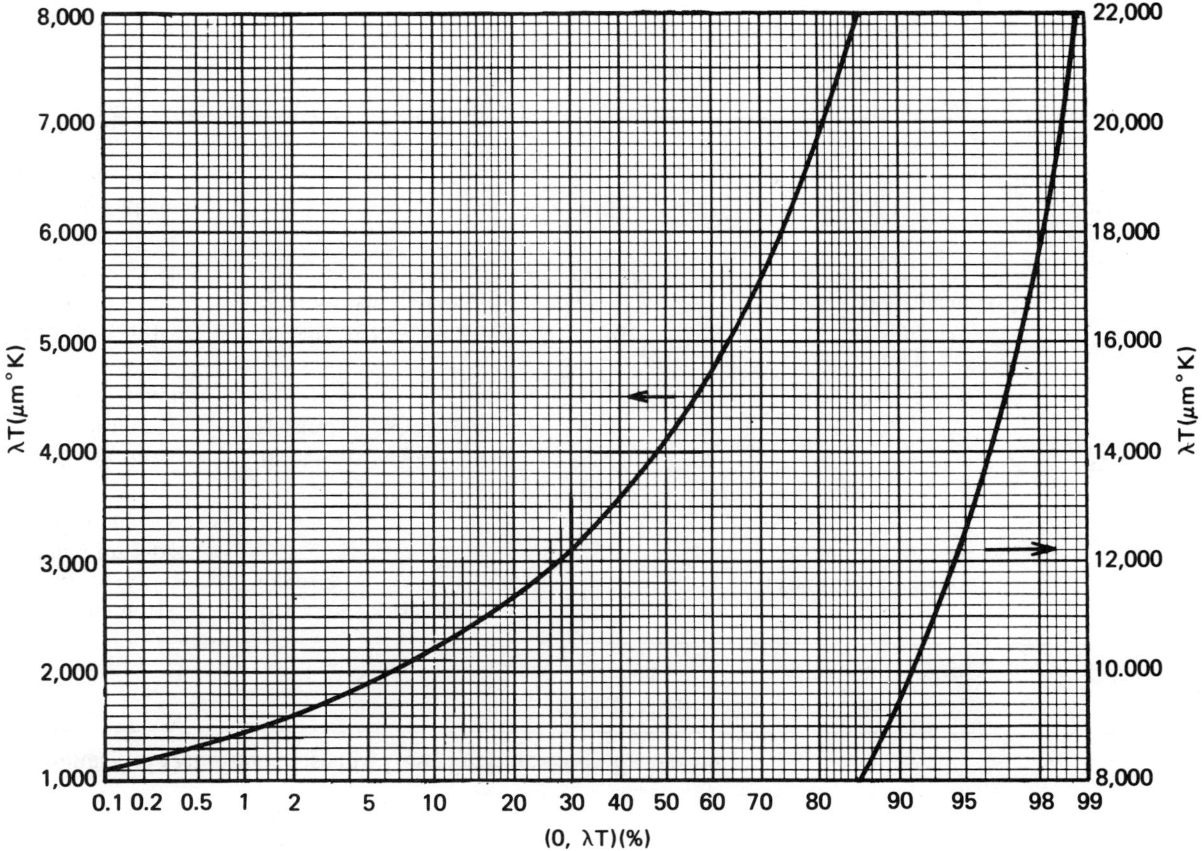

Figure 5.3. The function $\mathscr{F}(0, \lambda T)$ versus λT.

Table 5.1. Summary of Definitions Used to Describe Blackbody Radiation

Quantity	Name	Symbol Used Herein	Equation Number	Units
Energy emitted per unit area normal to the direction of propagation, per unit time, per unit wavelength, per unit solid angle	Blackbody spectral intensity	B_λ	(5.2) (The Planck function)	$W/m^2\ \mu m$
Energy emitted per unit surface area, per unit time, per unit wavelength (into a hemispherical half space)	Blackbody hemispherical spectral emissive power	πB_λ	—	$W/m^2\ \mu m$
Energy emitted per unit surface area, per unit time (into a hemispherical half space, in all wavelengths)	Blackbody hemispherical total emissive power	W_b	(5.4) (The Stefan–Boltzmann law)	W/m^2
Fraction of blackbody radiation contained in the wavelength interval 0 to λ_1	—	$\mathscr{F}(0 - \lambda_1)$	(5.5)	(dimensionless)

5.2. THE EQUATION OF RADIATIVE TRANSFER

We wish to describe the action of radiation as it traverses a medium which may partially absorb or emit radiation and which possesses spectral properties (i.e., absorption and emission coefficients) that are frequency dependent. An illustration is provided by considering infrared photons that originate at the sun and pass into the earth's atmosphere to be absorbed by molecules which also emit radiation of the same frequency, and other frequencies as well, both toward the earth and back to outer space. A quantitative description of this process is provided by the equation of radiative transfer.

We define the specific spectral intensity, I_ν, of radiation passing through space as the radiant energy (or number of photons) passing through a unit area normal to the direction of propagation, per unit time, in a unit solid angle in a specified direction, and at a specified frequency increment. The radiative transfer equation states that the gradient of intensity along a given path is the difference between the energy emitted per unit volume at the same frequency and direction, j_ν, and the energy absorbed per unit volume. The latter is the product of the absorption cross section per unit volume, k_ν, usually a highly frequency dependent quantity, and the magnitude of intensity itself. The equation of radiative transfer is then expressed as

$$\frac{dI_\nu}{ds} = -k_\nu I_\nu + j_\nu \tag{5.9}$$

The flux of radiation, F, passing through an element of surface area in a given direction, $\mathbf{s}$, is given by the integration of the specific spectral intensity over a hemispherical solid angle, and all frequencies,

$$F(\mathbf{s}) = \int_0^\infty \int_0^{2\pi} I_\nu \cos\theta \, d\omega \, d\nu \tag{5.10}$$

where θ is the angle between $\mathbf{s}$ and the direction of propagation. The above two relations are often tedious to solve in practice because the volumetric absorption coefficient requires a detailed knowledge of the spectroscopic properties of the gaseous species, each possessing many narrow absorption lines or series of lines known as bands. For example, a recent compilation lists over 107,000 lines for the seven most important absorbing-emitting species of the unpolluted atmosphere, which are CO_2, H_2O, O_3, N_2O, CH_4, O_2, and CO (Ref. 4).

If the gas is in thermodynamic equilibrium, then Kirchhoff's law applies and it can be shown that the emission coefficient is equal to $k_\nu B_\nu$ and the radiative transfer equation becomes

$$\frac{dI_\nu}{ds} = -k_\nu[I_\nu - B_\nu(T)] \tag{5.11}$$

The requirement of thermodynamic equilibrium, which demands thermal and radiative equilibrium, is completely fulfilled in an isothermal enclosure containing a gas at the temperature of the walls. The enclosed space would then be uniformly filled with isotropic radiation (i.e., I_ν is independent of direction) corresponding to the wall and gas temperature. The gas will be in thermodynamic equilibrium and the amount of energy per unit volume received at each point in space, $k_\nu B_\nu(T)$, is equal to the amount of energy emitted per unit volume at that point, j_ν.

Strict conformity to thermodynamic equilibrium is not found at any point in the atmosphere, where for example, during a cloudless day we may find the gas at 270°K, the radiation field to be of solar origin, highly anisotropic, and possessing a frequency distribution characteristic of a 6000°K blackbody source. Fortunately, the application of Kirchhoff's law and Eq. 5.11 requires only that the various energy levels of the atoms and molecules be determined by the local translational temperature of the gas. This is the condition of local thermodynamic equilibrium. In particular, the radiation intensity, isotropy, and spectral content are allowed to deviate from those of a blackbody corresponding to the local gas temperature. The condition of local thermodynamic equilibrium is fulfilled throughout the troposphere and in the lower stratosphere (Ref. 3), but is not fulfilled at higher altitudes where the very low densities result in much lower collision frequencies. For example, in the upper layers of the stratosphere, the concentration of ions and electrons is far in excess of the values appropriate for the local translational temperature given by thermodynamic equilibrium.

In some cases the spectral emission at the same frequency as the incident radiation is small enough to be ignored ($j_\nu \ll k_\nu I_\nu$). Equation 5.8 then may be integrated, if k_ν is constant, to yield the Lambert-Beer law,

$$\frac{I_\nu}{I_{\nu 0}} = \exp(-k_\nu l) \tag{5.12}$$

where l is the path length through which the radiation traverses, $I_{\nu 0}$ is the initial specific spectral intensity, and I_ν the emergent intensity. The quantity $k_\nu l$ is the optical depth, τ_ν, of the medium of specified property (k_ν) and physical extent. If τ_ν is small there is a weak decay of I_ν which is linear with distance and the medium is highly transparent to radiation. If the τ_ν is large then only a small fraction of the incident radiation traverses the medium, which would be described as nearly opaque. Both of these cases are illustrated by the atmosphere in different spectral regions even though

k_ν is usually highly altitude dependent in this case. In the visible region the transmission may be as high as 95% ($\tau_\nu \ll 1$) and in the ultraviolet the transmission is far less than 1% ($\tau_\nu \ll 1$). (Infrared frequencies often correspond to the case where the optical depth of the atmosphere is large but the emission coefficient j_ν may be comparable to $k_\nu I_\nu$ and the transmission equation does not apply. In these cases the full radiative transfer equation must be used.)

5.3. ATMOSPHERIC PHOTOCHEMISTRY AND THERMAL STRUCTURE

The atmospheric temperature profile, Fig. 5.4, reveals an interesting structure which is largely controlled by photochemical reactions. We discuss these different layers of the atmosphere starting at the highest altitude where the solar radiation has not yet experienced absorption by atmospheric species. More detailed discussions of atmospheric photochemistry are given in Refs. 6 and 7.

5.3.1. The Thermosphere

Energetic ultraviolet photons present at altitudes of 100 km and higher cause the photodissociation of molecular oxygen

$$O_2 + h\nu \longrightarrow 2O$$

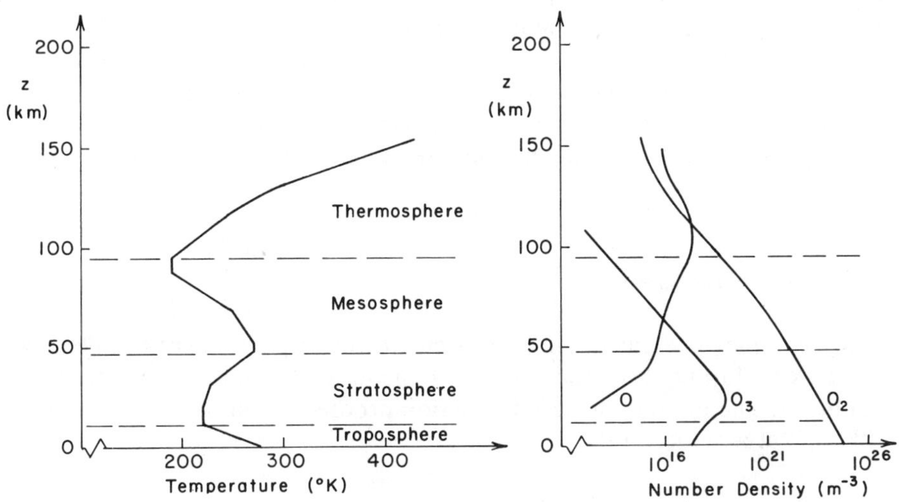

Figure 5.4. Atmospheric temperature and species concentration profiles.

At altitudes of about 200 km and higher the concentration of atomic oxygen far exceeds that of molecular oxygen. The atomic oxygen at these high altitudes is the dominant source of charged particles which are produced by the photoionization

$$O + h\nu \longrightarrow O^+ + e^-$$

The charged oxygen atoms can then exchange electrons with neutral oxygen molecules or react with neutral nitrogen molecules

$$O^+ + O_2 \longrightarrow O_2^+ + O$$

and

$$O^+ + N_2 \longrightarrow NO^+ + N$$

The first of these reactions is highly exothermic and partially contributes to the high temperatures of this region. Additional exothermic reactions are the dissociative recombination reaction

$$NO^+ + e^- \longrightarrow O + N$$

and

$$O_2^+ + e^- \longrightarrow 2O$$

The above series of reactions account for the high concentrations of charged particles in the thermosphere and also its high temperature, typically 1500°K at altitudes of 300 km and higher. As altitude decreases, the ultraviolet solar radiation is weakened; the photodissociation and ionization reactions become relatively weaker. Electron concentration and temperatures accordingly decrease until the mesopause is reached at the 80 to 90 km region.

5.3.2. The Mesosphere

Many complex chemical reactions occur in this region where the density has increased to the level that the simultaneous collision of three molecules can occur. The photodissociation reaction proceeds as in the thermosphere although at a reduced rate

$$O_2 + h\nu \longrightarrow 2O$$

The atomic oxygen can collide with molecular oxygen and a third atom in a three-body collision to form ozone, O_3, in a reaction represented by

$$O + O_2 + M \longrightarrow O_3 + M$$

The third body, M, is required in order to carry off excess energy and the absence of three body collisions at higher altitudes prevents O_3 formation by the above reaction. The ozone is decomposed by the exothermic reactions

$$O_3 + O \longrightarrow 2O_2$$

and

$$O_3 + h\nu \longrightarrow O_2 + O$$

The above three reactions occur at higher rates as altitude decreases owing to the increasing density that favors the ozone production. The high rates of the exothermic reactions bring about an increase in temperature with increasing ozone concentration as altitude decreases to the stratopause at about the 50 km level.

5.3.3. The Stratosphere

At the stratopause the ultraviolet radiation responsible for the exothermic photodecomposition of ozone has been depleted to the point that temperature again starts to fall with decreasing altitude. The production of ozone remains efficient because the gas density continues to increase. The concentration of ozone increases with decreasing altitude in part because of the lower rate of photodecomposition. Ozone concentration reaches its maximum level at 25 to 30 km and then decreases because the ultraviolet photons that are necessary for the O_2 dissociation reaction have been absorbed at high altitudes and consequently the atomic oxygen concentration decreases at a rapid rate. The formation of ozone by three-body collisions therefore proceeds at a very low rate and ozone concentration decreases with altitude. Temperature decreases until the tropopause is reached at an altitude that varies with latitude and season but is commonly considered to be 11 km.

The stratosphere is thus characterized as the layer in which the ozone production-depletion reactions result in a peak ozone concentration and the temperature increases with increasing altitude. The negative lapse rate results in a persistent stable stratification that tends to inhibit mixing and diffusion. For example, vertical velocities are estimated at one-tenth the vertical velocities in the troposphere and the residence time of foreign matter is estimated to be three years (Ref. 7).

5.3.4. The Troposphere

As altitude decreases still further, the natural ozone formation ceases because ultraviolet radiation that has been responsible for the production of atomic oxygen has been completely absorbed in the mesosphere and the stratosphere. Photoionization reactions are absent for the same reasons. Photochemical reactions are confined to the low energy, infrared spectrum and merely result in the excitation of the triatomic species—principally water vapor and carbon dioxide. The subsequent de-excitation results in the release of the energy of the absorbed photon as heat.

Temperature, which reached a minimum at the tropopause, increases continuously throughout the troposphere, and the lapse rate, although not as high as the adiabatic value, is positive and indicative of a weakly stable stratification. The temperature of the atmosphere near the surface is equal to the earth's temperature which is controlled by the processes of absorption and emission of radiant energy as described below. The relatively high density and weakly stable stratification combine to promote a strong large scale convective motion which is known to us as our weather activity.

We therefore define the troposphere by a set of symptoms which are:

1. The lowest layer in the atmosphere, whose height varies from 7 to 17 km according to the season and the latitude, in which the lapse rate is generally positive and slightly less than the dry adiabatic value.
2. The layer adjacent to the surface of the earth in which there is a high level of large scale convective activity. The earth's surface temperature influences and is influenced by the presence of this activity.
3. The lowest layer of the atmosphere in which photochemical reactions involve mainly absorption of infrared radiation; photodissociation and photoionization reactions are negligible in the unpolluted troposphere.

5.4. THE EARTH'S RADIATION BUDGET AND SECULAR TEMPERATURE TRENDS

The solar constant is the amount of radiant energy falling on a unit area that is normal to the solar rays and located at the mean earth-to-sun distance from the sun. The recent redetermination (Ref. 8) of the solar constant has been made from an aircraft flying at 10.7 km and is 1.353 kW/m^2 (equal to 1.940 cal/min cm^2 or 429.1 Btu/hr ft^2). About 58% (Ref. 8) of this energy is absorbed by gases in the earth's atmosphere and by the earth's surface and the remaining 42% is returned to space by scattering from atmospheric atoms, molecules, and particles and by reflection from the earth's surface.

The fraction of the total energy incident upon the earth's cross-sectional area that is returned by particulate scattering and surface reflection is called the albedo, a, of the planet.

The simplest theoretical model that one can use to calculate the earth's temperature is formulated by equating the fraction of the total solar radiation intercepted by the earth that is absorbed to the thermal energy emitted by the earth. The solar energy absorbed by the earth is the product of the solar constant, F_s, the cross-sectional area of the earth, and the fraction of the solar energy that is absorbed, which is $1 - a$. The total radiant energy emitted is given by the Stefan–Boltzmann law, Eq. 5.3, multiplied by the surface area of the earth. This radiation balance is expressed by

$$\pi R_e^2(1 - a)F_s = 4\pi R_e^2 \sigma T_e^4$$

where R_e is the radius of the earth and T_e is the effective temperature that is defined by this relationship. Solving this equation for the effective temperature, we find

$$T_e = \left(\frac{F_s(1 - a)}{4\sigma}\right)^{1/4} \tag{5.13}$$

In deriving this result we have assumed that the earth is a perfect emitter for the long wavelength radiation characteristic of objects at low temperatures even though it is not a perfect absorber-emitter for the short wavelength radiation emitted by the sun. The equilibrium temperature that is found with the above values of the solar constant and albedo is 242°K, which is somewhat lower than the observed mean surface temperature of 288°K.

The reason for this discrepancy is that the atmosphere surrounding the earth acts as a filter for radiant energy of distinctive optical properties. Referring to Fig. 5.5 we find that the atmosphere is relatively transparent to the sun's radiation and highly opaque to the earth's radiation. Thus about 43% of the incoming solar radiation is directly or diffusely transmitted to the earth's surface while only about 9% of the outgoing terrestrial radiation is transmitted by the atmosphere. The above calculation of the equilibrium temperature is therefore not adequate as a prediction of the earth's surface temperature because it fails to consider the radiative transfer within the atmosphere.

A more detailed calculation can be attempted in which the optical properties of the atmosphere are considered. The calculation becomes hopelessly complex unless simplifications are made. An elementary radiative model to determine the atmospheric and surface temperatures is based on the assumption that the atmosphere is quiescent and that each layer of the atmosphere is in radiative equilibrium with the surrounding layers.

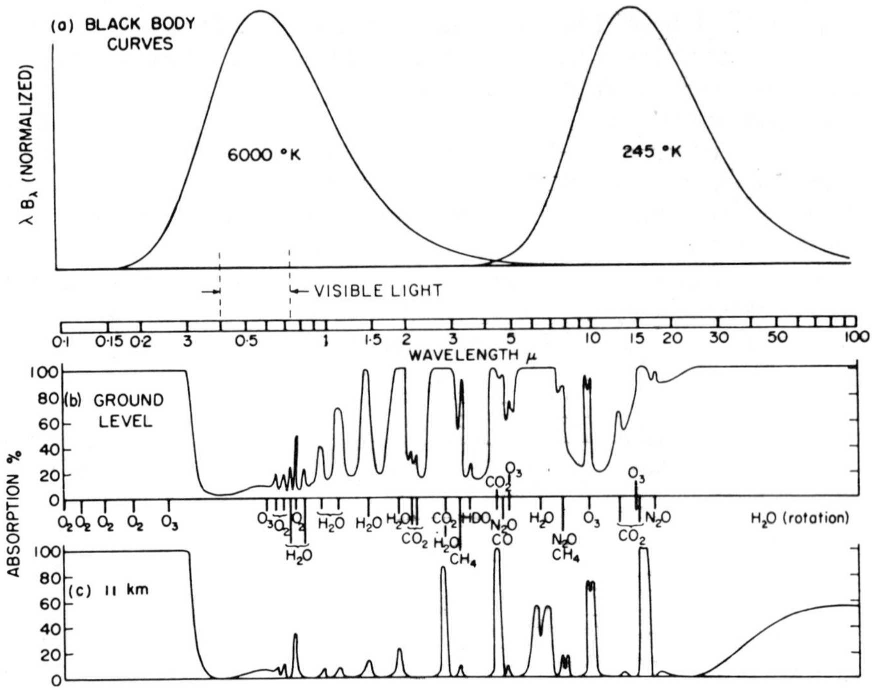

Figure 5.5. Atmospheric absorptions. (*a*) Blackbody curves for 6000°K and 245°K. (*b*) Atmospheric gaseous absorption spectrum for a solar beam reaching ground level. (*c*) The same for a beam reaching the temperate tropopause. The axes are chosen so that areas in (*a*) are proportional to radiant energy. Integrated over the earth's surface and over all solid angles the solar and terrestrial fluxes are equal; consequently, the two blackbody curves are drawn with equal areas beneath them. An absorption continuum has been drawn beneath the bands in (*b*). This is partly hypothetical because it is difficult to distinguish from the scattering continuum, particularly in the visible and near infrared spectrum. Conditions are typical of midlatitudes and for a solar elevation angle of 40° or diffuse terrestrial radiation. (From R. M. Goody, *Atmospheric Radiation*, Oxford Press, 1964. Reprinted with permission of the Oxford University Press and R. M. Goody.)

Simplified optical properties of the atmosphere are employed and the earth's surface temperature is calculated on the basis of its radiative equilibrium with the sun and the atmosphere. A calculation of this type is described in Ref. 5 and predicts the surface temperature of the earth to be 333°K, some 45°K higher than the observed value. In addition, this model predicts super-adiabatic lapse rates in the lower levels of the atmosphere. Thus a basic assumption of the model is not expected to be fulfilled; a quiescent atmosphere will not persist if the lapse rate is superadiabatic. We may reasonably conclude that the observed mean surface temperature results from the combined influences of atmospheric radiative transfer and the convective action of the moving atmosphere.

[The neighboring planets of the earth provide opportunities to test the ideas presented above (Ref. 5). Mars has a very thin atmosphere and its mean surface temperature is very near to the calculated equilibrium temperature of 216°K. Venus has a very dense, cloudy atmosphere and the surface temperature is about 700°K while the equilibrium temperature is only 244°K.]

The ability of the atmosphere to produce this surface warming because it transmits a large fraction of solar radiation and a small fraction of the earth's thermal radiation is called the "greenhouse effect." This term is also often criticized because the analogy to the actual greenhouse warming effect is imperfect. A greenhouse is enclosed by a thin layer of glass which is a solid material; the earth is enveloped by a thick layer of air which is a moving fluid. Nevertheless, both air and glass possess the ability to transmit visible light and absorb most ultraviolet and infrared radiation; at least to this extent the term greenhouse effect is appropriate. The important point is that the earth's surface temperature is determined by the magnitude of the solar constant, by the atmosphere and the absorbing and scattering species it contains, and by the reflective and absorptive properties of the surface itself.

There is widespread concern that the activities of industrialized nations will have an effect on the mean surface temperature on the earth and therefore a dramatic effect on its climate. This concern arises because man's activities have resulted in increased atmospheric concentrations of (a) carbon dioxide as a result of the burning of fossil fuels and (b) particulate material of a variety of compositions. Both of these species enter into the radiative calculation. For example, carbon dioxide absorbs strongly around 15 microns near the wavelength of peak terrestrial radiative emission. Particles scatter radiation and although this effect depends on their size and composition, it appears reasonable to believe that they would be more efficient scatterers of solar radiation than terrestrial radiation. The increase of concentration of both carbon dioxide and particles is well documented.

Careful measurements of the carbon dioxide concentration have been observed at the meteorological observatory at Mauna Loa, Hawaii since 1958 and are displayed in Fig. 5.6 from Ref. 9. These striking results show that the CO_2 concentration has not only a seasonal variation, but also a persistent increase in its mean value that recently amounts to 1.0 part per million by volume (ppm) per year. The actual production of CO_2 by burning of fossil fuels is estimated at about $18 \times 10^{+12}$ kg/yr. It is believed that the oceans, which contain 50 times as much CO_2 as the atmosphere, absorb

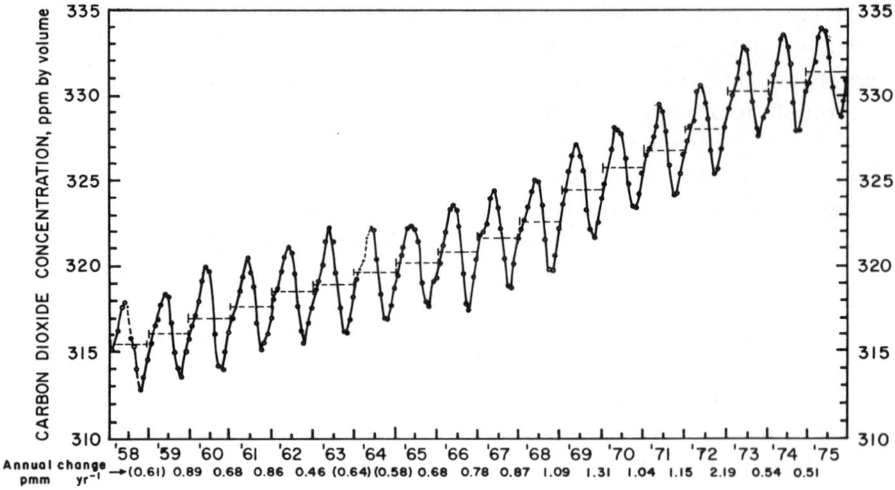

Figure 5.6. Mean monthly values of CO_2 concentration at Mauna Loa Observatory, Hawaii, for the period 1958 to 1975. [From C. D. Keeling, unpublished data, 1977, and Keeling et al., *Tellus*, **28**, 546 (1976), copyright The Swedish Geophysical Society. Reprinted with permission of the latter and C. D. Keeling.]

the difference between the total production and the atmospheric accumulation. An increase in the carbon dioxide concentration raises the mean temperature by increasing the optical depth of the atmosphere.

Particulate concentrations in the atmosphere are also increasing although the rate is more difficult to measure. Mitchell (Refs. 10 and 11) estimates that human activities are responsible for 30% of the total atmospheric particulate loading and that the human source strength is growing at a rate of 4% per year. The effects of the increased particle loading are also difficult to estimate. If the particles merely scatter radiation and thus directly return some solar radiation to space, then their effect is to lower the atmospheric temperature by increasing the earth's albedo. However, particles may partially absorb radiation in which case their presence in higher concentrations may have the effect of increasing the atmospheric mean temperature. This view is not widely accepted but is cited as evidence that our understanding of the influence of increased concentration of particles in the atmosphere is far from complete.

One tentative projection of the combined effect of the higher atmospheric concentrations of carbon dioxide and particles has been offered by Mitchell and is shown in Fig. 5.7. This estimate shows the warming effect by carbon dioxide and the estimated maximum cooling effect of particulates to be equal

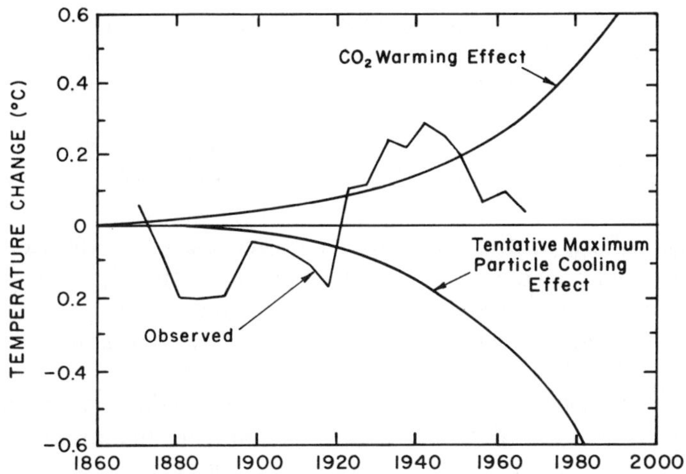

Figure 5.7. Mean temperature trends. The broken curve is the observed mean temperature. The smooth curves represent the estimated warming effect by CO_2 and a tentative cooling contribution by increased particle concentrations. [From J. M. Mitchell, Jr., *Quarternary Research*, **2**, 444 (1972). Reprinted with permission of the Academic Press and J. M. Mitchell, Jr.]

in magnitude. Mitchell concludes that the most likely net effect will be a modest warming of the climate by the dominant action of CO_2. Also in Fig. 5.7 is the observed trend in atmospheric temperature which is shown as a highly irregular curve that increased by 0.5°C from 1880 to 1940. Since 1940 the observed mean temperature has decreased by more than 0.2°C and this trend persists at this time. The causes of the fluctuations of the observed mean temperature are not known but they are presumed to be of natural origin. One possible natural cause is the volcanic dust loading in the atmosphere which is fluctuating according to volcanic activity. Thus, the question of man's impact on the mean surface temperature of the earth should take into consideration the observed natural fluctuation.

A more recent assessment of this problem is given in the study of the National Academy of Science (Ref. 12) which predicted that the CO_2 concentration in the year 2050 would be twice the value at the start of the industrial revolution if heavy reliance on fossil fuels continues. Over a time span of 200 years it was estimated that the world's average temperature could increase by more than 6°C and that polar temperatures could increase as much as three times this figure. The study concludes that the climatic impact of the release of CO_2 may be the primary limiting factor on the use of fossil fuels for the production of energy over the next few centuries. A worldwide research program was recommended to include topics such as the carbon cycle, climate, population trends, energy demands, and the mitigation of the impact of climatic changes on world food production.

5.5. RADIATIVE INVERSIONS

The surface of the earth, and material objects near the surface, radiate heat to the cool layers of the upper atmosphere when the sky is clear. During daylight hours these objects normally receive more radiant energy from the sun's rays than they radiate away. Consequently, they assume a temperature higher than the surrounding air. At night when the solar energy is absent, the surface of the earth continues to radiate to the upper atmosphere but no energy is received, except by convection from the atmosphere, and the temperature decreases. If the surface temperatures drop below the dewpoint temperature of the air, then condensation, commonly referred to as "dew" in this context, will form. During the spring or fall seasons of the year, hoarfrost may form when surface temperatures drop below freezing by this type of nocturnal radiative cooling even though air temperatures remain above freezing.

A parcel of air is subject to a similar type of radiative cooling. The process must be analyzed by solving the radiative transfer equation using a knowledge of the spectrally dependent absorption coefficient of carbon dioxide, water vapor, and other absorbing species. The air nearest the radiatively cooled ground loses heat to the latter by a combination of radiation and convection. The higher rate of cooling of the air near the ground is strong enough to result in the development of a negative lapse rate, or in other words, a temperature profile that is inverted from its usual shape, which is referred to as a radiative inversion. The development of radiative inversions is favored during clear nights when the wind speeds are low so that strong turbulent mixing does not develop.

Low winds and clear skies accompany the passage of anticyclones which independently may produce a subsidence inversion. Thus a radiative inversion may accompany the subsidence inversion. Since temperature inversions of any origin correspond to conditions of strong stable stratification, poor dispersal of air pollutants is observed during atmospheric conditions of this type. Radiative inversions are normally eliminated by the convective effects that result from the solar heating of the earth's surface during the morning hours. The formation of dew or hoarfrost and the rapid development of moderate pollution concentrations in populated areas during the early morning hours are often symptoms of the presence of a radiative inversion.

5.6. CHEMICAL THREATS TO STRATOSPHERIC OZONE

Foreign gases or particles that are deposited in the stratosphere possess a longer residence time there because of the stable stratification that exists in this layer. The concentration of ozone in the stratosphere influences plant, animal, and human life at the surface of the earth. In recent years there has been concern of possible chemical threats to the stratospheric ozone not only by pollutants emitted within the stratosphere but also by species released at sea level as well.

The component of solar radiation that directly causes sunburn (erythema) and suntan is a narrow spectral range centered at 0.297 μm in the ultraviolet region (see Fig. 5.8). The intensity of solar radiation above the atmosphere in the erythema-producing spectral region is about 100 times the sea level value. Attenuation of the erythemal ultraviolet is caused by stratospheric ozone; a reduction in its concentration will result in an increase in the sea-level intensity of the solar radiation in this spectral region. Skin cancer,

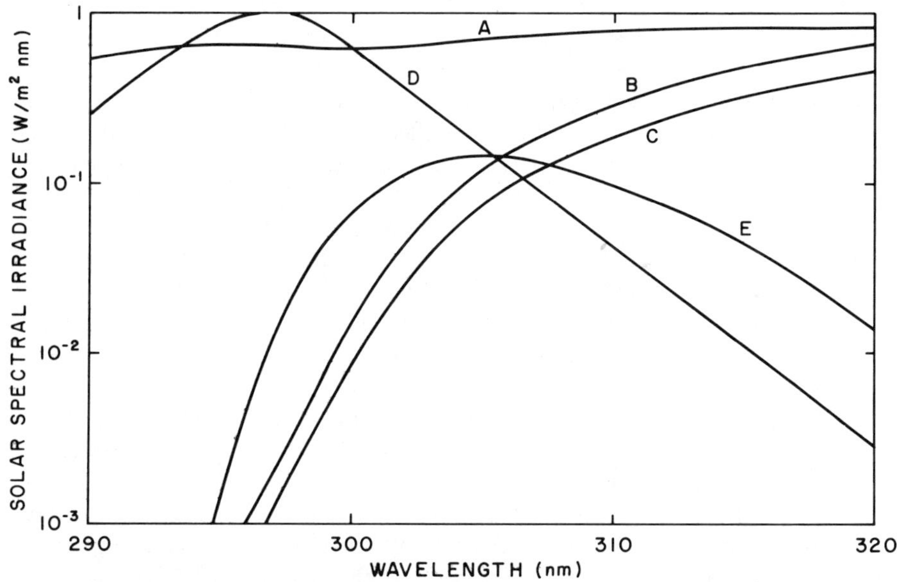

Figure 5.8. The erythemal effect of sunlight. Curve A, the distribution of energy in sunlight outside the atmosphere; curve B, sunlight filtered by 4 mm STP ozone; curve C, sunlight filtered by 4 mm STP ozone and 1 atm pure air plus scattered skylight; curve D, the relative effectiveness of ultraviolet radiation in producing erythema; and curve E, the relative contribution to erythema at various wavelengths due to sunlight filtered by 4 mm STP ozone and 1 atm of pure air plus scattered skylight. [From F. S. Johnson, *Aeronaut. and Astronaut.*, **11**, 14 (1973). Reproduced with permission of the American Institute of Aeronautics and Astronautics and F. S. Johnson.]

which can produce facial disfiguration if untreated and, in a less frequent form, can produce a fatal malignancy, is believed to be caused by ultraviolet radiation either in the erythemal spectral region or by nearby longer wavelength radiation which is strongly absorbed by the DNA molecule. Thus the reduction of the stratospheric ozone concentration could cause an increase in the incidence of skin cancer.

The combustion products from high altitude aircraft constitute one type of threat to the ozone shield. Contemporary long-range subsonic aircraft typically fly at altitudes of about 11 km which corresponds to the lower boundary of the stratosphere. The proposed extensive use of supersonic aircraft which operate economically well into the stratosphere at altitudes of 18 km is a cause of greater concern. The nitric oxide (NO) released by aircraft engines is the compound that appears to pose the most serious consequences. The dynamic balance between the chemical reactions that create and destroy ozone involves many species but Johnson (Ref. 13) describes a simplified reaction cycle between NO and O_3 as

$$NO + O_3 \longrightarrow NO_2 + O_2$$

$$NO_2 + O \longrightarrow NO + O_2$$

In this sequence the nitric oxide that is consumed in the first reaction is regenerated in the second and the net effect is the depletion of both ozone and atomic oxygen. Since NO is not consumed but remains to repeat the cycle, the reaction sequence is described as catalytic. Another catalytic reaction sequence may also develop with the hydroxyl (OH) molecule, which is also a product of combustion. A later study by the National Research Council (Ref. 14) provided a specific estimate of the reduction of ozone by supersonic aircraft based on a number of assumptions which include the emission rate of NO per engine and the size and duty cycle of the fleet. This report recommended additional research on both chemical and biological aspects of this problem and international monitoring to provide the basis for more confident projections of the impact of proposed supersonic flight and existing extensive subsonic air transport.

Another chemical threat to stratospheric ozone is posed by compounds that are used at the surface of the earth and which diffuse to the stratosphere. Examples are provided by halocarbons such as F-11 ($CFCl_3$) and F-12 (CF_2Cl_2). The chemical inertness and physical properties of these compounds made them commercially appealing as refrigerants and as materials to serve as pressure sources in the spray-can packing industry. The same chemical stability, which means that these compounds are nontoxic and noncombustible, also assures an extremely long risidence time when they are released in the atmosphere. Their dispersion throughout the troposphere and into the stratosphere is a consequence of the long atmospheric residence time. The ultraviolet intensity at stratospheric altitudes causes their photolytic decomposition and results in the release of atomic chlorine. The following reactions can then ensue

$$Cl + O_3 \longrightarrow ClO + O_2$$
$$ClO + O \longrightarrow Cl + O_2$$

In this instance the chlorine atom acts as a catalyst because it is regenerated in the second reaction, and the net effect again is the recombination of one O_3 molecule with one O atom to form two O_2 molecules. A study sponsored by the National Research Council (Ref. 15) examined this problem in detail and concluded that "the long term release of F-11 and F-12 at the present rates will cause an appreciable reduction in the amount of stratospheric ozone." The study also pointed out that hydrogen chloride (HCl) from the United States space shuttle and N_2O from the increasing use of fertilizers may also constitute threats to stratospheric ozone. Legislation by the United States Congress to restrict the use of halocarbons was a consequence of the study by the National Research Council.

REFERENCES

1. T. J. Love, *Radiative Heat Transfer* (C. E. Merrill Publishing Co., Columbus, Ohio, 1968).
2. R. Siegel and J. R. Howell, *Thermal Radiation Heat Transfer* (McGraw-Hill Book Company, New York, 1972).
3. R. M. Goody, *Atmospheric Radiation* (Oxford Press, London, 1964).
4. R. A. McLatchey, Personal communication, November 1973. Also R. A. McLatchey et al., "AFCRL Atmospheric Absorption Line Parameters Compilation," AFCRL-TR-73-0096, Air Force Cambridge Research Laboratories, January 26, 1973.
5. R. M. Goody and J. C. G. Walker, *Atmospheres* (Prentice-Hall, Inc., Englewood Cliffs, N.J., 1972).
6. R. D. Cadle and E. R. Allen, "Atmospheric Photochemistry," *Science*, **167**, 243–249 (1970).
7. R. E. Newell, "The Global Circulation of Atmospheric Pollutants," *Sci. Amer.*, **224**, 31–42 (January 1971).
8. J. L. Threlkeld, "Solar Radiation and Atmospheric Heat Balance," in *The Earth's Atmosphere*, Vol. XIII, AIAA Selected Reprint Series (A.I.A.A., New York, 1972).
9. C. D. Keeling, unpublished data (1977), and C. D. Keeling et al., *Tellus*, **28**, 538–551 (1976).
10. J. M. Mitchell, Jr., "The Natural Breakdown of the Present Interglacial and its Possible Intervention by Human Activities," *Quaternary Research*, **2**, 436–445 (1972).

11. J. M. Mitchell, Jr., "A Reassessment of Atmospheric Pollution as a Cause of Long-Term Changes of Global Temperature," in *Global Effects of Environmental Pollution* (D. Reidel Publishing Co., Holland, 1973).
12. National Research Council, "Energy and Climate," National Academy of Science, Washington, D.C., 1977.
13. F. S. Johnson, "SST Ozone and Skin Cancer" *Aeronaut. and Astronaut.*, 11, 16–21 (1973).
14. National Research Council, "Environmental Impact of Stratospheric Flight," National Academy of Science, Washington, D.C., 1975.
15. National Research Council, "Halocarbons: Effects of Stratospheric Ozone," National Academy of Science, Washington D.C., 1976.

PROBLEMS

5.1. Assume the earth's atmosphere could be approximated by a filter which transmits 80 % of the incoming solar radiation that lies between 0.3 and 1.5 μm. All other solar radiation is considered to be blocked by absorption lines or bands. What fraction of the total incoming solar radiation is incident upon the earth's surface?

5.2. Assume the earth's atmosphere to be opaque to radiation emitted by the earth's surface at 273°K except in the 8 to 14 μm wavelength interval. Within this interval we will assume the transmission is 60 % at all wavelengths. What fraction of the earth's radiation is transmitted through the atmosphere if the above approximations apply and the earth is a perfect emitter?

5.3. The following problem is an idealization of the "greenhouse effect." Suppose a radiation flux of 285 W/m^2 is incident upon a glassed-in enclosure. The glass has the property of a short-pass interference filter; that is, it transmits all light less than the cutoff wavelength and reflects all light beyond this wavelength which is at 12 μm. The incoming radiation originates at a blackbody source at 6000°K. The surface enclosed by the glass emits as a blackbody, but only the energy emitted at wavelengths less than 12 μm is transmitted through the window. Calculate the equilibrium temperature of the surface enclosed by the glass. This temperature will be such that the fraction of the energy emitted by the surface at $\lambda < 12\ \mu$m will equal the incident radiation. Convective effects are to be ignored.

5.4. Discuss the influence on the equilibrium temperature calculated in Problem 5.3(a) if the incoming radiative flux is reduced, and (b) if the cutoff wavelength is reduced.

6

THE ATMOSPHERIC BOUNDARY LAYER

The dynamics of the atmosphere have been discussed in the preceding chapters with a primary emphasis on some of the mechanisms that play an important role in the regions of the troposphere that are not adjacent to the ground. The surface of the earth affects the flow in the lowest edge of the troposphere, which is defined by the presence of this influence to be the atmospheric boundary layer. The surface exerts its influence through two different mechanisms. First, a highly chaotic or turbulent flow is produced over the roughness elements—vegetation, buildings—which are present in populated areas. Even over a very smooth plain the flow in the atmospheric boundary layer is turbulent. The second influence of the surface is to serve, during clear weather, as a heat source or a heat sink for the air immediately above it. For example, daytime incoming solar radiation heats the surface. which in turn heats the atmosphere and usually results in strong mixing action. The heat sink effect occurs during cloudless nights when the earth's surface is cooled by radiation to the upper atmosphere. The surface of the earth thus represents an important influence which was not considered in our earlier discussion.

The viscosity of air plays an important role in the flow in the lowest layer of the troposphere. The concept of viscosity is easily grasped by considering the simple shear flow which is established in the space between two parallel plates, one of which moves at constant speed while the other remains fixed (see Fig. 3.3). Experiments show that many fluids exert a shear force on the moving plate which varies directly as the plate velocity and inversely as the separation distance. When the retarding force is expressed as a force per unit area or shear stress τ, the relationship is expressed by

$$\tau = \mu \frac{U_0}{h}$$

where U_0 is the velocity of the moving plate and h is the plate separation distance. The constant of proportionality μ is called the absolute viscosity. Fluids which display this type of behavior include many liquids—most notably water—all ideal gases and mixtures of ideal gases such as air. Liquids and gases obeying this simplest law of viscosity are called Newtonian fluids. When the rate of strain, U_0/h, is not constant but varies with the distance from the surface as in a boundary layer, the above definition is generalized to

$$\tau = \mu \frac{du}{dz}$$

The absolute viscosity of liquids and ideal gases is a function of temperature. Often we choose to use the kinematic viscosity ν which is defined by

$$\nu = \frac{\mu}{\rho}$$

where ρ is the density of the fluid. The kinematic viscosity of gases and their mixtures depends on both pressure and temperature and its value for air is 14.8×10^{-6} m^2/s when $T = 15$°C and $p = 1013$ mb (see p. 312).

The importance of the viscosity of a fluid in controlling the character of the flow in a pipe was strikingly demonstrated by Osborne Reynolds. He showed, through the use of glass tubes and injected dye, that the flow was, on some occasions, in smooth layers or laminar while in other instances the flow was highly disordered and chaotic or turbulent. Reynolds concluded that the relevant physical quantities were the mean velocity U, the kinematic viscosity ν, and the pipe diameter D which could be formulated into the dimensionless number, UD/ν, that bears his name. The pipe flow was found to be turbulent if the Reynolds number was in excess of 2000, a value that continues to be accepted with but minor refinement.

In this chapter we introduce some concepts used to describe turbulent flows and discuss the influence of turbulence on the structure of the atmospheric boundary layer.

6.1. REYNOLDS EQUATIONS FOR ATMOSPHERIC FLOWS

Experience indicates that the flow near the surface of the earth is normally "gusty" or turbulent. A definition of a turbulent flow field that will meet widespread acceptance is difficult to formulate. At the suggestion of Stewart

(Refs. 1 and 2) we avoid attempting a precise definition in favor of presenting a set of symptoms which are common to turbulent flow fields:

1. A turbulent flow displays a highly disordered, chaotic character which never reproduces itself in full detail.
2. A turbulent flow displays a relatively strong propensity to promote mixing. The mixing action of turbulence serves to irradicate non-uniformities of scalar quantities such as temperature or the concentration of a particular chemical species. Clearly the ability of the atmosphere near the surface of the earth to promote mixing is of paramount importance in the dispersion of pollutants.
3. A turbulent flow field possesses a three-dimensional continuous distribution of vorticity which is also highly irregular and is not self-reproducing in time.

In view of the above description of turbulence it is not surprising that attempts to calculate turbulent flow fields are filled with complexities whose details in many cases are yet to be resolved. In this chapter some elements of turbulent flows are presented in an introductory fashion. A more extensive treatment of the topic is available in the standard fluid mechanics texts or, from a more advanced viewpoint, in the works of Tennekes and Lumley (Ref. 3), and Hinze (Ref. 4). Our immediate task will be to inquire into an appropriate form of the equations of motion of a turbulent flow field.

Because the value of any particular property in a turbulent flow possesses a complex time dependence, it is often advantageous to deal with the values of the various properties that are averaged over a suitable period. For example the time-averaged value of the x component of velocity, designated as $\bar{u}$, is defined by

$$\bar{u} = \frac{1}{T}\int_{t-T/2}^{t+T/2} u(t)dt \tag{6.1}$$

where T is the averaging time. In general the time-averaged value of a property may be time dependent in which case the value of T must be carefully chosen. In Fig. 6.1 we illustrate a turbulent velocity profile in which the earlier portion of the mean flow is time dependent but the later portion is constant. We use the term quasi-steady to describe the case of a turbulent flow whose time-averaged quantities are constant even though the instantaneous values of the properties are highly time dependent.

The difference between the instantaneous and the time-averaged values is the fluctuation

$$u'(t) = u(t) - \bar{u}$$

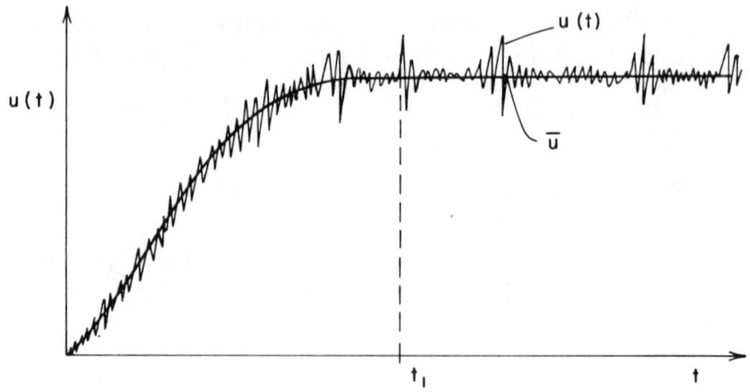

Figure 6.1. The x component of velocity in a turbulent flow. The flow is quasi-steady for $t > t_1$ when $\bar{u}$ is independent of time.

Substitution of the fluctuation into the time-averaging integral shows that its time average is zero, $\overline{u'(t)} = 0$. Special transducers of high frequency response are required to measure the fluctuating values of fluid properties in a turbulent flow. An example is hot wire anemometry that has been perfected for the measurement of the fluctuations of the components of the velocity vector.

We further explore the properties of the time-averaging process. By substituting the product of a constant scalar s and a scalar property $a(t)$ into Eq. 6.1, we find

$$\overline{(sa)} = s\bar{a} \tag{6.2}$$

By a similar process it is found that the time average of the sum of two scalar properties is the sum of the time-averaged values,

$$\overline{(a + b)} = \bar{a} + \bar{b} \tag{6.3}$$

The time average of a quantity that is already time-averaged is simply the time average

$$\overline{(\bar{a})} = \bar{a} \tag{6.4}$$

The order of integration and differentiation can be interchanged when the limits of integration are fixed and thus

$$\overline{\left(\frac{\partial a}{\partial x}\right)} = \frac{\partial \bar{a}}{\partial x} \quad \text{and} \quad \overline{\left(\frac{\partial a}{\partial t}\right)} = \frac{\partial \bar{a}}{\partial t} \tag{6.5}$$

In a quasi-steady flow the time derivative of the time-averaged values of the three velocity components are zero.

The equations which describe the motion of a fluid can be recast into forms that describe the mean velocity components of a turbulent flow by applying these properties. For example the continuity equation for a fluid which is incompressible—a fluid whose parcel densities remain constant with time, $D\rho/Dt = 0$—is the familiar expression

$$\frac{\partial u}{\partial x} + \frac{\partial v}{\partial y} + \frac{\partial w}{\partial z} = 0 \tag{6.6}$$

This equation that expresses the conservation of mass applies in a turbulent flow field when u, v, and w are understood to be instantaneous values of the velocity components. Time-averaging the entire equation and using the properties of Eqs. 6.3 and 6.4 we find

$$\frac{\partial \bar{u}}{\partial x} + \frac{\partial \bar{v}}{\partial y} + \frac{\partial \bar{w}}{\partial z} = 0 \tag{6.7}$$

Thus the continuity equation for the time-averaged or mean velocity components is identical to the continuity equation for the instantaneous velocity components. Further we note, since $u = \bar{u} + u'$, that

$$\frac{\partial u}{\partial x} = \frac{\partial \bar{u}}{\partial x} + \frac{\partial u'}{\partial x}$$

and similarly for the other components. Equation 6.6 can then be written in the expanded form

$$\frac{\partial \bar{u}}{\partial x} + \frac{\partial u'}{\partial x} + \frac{\partial \bar{v}}{\partial y} + \frac{\partial v'}{\partial y} + \frac{\partial \bar{w}}{\partial z} + \frac{\partial w'}{\partial z} = 0 \tag{6.8}$$

Subtracting Eq. 6.7 from Eq. 6.8, we find

$$\frac{\partial u'}{\partial x} + \frac{\partial v'}{\partial y} + \frac{\partial w'}{\partial z} = 0 \tag{6.9}$$

and thus the velocity fluctuations satisfy a continuity equation of the customary form.

The equations for the conservation of momentum yield an interesting result when recast in the form applicable for the time-averaged values. We return to Eqs. 2.6 which are essentially Euler's equations written in tangent plane coordinates (x east, y north, and z parallel and oppositely directed to the acceleration-of-gravity vector). The important Coriolis terms appear only in the x and y components; the gravitational acceleration appears in the z equation. Expanding the particle derivative in the usual fashion, we write

$$\frac{\partial u}{\partial t} + u\frac{\partial u}{\partial x} + v\frac{\partial u}{\partial y} + w\frac{\partial u}{\partial z} = -\frac{1}{\rho}\frac{\partial p}{\partial x} + fv \tag{6.10}$$

$$\frac{\partial v}{\partial t} + u\frac{\partial v}{\partial x} + v\frac{\partial v}{\partial y} + w\frac{\partial v}{\partial z} = -\frac{1}{\rho}\frac{\partial p}{\partial y} - fu \tag{6.11}$$

$$\frac{\partial w}{\partial t} + u\frac{\partial w}{\partial x} + v\frac{\partial w}{\partial y} + w\frac{\partial w}{\partial z} = -\frac{1}{\rho}\frac{\partial p}{\partial z} - g \tag{6.12}$$

We proceed to time-average the above equations and, for example, Eq. 6.10 becomes

$$\frac{\partial \bar{u}}{\partial t} + \overline{\left(u\frac{\partial v}{\partial x}\right)} + \overline{\left(v\frac{\partial u}{\partial y}\right)} + \overline{\left(w\frac{\partial u}{\partial z}\right)} = -\frac{1}{\rho}\frac{\partial \bar{p}}{\partial x} + f\bar{v}$$

A convective term may be expanded by noting

$$\begin{aligned}
\overline{\left(u\frac{\partial u}{\partial x}\right)} &= \overline{(\bar{u} + u')\frac{\partial(\bar{u} + u')}{\partial x}} \\
&= \overline{\bar{u}\frac{\partial \bar{u}}{\partial x} + u'\frac{\partial \bar{u}}{\partial x} + \bar{u}\frac{\partial u'}{\partial x} + u'\frac{\partial u'}{\partial x}} \\
&= \bar{u}\frac{\partial \bar{u}}{\partial x} + \overline{\left(u'\frac{\partial u'}{\partial x}\right)}
\end{aligned}$$

since $\overline{u'} = 0$. It is apparent that the application of the time-averaging process to the convective terms gives rise to three new terms which are in fact non-zero. To recast these terms we note the following identity,

$$\begin{aligned}
\frac{\partial(u'u')}{\partial x} + \frac{\partial(v'u')}{\partial y} + \frac{\partial(w'u')}{\partial z} &= u'\frac{\partial u'}{\partial x} + v'\frac{\partial u'}{\partial y} + w'\frac{\partial u'}{\partial z} \\
&\quad + u'\left(\frac{\partial u'}{\partial x} + \frac{\partial v'}{\partial y} + \frac{\partial w'}{\partial z}\right)
\end{aligned}$$

The three terms in parenthesis on the right side of the equation vanish by virtue of the continuity equation for the fluctuating components. The resulting expression can be time-averaged to yield

$$\overline{u'\frac{\partial u'}{\partial x}} + \overline{v'\frac{\partial u'}{\partial y}} + \overline{w'\frac{\partial u'}{\partial z}} = \frac{\partial \overline{u'u'}}{\partial x} + \frac{\partial \overline{v'u'}}{\partial y} + \frac{\partial \overline{w'u'}}{\partial z}$$

The three new terms in the x component of the momentum equation are written on the right-hand side with the appropriate change of sign. This process is repeated for the other two components of momentum and we finally obtain

$$\frac{\partial \bar{u}}{\partial t} + \bar{u}\frac{\partial \bar{u}}{\partial x} + \bar{v}\frac{\partial \bar{u}}{\partial y} + \bar{w}\frac{\partial \bar{u}}{\partial z} = -\frac{1}{\rho}\frac{\partial \bar{p}}{\partial x} + f\bar{v} - \left(\frac{\partial \overline{u'u'}}{\partial x} + \frac{\partial \overline{u'v'}}{\partial y} + \frac{\partial \overline{u'w'}}{\partial z}\right) \tag{6.13}$$

$$\frac{\partial \bar{v}}{\partial t} + \bar{u}\frac{\partial \bar{v}}{\partial x} + \bar{v}\frac{\partial \bar{v}}{\partial y} + \bar{w}\frac{\partial \bar{v}}{\partial z} = -\frac{1}{\rho}\frac{\partial \bar{p}}{\partial y} - f\bar{u} - \left(\frac{\partial \overline{v'u'}}{\partial x} + \frac{\partial \overline{v'v'}}{\partial y} + \frac{\partial \overline{v'w'}}{\partial z}\right) \tag{6.14}$$

$$\frac{\partial \bar{w}}{\partial t} + \bar{u}\frac{\partial \bar{w}}{\partial x} + \bar{v}\frac{\partial \bar{w}}{\partial y} + \bar{w}\frac{\partial \bar{w}}{\partial z} = -\frac{1}{\rho}\frac{\partial \bar{p}}{\partial z} - g - \left(\frac{\partial \overline{w'u'}}{\partial x} + \frac{\partial \overline{w'v'}}{\partial y} + \frac{\partial \overline{w'w'}}{\partial z}\right) \tag{6.15}$$

These expressions are Reynolds equations for a turbulent flow here written for the rotating tangent plane coordinate system. These equations differ from the corresponding instantaneous flow equations, Eqs. 6.10 to 6.12, by the appearance of nine new terms on the right-hand side which, if multiplied by density, are called Reynolds stress terms. The appearance of these terms has major consequences which will be examined below.

For example, certain classes of laminar, constant density flow problems can be described mathematically by equations similar to Eqs. 6.6, 6.10, 6.11, and 6.12. These equations together with appropriate boundary conditions can be solved since they represent a system of four equations in the four dependent variables u, v, w, and p. The corresponding expressions for the mean velocity components are Eqs. 6.7, 6.13, 6.14, and 6.15, but these equations contain, if ρ is treated as constant, the four mean variables $\bar{u}$, $\bar{v}$, $\bar{w}$, and $\bar{p}$ plus the nine components that represent the Reynolds stress terms. The equations for the mean flow are thus not a closed or determinant system. The major problem in the analytic description of turbulent flows is the lack of information that is necessary to write a complete set of equations for the mean flow. This is the so-called closure problem of turbulence.

The various components of the Reynolds stress terms can be arranged in an array which is the turbulent stress tensor τ_{ij}^T composed of the nine terms contained on the right side of Eqs. 6.13, 6.14, and 6.15.

$$\tau_{ij}^T = \begin{vmatrix} \tau_{xx}^T & \tau_{xy}^T & \tau_{xz}^T \\ \tau_{yx}^T & \tau_{yy}^T & \tau_{yz}^T \\ \tau_{zx}^T & \tau_{zy}^T & \tau_{zz}^T \end{vmatrix} = \begin{vmatrix} -\rho\overline{u'u'} & -\rho\overline{u'v'} & -\rho\overline{u'w'} \\ -\rho\overline{v'u'} & -\rho\overline{v'v'} & -\rho\overline{v'w'} \\ -\rho\overline{w'u'} & -\rho\overline{w'v'} & -\rho\overline{w'w'} \end{vmatrix} \tag{6.16}$$

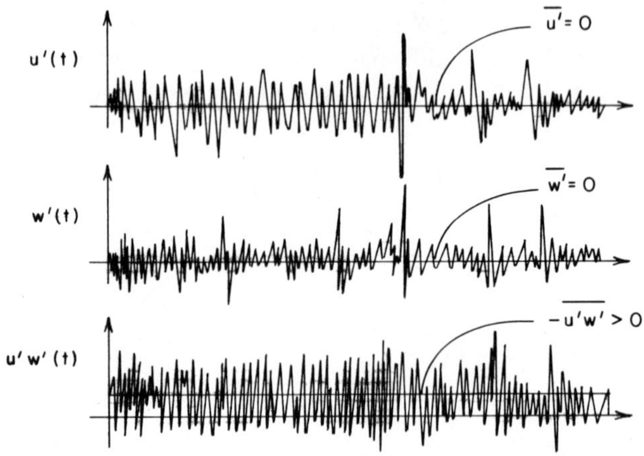

Figure 6.2. The velocity fluctuations in the x and z directions produce nonzero time-averaged correlations. Although $\bar{u}' = \bar{w}' = 0$, the correlation $-\overline{u'w'} > 0$ in a boundary layer (schematic drawing).

The notation τ_{ij}^T designates the turbulent stress on the ith face in the jth direction.

The turbulent stress tensor is clearly symmetric about the diagonal elements because $\overline{v'u'} = \overline{u'v'}$, $\overline{u'w'} = \overline{w'u'}$, and $\overline{w'v'} = \overline{v'w'}$. Thus the nine components of the Reynolds stress tensor are made up of six independent quantities. A diagonal term of the array given by Eq. 6.16 is nonzero because the square of a fluctuating quantity is a rectified fluctuation which has a nonzero average value. An off-diagonal term such as $-\overline{u'w'}$ will be nonzero if the negative product of u' and w', each of which possesses a zero time-averaged value, is usually positive as depicted in Fig. 6.2. The quantity $-\overline{u'w'}$ is then a nonzero positive quantity even when $u'w' = 0$ because either u' or w' is instantaneously equal to zero. Two fluctuating quantities whose time-averaged product is nonzero are said to be correlated. If the time-averaged product of two fluctuations is zero, the quantities are said to be uncorrelated. This concept is made quantitative through the use of a correlation coefficient. For example, the correlation coefficient for the x and z directed velocity fluctuations is given by

$$R_{uw} = \frac{-\overline{u'w'}}{(\overline{u'^2})^{1/2}(\overline{w'^2})^{1/2}} \tag{6.17}$$

In the improbable event that $u' = -\overline{w'}$ over a certain time interval, then $R_{uw} = 1.0$ and a perfect correlation would exist. A value $R_{uw} = 0$ would indicate the fluctuations are uncorrelated. A turbulent flow field whose properties are independent of direction, $\overline{u'^2} = \overline{v'^2} = \overline{w'^2}$, is said to be *isotropic*. If the properties of a turbulent flow field are independent of position, then the turbulence is said to be *homogeneous*. Finally if the statistical properties are independent of time the turbulence is said to be *stationary*. The turbulence produced in the wake of the grid of closely spaced perpendicular wires is stationary and homogeneous in planes parallel to the grid.

6.2. TRANSPORT CONCEPTS IN A TURBULENT FLOW

Momentum and heat may be transported in a moving fluid, and the presence of these processes can have a major influence on the flow field. For example, the moving plate of the laminar shear flow that was discussed earlier imparts momentum to the fluid adjacent to it as a result of the shear stress which can be sustained by a viscous fluid. Momentum is imparted through the successive layers of the fluid by the action of viscosity to the fixed plate where a shear stress is imparted. If a temperature difference exists between the two plates, heat is conducted in a manner entirely similar to the conduction of heat in a solid material. On the other hand, if the flow between the plates is turbulent rather than laminar, then the transport processes are drastically altered. In the following sections we examine transport concepts in a turbulent flow near a surface and consider the influence of the interaction on the flow itself.

6.2.1. Momentum and Heat Transport in a Turbulent Boundary Layer

To illustrate turbulent transport processes, we consider the quasi-steady turbulent flow over a surface in the absence of the Coriolis force. In this case the time-averaged velocity near the surface is in the direction of the mean velocity vector at the outer edge of the boundary layer and the magnitude of the mean velocity vector is a function only of distance from the surface. We designate the mean velocity as $\bar{u}$ and the distance from the surface as z.

Referring to Fig. 6.3*a*, we visualize a fluid parcel at distance z_1 from the surface traveling at the mean velocity $\bar{u}_1$ when there occurs a positive fluctuation of velocity w' in the direction away from the wall. The parcel of fluid arrives at z_2 where the x component of velocity is $\bar{u}_2$. If we assume that the parcel essentially retains its original velocity $\bar{u}_1$, then it arrives at z_2 with a decrement of x directed velocity, which approaches $\bar{u}_1 - \bar{u}_2$ and is therefore negative. While individual parcels may deviate from this picture, the description will be statistically correct for a large number of parcels. Thus a positive

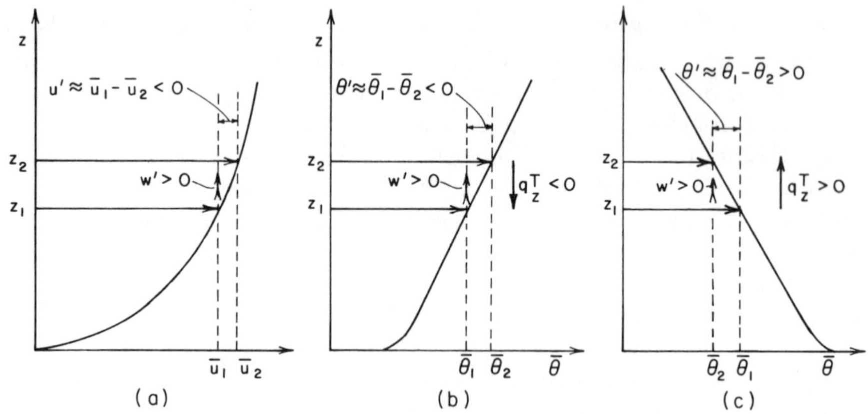

Figure 6.3. The transport of momentum and heat in a turbulent stratified boundary layer. (*a*) The fluctuations u' and w' usually have opposite signs and the turbulent stress $-\rho\overline{u'w'} > 0$. (*b*) When the lapse rate is subadiabatic, the fluctuations θ' and w' usually have opposite signs. The turbulent heat flux $q_t^T = +\rho c_p \overline{\theta' w'}$ is negative and directed toward the surface. (*c*) If the lapse rate is superadiabatic, the fluctuations θ' and w' usually have the same sign. The turbulent heat flux is then positive and directed away from the surface.

value of w' will be associated with a negative value of u' and consequently the product $-u'w'$ will usually be positive, while both the velocity correlation $-\overline{u'w'}$ and the Reynolds stress $-\rho\overline{u'w'}$ will always be positive. Furthermore the parcel, in moving from the z_1 plane to the z_2 plane, carries a deficiency of x directed momentum per unit volume equal to $+\rho u'$ where $u' < 0$. Since $w' > 0$ this deficit of momentum per unit volume is transported from the region of lower momentum to the region of higher momentum at the instantaneous rate of $\rho u'w'$. Thus the turbulent shear stress $-\rho\overline{u'w'}$ represents both the x directed shear force exerted and the average flow of x momentum passing through a unit area parallel to the surface.

The above argument also applies when the turbulent fluctuation of the parcel is toward the surface. A parcel initially at z_2 travels at a mean velocity $\bar{u}_2$ when a velocity fluctuation $w' < 0$ occurs which may carry it to z_1. There the parcel will be surrounded by fluid traveling at a lower velocity and at least momentarily will possess a fluctuating x component u' approaching $\bar{u}_2 - \bar{u}_1$ which is greater than zero. The negative value of w' will be associated with a positive u' and $-u'w'$, will usually be positive while $-\overline{u'w'}$ and $-\rho\overline{u'w'}$ will always be positive quantities. The region of higher momentum supplies an increment of x directed momentum to the region of lower momentum. The momentum flux remains, as in the previous case, directed toward the surface.

The ability of a turbulent flow to transport heat in the presence of a gradient of the mean temperature is demonstrated in a similar manner. We visualize a temperature gradient to exist in the presence of a gravitational field and therefore a density stratified turbulent flow is present. For this reason we deal with potential temperature rather than actual temperature because the former remains unchanged when the fluid parcel changes its height in the idealization that this change is accomplished adiabatically. Of course the changes in actual and potential temperature are equal to one another for small changes in height. The potential temperature in a turbulent flow field can be expressed as the sum of a time-averaged or mean value $\bar{\theta}$ and a fluctuation θ'. If $d\bar{\theta}/dz > 0$, corresponding to the statically stable case, as depicted in Fig. 6.3*b*, then a parcel which moves from z_1 to z_2 as a result of a positive value of w' will find itself surrounded by air of higher potential temperature and will possess, on the average, a negative value of θ' which approaches $\bar{\theta}_1 - \bar{\theta}_2$. The product $+w'\theta'$ is generally negative and the velocity-temperature correlation $+\overline{w'\theta'}$ is always negative. The parcel arrives at z_2 with a defect of temperature, $\theta' < 0$, and therefore a deficiency of enthalpy per unit volume equal to $\rho c_p \theta'$. The instantaneous rate of volumetric heat transfer per unit area parallel to the surface is $+\rho c_p w'\theta'$ and its average value is $+\rho c_p \overline{w'\theta'}$. We define the latter to be the turbulent heat flux q_z^T. Thus, in the case examined, the turbulent flow interacts with the gradient of potential temperature to produce a negative turbulent heat flux—a heat flow that is directed along the negative z axis.

The argument holds for an excursion of a parcel from z_2 to z_1 as a result of a momentary fluctuation $w' < 0$. In this case the potential temperature fluctuation θ' approaches the value $\bar{\theta}_2 - \bar{\theta}_1$ and is generally positive. The product $+w'\theta'$ is generally negative and the velocity–temperature correlation $+\overline{w'\theta'}$ is always negative. The average rate of turbulent heat transfer per unit area, $+\rho c_p \overline{w'\theta'}$, is again negative representing turbulent heat transfer in the negative z direction.

A superadiabatic lapse rate, as depicted in Fig. 6.3*c* will result in a turbulent heat flux in the positive z direction. (A superadiabatic lapse rate corresponds to an unstable stratification and is less frequently observed. A ground level superadiabatic lapse rate of limited vertical extent must be present during daylight hours of strong insolation when the surface of the earth is hotter than the air above it.) In this case we visualize a parcel originally at z_1 which experiences a fluctuation $w' > 0$ that carries it to z_2 with an excess of enthalpy per unit volume of $\rho c_p \theta'$ where θ' approaches $\bar{\theta}_1 - \bar{\theta}_2$ on the average and is therefore usually positive. The instantaneous rate of heat transfer per unit volume across a unit of surface area is $+\rho c_p \theta' w'$ and its average value is $+\rho c_p \overline{\theta' w'} = q_z^T$. In this case the signs of both the velocity and temperature fluctuations are positive, and the turbulent heat flux is directed in the positive z direction which is defined to be the positive sense.

We obtain the same result in the case of the superadiabatic lapse rate if the parcel is initially at z_2. A negative velocity fluctuation $w' < 0$ can transport the parcel to z_1 where θ' will approach $\bar{\theta}_2 - \bar{\theta}_1$ and will generally be negative. The product $+w'\theta'$ will be generally positive and the corresponding correlation will always be positive. Accordingly the turbulent heat flux, $q_z^T = +\rho c_p \overline{w'\theta'}$, is positive which signifies turbulent heat transport in the positive z direction or away from the surface.

6.2.2. The Mixing Length

The above discussion alludes to the concept that a turbulent flow consists of parcels of fluid or "eddies" of unspecified size that usually travel at the mean velocity but which undergo turbulent excursions to a new location giving rise to a new local velocity fluctuation. This oversimplified picture would be more realistic if it is recognized that the parcels are in fact of a wide variety of sizes, that their turbulent excursions are over various lengths, and that each parcel is subject to the mixing action of turbulence and therefore will eventually lose its identity. The mixing length l_m is the distance which the "eddy" travels calculated on the basis of an averaging procedure which we do not specify, but which takes the above factors into consideration. The mixing length in a turbulent flow is an extension of a more clearly definable concept from the kinetic theory of gases. The motion of any one molecule in a gas is very complicated as it undergoes successive collisions with other particles moving at various velocities. The average distance that a particle can travel between two successive collisions, the mean free path of the molecule, can be accurately calculated and related to macroscopic gas properties. Although the mixing length for turbulent eddies is not subject to a similar type of exact mathematical analysis, nevertheless it has proved to be a useful concept.

The argument of Section 6.2.1 leads us to an estimate of the mixing length since we note that the average turbulent fluctuation $(\overline{u'^2})^{1/2}$ is roughly equal to the mean gradient $d\bar{u}/dz$ times the mixing length,

$$\left(\overline{u'^2}\right)^{1/2} \approx l_m \frac{d\bar{u}}{dz}$$

The value of the mean velocity fluctuations in the x and z directions will be of comparable magnitude, $(\overline{u'^2})^{1/2} \approx (\overline{w'^2})^{1/2}$, hence

$$\left(\overline{u'^2}\right)^{1/2}\left(\overline{w'^2}\right)^{1/2} \approx l_m^2\left(\frac{d\bar{u}}{dz}\right)^2 \tag{6.18}$$

Combining Eqs. 6.17 and 6.18, we obtain

$$-\rho\overline{u'w'} = \rho(R_{uw}^{1/2}l_m)^2\left(\frac{d\bar{u}}{dz}\right)^2$$

The value of $R_{uw}^{1/2}$ is not very much less than unity in a boundary layer and can be absorbed within l_m. The shear stress τ_{zx}^T is then given by

$$\tau_{zx}^T = -\rho\overline{u'w'} = \rho l_m^2\left(\frac{d\bar{u}}{dz}\right)^2 \tag{6.19}$$

The use of the above expression, which is called the mixing length formula of Prandtl who formulated this concept, is illustrated below.

6.2.3. The Velocity Distribution near a Surface in the Absence of Buoyancy Effects

For the flow near a surface, Prandtl argued on dimensional reasoning that the mixing length is proportional to the distance from the surface. Therefore

$$l_m = \kappa z \tag{6.20}$$

where κ is a universal constant called the von Karman constant.

Combining Eqs. 6.19 and 6.20, we obtain

$$\tau_{zx}^T = \rho\kappa^2 z^2\left(\frac{d\bar{u}}{dz}\right)^2 \tag{6.21}$$

Equation 6.21 provides the basis for determining the velocity profile if it is recognized that near the surface the shear stress is approximately constant independent of z. We designate the value of shear stress near the surface as τ_0 and, since τ_0/ρ has the dimensions of the square of a velocity, we define the friction velocity as

$$u_* \equiv \sqrt{\frac{\tau_0}{\rho}} \tag{6.22}$$

With this definition Eq. 6.21 can be written as

$$\frac{1}{u_*}\frac{d\bar{u}}{dz} = \frac{1}{\kappa z} \tag{6.23}$$

Equation 6.23 becomes dimensionless when we define $u^+ = \bar{u}/u_*$ and $z^+ = zu_*/\nu$,

$$\frac{du^+}{dz^+} = \frac{1}{\kappa z^+} \tag{6.24}$$

Integration of this expression gives the logarithmic velocity profile

$$u^+ = \frac{1}{\kappa} \ln z^+ + c_1 \tag{6.25}$$

including an integration constant which must be determined. Experiments show that the flow immediately adjacent a smooth wall is essentially laminar. The flow in this so-called laminar sublayer is dominated by the shear stress delivered to it by the turbulent outer region. The velocity profile is that of the simple shear flow described above,

$$u = \frac{U_s}{\delta_s} z \tag{6.26}$$

where U_s is the velocity at the outer edge of the sublayer possessing the thickness δ_s. Considering that $\tau = \rho \nu U_s/\delta_s = u_*^2$, we rewrite Eq. 6.26 in the dimensionless form

$$u^+ = z^+ \tag{6.27}$$

For pipe flow Nikuradse (see Schlichting, Ref. 5) found Eq. 6.25 applicable for $z^+ > 70$ with $\kappa = 0.40$ and $c_1 = 5.5$. Equation 6.27 was found applicable for $z^+ \lessapprox 5$ which represents the thickness of the laminar sublayer. For the intermediate values of wall distance, the data of Nikuradse blended smoothly between these two curves which are depicted in Fig. 6.4*a*.

The above described velocity profile applies equally well to the turbulent boundary layer which fills a pipe or which is found adjacent to a flat surface or plate. For this reason the profile is termed a "universal" velocity profile. Velocity profiles are sometimes found to be described by the logarithmic law well beyond the constant shear layer. Thus the observation that velocity varies logarithmically with distance should not be construed to imply that shear stress is constant.

For our purposes the flow over rough surfaces, which was also investigated by Nikuradse, is of greater interest. The roughness elements are described by a single height scale h_r often recommended as the root-mean-square surface roughness. Experiments indicate that the roughness has no influence on the flow in the boundary layer if the roughness height is smaller than the thickness of the laminar sublayer, namely for $h_r < 5\nu/u_*$. The rough surface is termed hydraulically smooth and, in this case, the drag force on the rough surface is communicated through viscous shear stress. On the other hand, for the fully rough regime when the roughness scale is such that $u_* h_r/\nu > 70$, the roughness elements protrude into the turbulent region and resistance to the flow is caused mainly by the drag force which they experience. This force results largely from the differences of pressure over each individual surface obstacle. These obstacles are frequently of such a shape that a flow separation occurs and, in this case, pressure drag (form drag) dominates over viscous shear drag (skin friction).

To describe the velocity distribution over a rough surface, we return to the logarithmic velocity profile. The roughness height h_r is an obvious length with which to nondimensionalize the height z. We may obtain the velocity profile over the rough wall in a manner that will include the smooth wall as a special case by noting

$$\frac{zu_*}{v} = \frac{z}{h_r}\frac{h_r u_*}{v}$$

hence Eq. 6.25 can be recast as

$$\frac{\bar{u}}{u_*} = \frac{1}{\kappa}\ln\frac{z}{h_r} + B \tag{6.28}$$

where

$$B = \frac{1}{\kappa}\ln\left(\frac{h_r u_*}{v}\right) + c_2 \tag{6.29}$$

and c_2 is itself a function of the roughness Reynolds number $h_r u_*/v$. For example, the velocity profile over a smooth wall, $h_r u_*/v < 5$, is given by

$$B = 5.5 + 2.5\ln\left(\frac{h_r u_*}{v}\right) \tag{6.30}$$

Nikuradse (Schlichting, Ref. 5) made systematic tests on pipes whose inner surface was coated with sand of varying grain size to determine B as a function of $h_r u_*/v$. These results are depicted in Fig. 6.4*b* and show, for $\kappa = 0.4$, that $B = 8.5$ for $h_r u_*/v > 66$. Thus for "high" roughness elements in which the drag at the surface is mainly pressure drag the velocity profile becomes independent of viscosity,

$$\frac{\bar{u}}{u_*} = \frac{1}{\kappa}\ln\frac{z}{h_r} + 8.5 \tag{6.31}$$

This expression is often rewritten by combining the constant with the roughness height. We find, accordingly,

$$\frac{\bar{u}}{u*} = \frac{1}{\kappa}\ln\frac{z}{z_0} \tag{6.32}$$

where the aerodynamic roughness z_0 is related to the physical roughness when $\kappa = 0.40$ by

$$z_0 = \frac{h_r}{30} \tag{6.33}$$

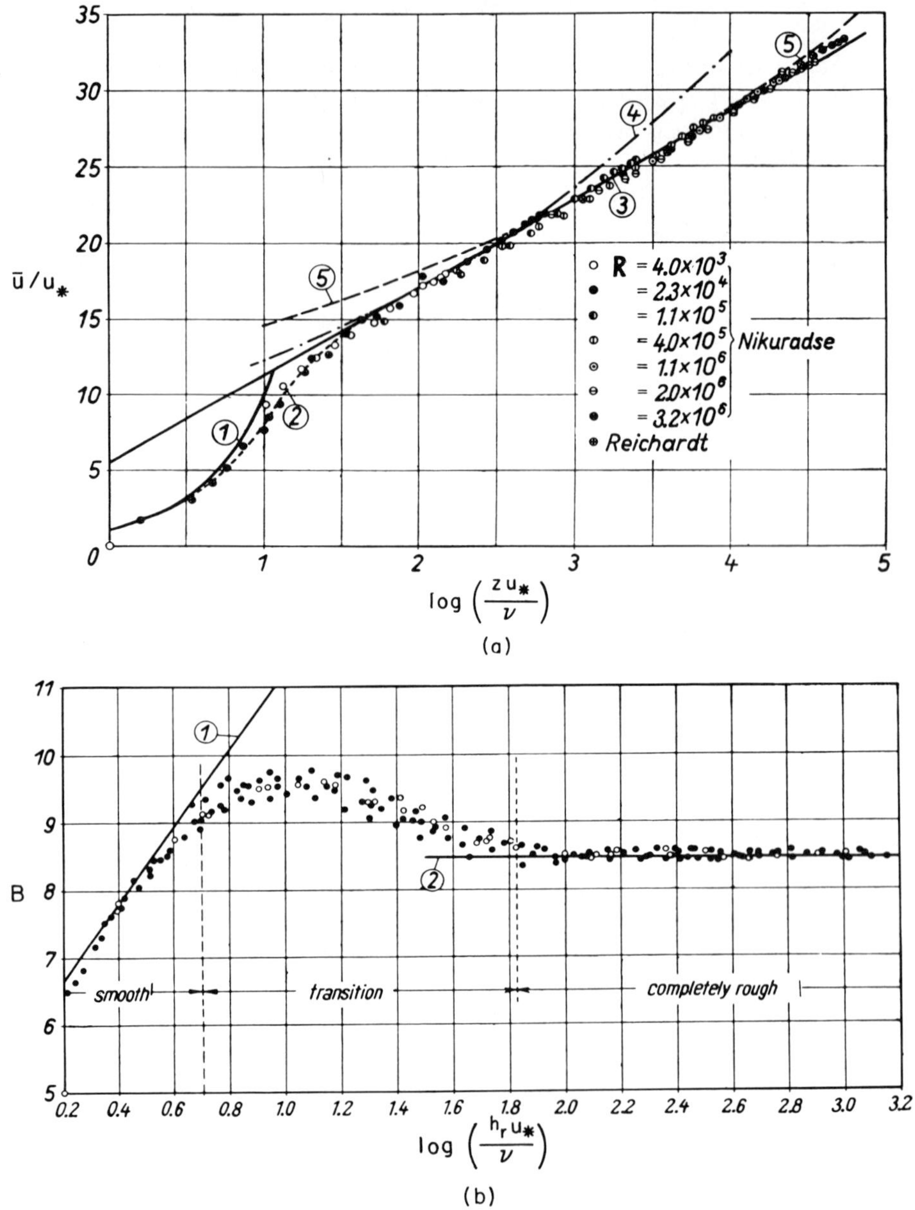

Figure 6.4. The turbulent flow in a boundary layer over walls of various roughness height. (*a*) The universal velocity distribution for smooth pipes. Curve 1 refers to Eq. 6.27, and curve 3 refers to Eq. 6.25, with $\kappa = 0.40$ and $c_1 = 5.5$. (*b*) The function B versus the roughness Reynolds number. Curve 1 refers to Eq. 6.30, and curve 2 refers to B = 8.5. (From H. Schlichting, *Boundary Layer Theory*, 6th Edition, 1968. Used with permission of McGraw-Hill Book Company and the author.)

The results expressed by Eqs. 6.32 and 6.33, originally derived from pipe flow experiments, are often used to describe the mean flow near the earth's surface.

Equation 6.32 must be understood to be applicable above the roughness elements or for $z > h_r$. For values of $z \lesssim h_r$ the flow will depend on the specific geometry of the roughness elements and the value of u_*. This flow is normally sufficiently complex to defy a detailed description by analytical methods. For this reason wind tunnel testing of the flow over scale models of urban terrain plays a leading role in the elucidation of the complex turbulent flow fields in the space immediately adjacent to roughness elements of complex shape and spatial distribution.

6.2.4. The Enhancement of Turbulence by Effects of Buoyancy and the Richardson Numbers

The stability of a quiescent atmosphere was discussed in Chapter 1 and the lapse rate and its relationship to the adiabatic lapse rate was determined. We now consider the factors which determine whether the atmospheric temperature profile acts to reinforce the development of the turbulence that is present, for example, in the atmospheric boundary layer. We consider a fluid parcel of unit volume which experiences a vertical velocity fluctuation $w' > 0$. The instantaneous vertical component of kinetic energy per unit volume of the parcel is $\bar{\rho}^* w'^2/2$. The potential density $\bar{\rho}^*$ is used here because this quantity is invariant with vertical displacements under the adiabatic idealization discussed earlier. Since $u' \approx w'$ in neutral stability, we may replace the latter by $l_{mn}\, d\bar{u}/dz$ in view of the mixing length hypothesis. Accordingly we represent the instantaneous kinetic energy of the fluid parcel or "eddy" as

$$E_K = \frac{\rho^*}{2}\left(\frac{d\bar{u}}{dz}\right)^2 l_{mn}^2 \tag{6.34}$$

where l_{mn} is the mixing length in the absence of stratification effects. A change in potential energy of the parcel results if it undergoes a vertical excursion from height z_1 to z_2. Again we use the potential density to formulate this increment of potential energy as

$$E_p = -g(\bar{\rho}_2^* - \bar{\rho}_1^*)(z_2 - z_1) = -g\frac{d\bar{\rho}^*}{dz}(z_2 - z_1)^2 \tag{6.35}$$

Equating the potential energy to the kinetic energy of the eddy, we find the relationship between the change in height and l_{mn} as

$$\frac{1}{2}\left[\frac{l_{mn}}{z_2 - z_1}\right]^2 = \frac{-\dfrac{g}{\bar{\rho}^*}\dfrac{d\bar{\rho}^*}{dz}}{\left(\dfrac{d\bar{u}}{dz}\right)^2} \tag{6.36}$$

where the quantity on the right side is the gradient Richardson number Ri_g named for the English meteorologist. In view of Eqs. 1.14 and 1.20 we write the gradient Richardson number in its equivalent, more usual form

$$Ri_g = \frac{\left(\dfrac{g}{\bar{\theta}}\dfrac{d\bar{\theta}}{dz}\right)}{\left(\dfrac{d\bar{u}}{dz}\right)^2} \tag{6.37}$$

hence

$$z_2 - z_1 = \frac{l_{mn}}{[2Ri_g]^{1/2}} \tag{6.38}$$

With the latter relation we are able to conclude:

1. Existing turbulence will be suppressed if $Ri_g \gg \frac{1}{2}$. Buoyancy effects convert vertically oriented kinetic energy into potential energy.
2. Existing turbulence will be unaffected by stratification if Ri_g is a critical value $\approx \frac{1}{2}$.
3. Existing turbulence is amplified if $Ri_g \ll \frac{1}{2}$. In this case the buoyancy effects act to reinforce the existing kinetic energy in the turbulent motion.

The critical value of the gradient Richardson number, here found to be one half, has been the subject of controversy and values ranging from 0.1 to 1.0 have been suggested. A numerically precise criterion for the amplification of existing turbulence by buoyancy effects would seem to require additional information on the statistical characteristics of the turbulence of the particular flow field. Lacking this detailed information it seems appropriate to consider the critical value of the Richardson number to be a small positive number whose precise value is determined by the structural details of the turbulent flow.

(The development of turbulence in an initially laminar flow possessing vertical velocity and density gradients has been the object of both theoretical and experimental study. The detailed analysis applicable in this case indicates that a corresponding gradient Richardson number is the appropriate criterion for the development of turbulence. In this case the flow is found to be stable for $Ri > \frac{1}{4}$ and recent experimental results support this conclusion. This criterion is not directly related to the question of the critical value of the gradient Richardson number for the amplification of preexisting turbulence.)

The argument offered above implies that turbulence will be suppressed if

$$\frac{d\bar{\theta}}{dz} \geq \frac{1}{2}\frac{\bar{\theta}}{g}\left(\frac{d\bar{u}}{dz}\right)^2 \tag{6.39}$$

or, in other words, if $d\bar{\theta}/dz$ exceeds a critical value whose magnitude is dependent upon the vertical velocity gradient. Near the earth's surface the vertical velocity gradients are high, and accordingly, the value of $d\bar{\theta}/dz$ must be relatively high to suppress turbulence by the stabilizing influence of a positive gradient of potential temperature. At higher altitudes the value of $d\bar{u}/dz$ decreases and turbulence will be suppressed by a relatively smaller gradient of potential temperature. Finally, as the value of $d\bar{u}/dz$ goes to zero the criterion for suppression of turbulence becomes $d\bar{\theta}/dz > 0$, which coincides with the criterion for the static stability for a quiescent atmosphere as discussed in Chapter 1.

A second dimensionless ratio, which also is associated with the name of Richardson, arises from considering the rate of change of turbulent energy in a shear flow when buoyancy effects are present. An equation for the rate of change of turbulent kinetic energy per unit mass contained by a specified fluid parcel may be obtained from the equations of motion that include a z directed buoyancy term. This equation, which is derived in detail by Plate (Ref. 6),

$$\frac{D}{Dt}\overline{\left(\frac{q'^2}{2}\right)} = -\frac{\partial}{\partial z}\overline{\left[w'\left(\frac{p'}{\rho} + \frac{q'^2}{2}\right)\right]} - \overline{u'w'}\frac{\partial \bar{u}}{\partial z} + \frac{g}{T}\overline{w'\theta'} - \varepsilon \tag{6.40}$$

where

$$q'^2 = u'^2 + v'^2 + w'^2 \tag{6.41}$$

represents a balance among terms that represent production or destruction of turbulent kinetic energy. The term on the left side of Eq. 6.40 is the rate of change of turbulent kinetic energy per unit mass of the fluid parcel itself. The right side represents a series of source or sink terms whose sum will determine whether the turbulent energy increases or decreases. The first term is the rate of vertical transport of turbulent energy by the turbulence itself; it is called the flux divergence term and is composed purely of the vertical transport of kinetic energy plus the transport of energy contained in the pressure–velocity $\overline{w'p'}$ correlation. The second term $-\overline{u'w'}\partial\bar{u}/\partial z$ represents the extraction of energy from the mean flow by the Reynolds stress and is called mechanical or shear production. Mechanical production is always an important source term in a neutrally stable turbulent boundary layer. The term $(g/T)\overline{w'\theta'}$ is the production of turbulent kinetic energy by the interaction of the buoyancy term with the gravitational field and is called the buoyant production. Since its sign may be positive or negative, it acts either as a source or a sink for turbulent energy. Finally, ε, which is intrinsically positive because it is made up of the squares of various velocity gradients, is always a sink term and represents the irreversible conversion of kinetic energy into heat by the action of viscosity. This term is called the viscous dissipation.

We illustrate the application of the above equation in the case of quasi-steady flow when the turbulent kinetic energy is constant so the left side of Eq. 6.40 vanishes. The difficulty in making measurements of all terms prevents an entirely satisfactory accounting. The following description is provided by Wyngaard and Coté (Ref. 7). For nearly neutral stratification the buoyancy production is absent and vertical transport through the flux divergence term appears to be small. The entire energy balance reduces to an equivalence between mechanical production and viscous dissipation. With unstable stratification the vertical convective activity causes the buoyancy term to act as a strong source, while the $\overline{w'q'^2}/2$ portion of the vertical transport acts as a sink term. Mechanical production is less important and dissipation offsets any net imbalance. With weakly stable stratification the buoyancy term is present as a sink for turbulent kinetic energy but does not play a role as important as in the unstable case. Again mechanical production and viscous dissipation are the important terms.

The flux Richardson number is defined as the negative of the ratio of the buoyancy and mechanical production terms,

$$Ri_f = \frac{\dfrac{g}{T}\overline{w'\theta'}}{+\overline{u'w'}\dfrac{\partial \bar{u}}{\partial z}} \tag{6.42}$$

This ratio is of particular interest because it represents the relative importance of the two strong terms which are, in a sense, external to the turbulence itself. The value $Ri_f = 1$ corresponds to the case where the mechanical production is equal to the destruction of turbulent energy by buoyancy effects.

(In general the flux and gradient Richardson numbers are not simply related to one another. If we assume that the turbulent fluxes of momentum and heat are simply proportional to the gradients of the mean velocity and temperature profiles,

$$-\overline{u'w'} = K_m \frac{\partial \bar{u}}{\partial z} \qquad \text{and} \qquad -\overline{w'\theta'} = K_h \frac{\partial \bar{\theta}}{\partial z} \tag{6.43}$$

then

$$Ri_f = \frac{K_h}{K_m} Ri_g \tag{6.44}$$

The constants of proportionality K_m and K_h are the eddy diffusivities of heat and momentum. The motivation of the assumptions of Eq. 6.43 are discussed below. Plate (Ref. 6) discusses the relationship between Ri_f, Ri_g, and K_h/K_m that has been proposed in the literature.)

Near the surface of the earth where the shear stress is constant, we use the definition of u_* and Eq. 6.23 for $\partial\bar{u}/\partial z$ to express the flux Richardson number as

$$Ri_f = -\frac{\kappa g z q_z^T}{\rho c_p T u_*^3} = +\frac{z}{L_{mo}} \tag{6.45}$$

where q_z^T is the turbulent heat flux and L_{mo} is the Monin–Obukhov length scale defined as

$$L_{mo} = \frac{-\rho c_p T u_*^3}{\kappa g q_z^T} \tag{6.46}$$

With this definition, the sign of L_{mo} will be positive when q_z^T is negative as in the case when the temperature profile corresponds to a statically stable stratification. Within the constant stress layer a value of $z = L_{mo}$ defines the altitude at which the mechanical production is balanced by its consumption through the action of buoyancy. For $z < L_{mo}$ within the constant stress layer the mechanical generation of turbulent energy is larger than its consumption by buoyancy. The Monin–Obukhov length scale provides a convenient parameter to use as a measure of the buoyancy production.

6.3. THE NEUTRAL BAROTROPIC ATMOSPHERIC BOUNDARY LAYER

The flow in the atmospheric boundary layer is more easily described under a highly restrictive set of circumstances—a set of conditions that are only occasionally observed to be fulfilled in the atmosphere. The detailed examination of the simplified case affords important insight into the flow near the earth's surface and also provides the basis for the more detailed analyses in which certain restrictions are removed.

6.3.1. The Equations of Motion of the Neutral, Barotropic Atmospheric Boundary Layer

The following assumptions are traditionally imposed in order to simplify the Reynolds equations as they apply to the atmospheric boundary layer:

1. The flow is assumed to be quasisteady. This condition will be fulfilled if the geostrophic wind vector is constant for several hours. Turbulent transfer of momentum is effective in establishing a quasisteady flow adjacent to the earth's surface.
2. The stability is taken to be neutral. Convective activity resulting from solar heating of the surface is absent.
3. The surface of the earth is presumed to possess a uniform scale of roughness and all flow properties are assumed to be constant on horizontal planes. The boundary layer is then said to possess horizontal homogeneity. These assumptions are normally best fulfilled over a vast plain such as in the central United States, certain regions in Australia and Russia, and also over the oceans. Their appropriateness over populated areas is difficult to assess.
4. The flow is assumed to be barotropic, $p = p(\rho)$, and therefore the geostrophic wind speed and direction are independent of altitude (see Section 2.5). The horizontal pressure force is a constant vector throughout the atmospheric boundary layer.
5. The term $\partial(\overline{w'^2})/\partial z$ is small compared with the acceleration of gravity. The term is clearly nonzero because w' vanishes near the surface, but it will be finite as z increases. The term is nevertheless many magnitudes smaller than the acceleration of gravity and can be ignored.

These assumptions result in a vast simplification of the Reynolds equations which apply more generally within the atmospheric boundary layer. According to assumptions 1, 2, and 3 the left sides of Eqs. 6.13, 6.14, and 6.15 vanish; the horizontal derivatives of the Reynolds stress vanish as a result of assumption 3; Eq. 6.15 reduces to the hydrostatic equation as a result of assumptions 2 and 5. The Reynolds equations reduce to

$$\frac{d\overline{(w'u')}}{dz} = -\frac{1}{\rho}\frac{\partial \bar{p}}{\partial x} + fv_b \tag{6.47}$$

$$\frac{d\overline{(w'v')}}{dz} = -\frac{1}{\rho}\frac{\partial \bar{p}}{\partial y} - fu_b \tag{6.48}$$

$$g = -\frac{1}{\rho}\frac{\partial \bar{p}}{\partial z} \tag{6.49}$$

These equations remain in the tangent plane coordinate system—x east, y north, and z vertically upward. The subscript b is added to the time-averaged velocity components to designate the applicability of these equations within the atmospheric boundary layer. The primed quantities designate the fluctuations of the respective velocity components.

The vector form of the above equations is particularly useful. We define the horizontal shear stress vector,

$$\boldsymbol{\tau}_z^T = \tau_{zx}^T \mathbf{i} + \tau_{zy}^T \mathbf{j} \tag{6.50}$$

where, as previously,

$$\frac{\tau_{zx}^T}{\rho} = -\overline{w'u'} \qquad \frac{\tau_{zy}^T}{\rho} = -\overline{w'v'} \tag{6.51}$$

Equations 6.47 and 6.48 are multiplied by the unit vectors $\mathbf{i}$ and $\mathbf{j}$ corresponding to the x and y directions, respectively. The addition of these two equations gives the following form,

$$-\frac{1}{\rho}\nabla_h \bar{p} - \mathbf{f} \times \mathbf{q}_b + \frac{d}{dz}\left(\frac{\boldsymbol{\tau}_z^T}{\rho}\right) = 0 \tag{6.52}$$

where

$$\nabla_h \bar{p} = \frac{\partial \bar{p}}{\partial x}\mathbf{i} + \frac{\partial \bar{p}}{\partial y}\mathbf{j}$$

and

$$\mathbf{q}_b = u_b \mathbf{i} + v_b \mathbf{j}$$

This equation asserts that the vector sum of the pressure, coriolis, and frictional forces is equal to zero. We recall from the discussion of the geostrophic flow approximation, Section 2.4.4, that above the atmospheric boundary layer the condition of geostrophic balance is expressed by

$$\mathbf{f} \times \mathbf{q}_G = -\frac{1}{\rho}\nabla_h p \tag{6.53}$$

With this relationship we eliminate $\nabla_h \bar{p}$ in Eq. 6.52 and express the result as

$$\mathbf{f} \times (\mathbf{q}_b - \mathbf{q}_G) = \frac{d}{dz}\left(\frac{\boldsymbol{\tau}_z^T}{\rho}\right) \tag{6.54}$$

We now abandon the tangent plane coordinate system in favor of coordinates chosen so that the x direction coincides with the direction of the geostrophic wind. We choose the unit vector $\mathbf{i}'$ parallel to $\mathbf{q}_G$, and $\mathbf{j}'$ perpendicular and to the left of $\mathbf{q}_G$. Thus

$$\mathbf{q}_G = q_G \mathbf{i}' \quad \text{and} \quad \mathbf{q}_b = u_b \mathbf{i}' + v_b \mathbf{j}'$$

and

$$\boldsymbol{\tau}_z^T = \tau_{zx}^T \mathbf{i}' + \tau_{zy}^T \mathbf{j}'$$

The unit vector **k** remains normal to the horizontal plane. It is understood that these components of velocity and shear stress, which properly should be designated as the primed values, are not equal to the corresponding components in the tangent plane coordinates but their vector sums are identical in both cases. Performing the cross product indicated in Eq. 6.54 and reverting to component form, we find

$$\frac{d}{dz}\left(\frac{\tau_{zx}^T}{\rho}\right) + fv_b = 0$$

and (6.55)

$$\frac{d}{dz}\left(\frac{\tau_{zy}^T}{\rho}\right) + f(q_G - u_b) = 0$$

Equations 6.52, 6.54, and 6.55 clearly contain the same information, and they represent useful alternate ways of expressing the conservation of momentum in the atmospheric boundary layer which is quasi-steady, neutrally stable, horizontally homogeneous, and barotropic. Of the six independent components of the Reynolds stress tensor, four do not appear in Eq. 6.55 because their derivatives are small. In spite of the simplified form of the two Eqs. 6.55, they contain four unknowns—the two mean velocity components u_b and v_b and the two Reynolds stress components τ_{zx}^T and τ_{zy}^T—and cannot be solved in their present form. The closure problem of turbulence that was associated with the Reynolds equations persists in these drastically simplified expressions.

6.3.2. The Ekman Solution

In our discussion above we noted that the shear stress in a laminar boundary layer is related to the velocity derivative by

$$\tau_{zx} = \rho \nu \frac{\partial u}{\partial z} \tag{6.56}$$

where ν is the kinematic viscosity of the particular fluid. A similar relationship is the Fourier heat conduction law

$$q_z = -K \frac{\partial T}{\partial z}$$

where q_z is the heat flux in the z direction, and K is the constant of proportionality which is called the thermal conductivity and is also a property of the material. We may recast the above expression by means of the thermal diffusivity α_h which is defined by $\alpha_h = K/\rho c_p$, where c_p is the specific heat at constant pressure. The Fourier heat conduction law can then be expressed as

$$q_z = -\rho\alpha_h \frac{\partial(c_p T)}{\partial z} \tag{6.57}$$

if the specific heat is constant. The negative sign appears in this expression in order that the direction of the flow of heat will be toward the region of lower temperature as is required by the second law of thermodynamics. Equations 6.56 and 6.57 represent two constitutive relations which relate two fluxes to local gradients. Regarding τ_{zx} as the flux of x directed momentum in the z direction, the quantity ν represents the "diffusivity of momentum" just as α_h represents the diffusivity of heat. For given magnitudes of the gradients, the respective diffusivities represent the effectiveness of the fluid in promoting transfer of momentum and heat. Finally the ratio ν/α_h is defined to be the Prandtl number, which for gases is weakly dependent of pressure and is only slightly dependent on temperature. The Prandtl number of air is 0.72 at a temperature of 273°K and thus air in laminar flow will transfer momentum and heat with nearly equal effectiveness.

The crudest method of coping with the closure problem of turbulence is to define diffusivities that are applicable in turbulent flow by analogy to laminar flow. Thus Boussinesq suggested the use of an eddy kinematic viscosity K_m relating the turbulent shear stress to the gradient of the velocity

$$\tau_{zx}^T = \rho K_m \frac{d\bar{u}}{dz} \tag{6.58}$$

The corresponding eddy diffusivity of heat K_h is given by

$$q_z^T = -\rho K_h \frac{d(c_p \overline{T})}{dz} \tag{6.59}$$

where q_z^T is the turbulent heat flux in the z direction and $\overline{T}$ is the mean temperature. Since c_p is practically constant for air, we temporarily choose to write it within the derivative to display more clearly the close relationship between the two turbulent closure relationships. The diffusivities ν and α_h are properties of the particular fluid and can be determined by the methods of kinetic theory. The corresponding turbulent diffusivities K_m and K_h are, in contrast, properties of the flow field and are determined not only by the fluid and its properties, but also by the velocity components, their derivatives, the flow geometry, and probably the history of these quantities for any particular fluid parcel. Viewed from this standpoint Eqs. 6.58 and 6.59 seem not particularly useful since the crucial constants of proportionality—the turbulent diffusivities—are unknown. A possible justification for this approach might be that the determination of the appropriate K_m and K_h in one particular steady flow may also apply, if properly scaled, to a geometrically similar flow.

Ekman used the concept of an eddy kinematic viscosity as a closure scheme to solve Eq. 6.55 in a manner that provides valuable insight into the nature of the flow in the boundary layer. He assumed that the turbulent shear stresses were related to the mean velocity gradients by a common eddy kinematic viscosity,

$$\frac{\tau_{zx}^T}{\rho} = K_m \frac{du_b}{dz} \qquad \frac{\tau_{zy}^T}{\rho} = K_m \frac{dv_b}{dz} \tag{6.60}$$

When the above equations are combined with Eq. 6.55 and K_m is taken to be a constant independent of z, we find

$$K_m \frac{d^2 u_b}{dz^2} + f v_b = 0$$

and (6.61)

$$K_m \frac{d^2 v_b}{dz^2} + f(q_G - u_b) = 0$$

These equations now contain only two unknowns and therefore can be solved if the proper boundary conditions are to be applied.

Traditionally it is assumed that the two mean velocity components, u_b and v_b, will vanish at the surface where $z = 0$. Thus the usual no-slip condition is applied as two boundary conditions at the surface. The upper boundary conditions are that $u_b \to q_G$ and $v_b \to 0$ as $z \to \infty$. The solution obtained with constant K_m and the no-slip boundary conditions at the surface is sometimes termed as a laminar flow solution.

Equations 6.61 are simultaneous, linear, homogeneous equations of the type commonly encountered in electrical or mechanical systems possessing two degrees of freedom. The general solution is then found to be of the form

$$u_b - q_G = c_1 e^{+az} \cos(az + \phi_1) + c_2 e^{-az} \cos(az + \phi_2)$$
$$v_b = c_1 e^{+az} \sin(az + \phi_1) + c_2 e^{-az} \sin(az + \phi_2)$$

Since $u_b - q_G$ and $v_b \to 0$ as $z \to \infty$, c_1 must be zero. As $z \to 0$, $u_b - q_G \to -q_G$ and $v_b \to 0$, hence $\phi_2 = 0$ and $c_2 = -q_G$. The final result is

$$u_b = q_G(1 - e^{-az} \cos az) \tag{6.62}$$

$$v_b = q_G e^{-az} \sin az \tag{6.63}$$

where

$$a = \left(\frac{f}{2K_m}\right)^{1/2} \tag{6.64}$$

The values of u_b and v_b versus z are depicted in Fig. 6.5*a* where it is apparent that both velocity components satisfy their outer boundary conditions through an oscillatory behavior. Another interesting presentation of the Ekman velocity profile is through the hodagraph—a graph of v_b versus u_b in which z is an implicit parameter depicted in Fig. 6.5*b*. The geometric figure shown in Fig. 6.5*b* is the well known Ekman spiral, a so-called equi-angular spiral (see Problem 6.1).

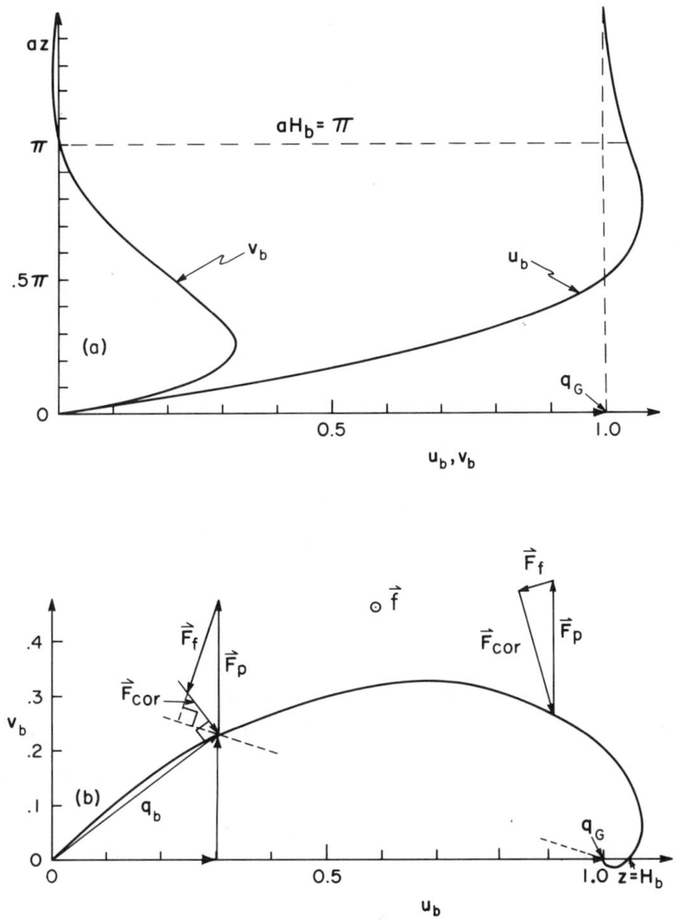

Figure 6.5. The Ekman solution for the flow in the atmospheric boundary layer. This solution is obtained if eddy kinematic viscosity is assumed to be constant. (*a*) The geostrophic component of velocity u_b and transverse component of velocity v_b versus altitude. (*b*) The hodagraph showing the force triangles at two values of altitude.

The height of the Ekman layer is defined as the level at which the value of v_b has its first zero for $z > 0$. This will correspond to the height H_b where $\sin aH_b = 0$ or $aH_b = \pi$ and therefore

$$H_b = \pi\sqrt{\frac{2K_m}{f}} \tag{6.65}$$

The spiral shape of the form $z > H_b$ is not observed and results from two artificialities that were imposed by the method of solution. The first is that K_m was prescribed to be constant for all values of z. The second flaw is that the imposition of the concept of an eddy kinematic viscosity requires that the vectors $\boldsymbol{\tau}_z$ and $d\mathbf{q}_b/dz$ are parallel at all altitudes. There is evidence that these vectors are not parallel in the upper portion of the Ekman layer, and therefore the concept of a scalar eddy kinematic viscosity may be invalid in this region.

The above solution can be readily interpreted by means of the force diagram expressed by the vector relationships Eqs. 6.52, 6.53, or 6.54. In the northern hemisphere where $\mathbf{f}$ is out of the plane of the paper with reference to Fig. 6.5*b*, the pressure force $\mathbf{F}_p = \mathbf{f} \times \mathbf{q}_G$ is in the transverse direction of $+v_b$; the Coriolis force $\mathbf{F}_{\text{cor}}$ is perpendicular to $\mathbf{q}_b$; and, by Eq. 6.54, the frictional force $\mathbf{F}_f = d/dz(\boldsymbol{\tau}_z/\rho)$ is perpendicular to $\mathbf{q}_b - \mathbf{q}_G$. The force diagram associated with several values of z for an Ekman spiral is depicted in Fig. 6.5*b*.

According to meteorological convention the wind direction is defined as the angle measured clockwise from true north to the direction from which the wind originates. The important conclusion from the above analysis is that the wind direction increases monotonically with z in the interval $0 \le z \le H_b$. Wind speed also increases monotonically to a level nearly equal to H_b and then decreases slightly. These trends are sometimes observed in measurements of winds by radiosonde observations. In many instances, however, the influence of one or more of the several factors not included in the above analysis will invalidate the assumptions that yield the Ekman solution and will prevent its predictions from being observed.

An alternative description of the steady, neutral, barotropic atmospheric boundary layer is given in Section 6.3.4. The method given there is free from the imposition of a scalar eddy kinematic viscosity and can be extended to describe the nonneutral and baroclinic boundary layers as well.

6.3.3. Ekman Layer Pumping and Subsidence in the Boundary Layer

The occurrence of a transverse wind component v_b in the Ekman layer results in a secondary flow pattern in the atmospheric boundary layer which

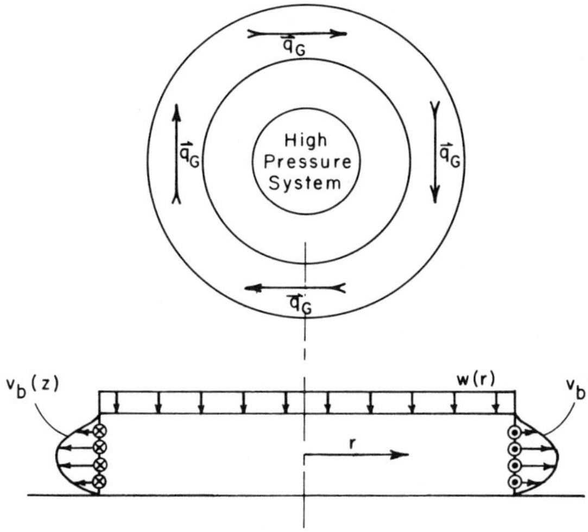

Figure 6.6. Ekman pumping and subsidence in the atmospheric boundary layer. A net outflow of air occurs from the boundary layer under an anticyclone because the transverse component of the boundary layer wind is directed away from the center of the pressure system. A downward component of velocity results at the top of the boundary layer.

has important consequences. If we again idealize the anticyclonic wind-pressure system as consisting of circular isobars, as depicted in Fig. 6.6, it is apparent that the transverse boundary layer wind component given by Eq. 6.63 causes a radially outward flow within the boundary layer. This flow requires that an equal mass flow must occur into the top of the boundary layer. A simple model of this secondary flow can be formulated by equating the two components of mass flux,

$$2\pi r\rho \int_0^{H_b} v_b(z)dz = -2\pi\rho \int_0^r w(r)rdr \tag{6.66}$$

where r is the radial distance from the center of the anticyclone. In the above mass balance v_b is positive if directed away from the center of the pressure system and w is positive if directed upward. We treat ρ as a constant for the limited range of z of interest and H_b as a constant for all r in this approximate analysis. By inserting $v_b(z)$ from Eq. 6.63 into the left side and evaluating the integral (see Problem 6.5 for more detail), we are able to write an elementary integral equation for $w(r)$

$$\int_0^r w(r)rdr = \frac{1 + e^{-\pi}}{2\pi} q_G H_b r$$

This equation is solved for $w(r)$ by differentiating both sides with respect to r,

$$w(r) = \frac{1 + e^{-\pi}}{2\pi} \left[\frac{1}{r} \frac{d}{dr} (q_G r) \right] H_b$$

The quantity on the right side in the square bracket represents the relative vorticity associated with the geostrophic wind ζ_G expressed in cylindrical coordinates,

$$\zeta_G = (\nabla \times \mathbf{q}_G)_z = \frac{1}{r} \frac{\partial}{\partial r} (q_G r)$$

Accordingly we find the vertical component of velocity is given by

$$w(r) = \frac{1 + e^{-\pi}}{2\pi} \zeta_G H_b \approx \tfrac{1}{6} \zeta_G H_b$$

In an anticyclonic wind-pressure system the relative vorticity is negative and the sign of w is negative indicating a downward flow or subsidence. This subsidence will act to extend the influence of the subsidence which is normally associated with the developing high pressure region.

A numerical estimate of the magnitude of $w(r)$ is obtainable from the above equation. For example, the relative vorticity associated with an anticyclone is about $-8 \times 10^{-5}\ \mathrm{s}^{-1}$, and assuming a boundary layer height of 1000 m the value of w is estimated at 50 m/hr. This velocity, although small, frequently combines with other factors which adversely affect the ability of the atmosphere to disperse pollutants (see discussion pp. 67 and 68). The strength of the subsidence is directly proportional to the magnitude of the geostrophic vorticity.

Certain types of atmospheric events can be detected by means of a recently devised experimental method which is referred to as acoustic sounding. Short pulses of audible sound beamed vertically upward are partially reflected by weak temperature nonuniformities that may be present in the atmosphere. The strength of the return signal is plotted by a recorder in which the record advances transversely at a very slow rate. The resulting graph provides a history of certain atmospheric events as a function of altitude and time. Figure 6.7 shows an acoustic sounding obtained from N. A. Shaw (Ref. 8) while at the University of Melbourne, Australia. This sounding shows an elevated thermal inversion descending toward the surface as a result of the subsidence that accompanies the passage of an anticyclonic wind-pressure system. As noted earlier (p. 68), a subadiabatic lapse rate tends to become more strongly stabilizing in a subsidence.

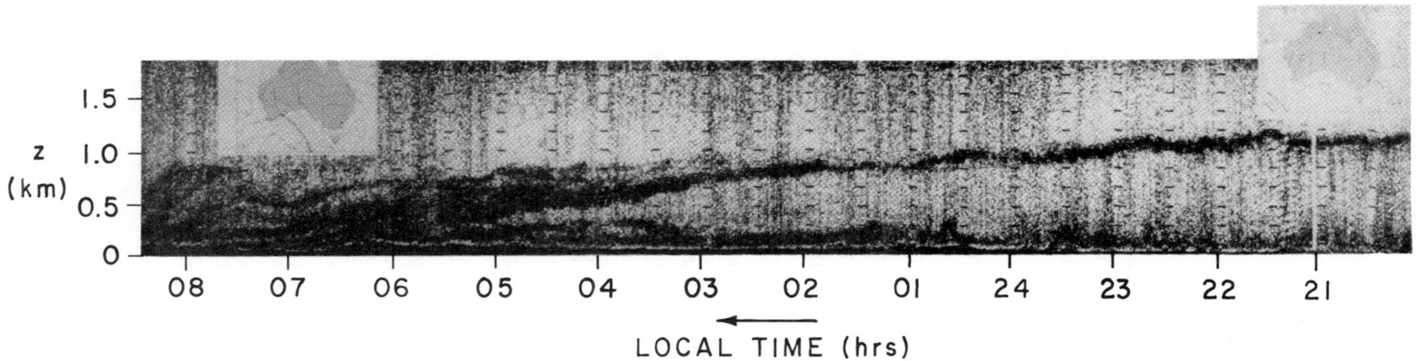

Figure 6.7. An acoustic sounding obtained near Melbourne, Australia, on August 8, 1970. An elevated inversion is carried toward the surface by the subsidence associated with the nighttime passage of an anticyclone. (From Ph.D. Thesis of N. A. Shaw, University of Melbourne, 1971. Reproduced with consent of N. A. Shaw.)

6.3.4. The Similarity Theory for the Neutral, Barotropic Boundary Layer

A more realistic description of the neutral atmospheric boundary layer is afforded by the Rossby number similarity theory. The following brief discussion of this theory takes its origin from asymptotic analysis and emphasizes the important physical concepts; further details are given by Tennekes (Ref. 9) and in the references that he cites.

As suggested by the previous discussion, the similarity theory for the steady, neutral, horizontally homogeneous atmospheric boundary layer arrives at a two-layer structure. In the outer layer, termed the region of geostrophic adjustment or simply the Ekman layer, the turbulent shear stress is reoriented to coincide with the direction of the surface shear stress through an interaction with the Coriolis and pressure forces. The inner layer is a region in which the flow responds to the applied shear stress on its upper face and which matches the boundary condition at the surface whose roughness is characterized by the aerodynamic roughness z_0. Both mathematical solutions for the inner and outer layers apply at an intermediate altitude which we term the interface. The similarity theory takes advantage of this physical description to deduce certain overall features of the boundary layer without specifying the full detail of the mathematical solution.

In the outer layer, Eq. 6.54 is applicable and is resolved into components parallel and perpendicular to the surface shear stress. The results are

$$\frac{d}{dz}(-\overline{u'w'}) - f(V_G - v_b) = 0 \tag{6.67}$$

$$\frac{d}{dz}(-\overline{v'w'}) + f(U_G - u_b) = 0 \tag{6.68}$$

where u_b and U_G are the velocity components parallel to the surface shear stress and v_b and V_G are the corresponding transverse components. The velocity defects $U_G - u_b$ and $V_G - v_b$ must be functions of the height z, the Coriolis parameter f, and the friction velocity u_* which represents the value of shear stress at the bottom of the layer. Thus

$$u_b - U_G = \mathscr{F}_u(z, f, u_*) \qquad v_b - V_G = \mathscr{F}_v(z, f, u_*)$$

These equations are readily nondimensionalized using u_* as the characteristic velocity and u_*/f as the characteristic length. Accordingly the dimensionless form of the Ekman layer solution is

$$\frac{u_b - U_G}{u_*} = \mathscr{F}_u\left(\frac{zf}{u_*}\right) \tag{6.69}$$

$$\frac{v_b - V_G}{u_*} = \mathscr{F}_v\left(\frac{zf}{u_*}\right) \tag{6.70}$$

The velocity in the inner layer is in the direction of the surface shear stress and must depend on z, u_*, and z_0,

$$u_b = f_u(z, z_0, u_*)$$

and

$$v_b = 0$$

In the nondimensional form, we find

$$\frac{u_b}{u_*} = f_u\left(\frac{z}{z_0}\right) \tag{6.71}$$

$$\frac{v_b}{u_*} = 0 \tag{6.72}$$

At the interface where $zf/u_* \ll 1$ and $z/z_0 \gg 1$ the two solutions must be in agreement, and we accomplish their matching by requiring that the derivatives of Eqs. 6.69 and 6.71 are equal at this level. The matching of the velocity gradients is equivalent to matching shear stress at the interface. The shear stresses are evaluated in the dimensionless forms as

$$\frac{z}{u_*}\frac{du_b}{dz} = \frac{z}{z_0}\frac{df_u}{d\left(\dfrac{z}{z_0}\right)} = \phi_m\left(\frac{z}{z_0}\right) \tag{6.73}$$

$$\frac{z}{u_*}\frac{du_b}{dz} = \frac{fz}{u_*}\frac{d\mathscr{F}_u}{d\left(\dfrac{zf}{u_*}\right)} = \Phi_m\left(\frac{zf}{u_*}\right) \tag{6.74}$$

The matching condition at the interface is expressed by

$$\phi_m(\infty) = \Phi_m(0)$$

We conclude, since ϕ_m and Φ_m are functions of different dependent variables, that in the indicated limit both quantities must equal a constant. By setting this constant equal to $1/\kappa$ where κ is the von Karman constant, we recover the logarithmic velocity profile in the surface layer

$$\frac{u_b}{u_*} = f_u = \frac{1}{\kappa}\ln\left(\frac{z}{z_0}\right) + c_1 \tag{6.75}$$

A logarithmic profile is also obtained for one component of the velocity defect in the Ekman layer,

$$\frac{u_b - U_G}{u_*} = \mathscr{F}_u = \frac{1}{\kappa}\ln\left(\frac{zf}{\mu_*}\right) + c_2 \tag{6.76}$$

Subtracting Eq. 6.76 from Eq. 6.75 and replacing $c_2 - c_1$ by A/κ, where A is an arbitrary constant, we obtain

$$\frac{U_G}{u_*} = \frac{1}{\kappa}\ln\left(\frac{u_*}{fz_0}\right) - \frac{A}{\kappa} \tag{6.77}$$

The transverse velocity and velocity defect yield, at the interface where these solutions are matched,

$$\frac{v_b}{u_*} = f_v\left(\frac{z}{z_0}\right) = 0$$

and

$$\frac{v_b - V_G}{u*} = \mathscr{F}_v\left(\frac{zf}{u_*}\right) = c_3$$

Subtracting the second from the first, we find

$$\frac{V_G}{u_*} = -\frac{B}{\kappa}$$

where c_3 has been replaced by the arbitrary constant B divided by κ. For α_0 defined positive in the counterclockwise direction, $V_G = -q_G \sin \alpha_0$ and therefore

$$\sin \alpha_0 = \frac{B}{\kappa}\frac{u_*}{q_G} \tag{6.78}$$

We combine Eqs. 6.77 and 6.78 to obtain, with the aid of $U_G^2 = q_G^2 - V_G^2$,

$$\ln\frac{q_G}{fz_0} = A - \ln\frac{u_*}{q_G} + \left[\left(\frac{\kappa q_G}{u_*}\right)^2 - B^2\right]^{1/2} \tag{6.79}$$

Equations 6.78 and 6.79 are two equations relating $\sin \alpha_0$ and u_*/q_G to q_G/fz_0 through the constants A, B, and κ. The ratio of u_*/q_G is called the geostrophic drag coefficient since its square is the dimensionless shear stress at the interface between the surface and the Ekman layers. Once the geostrophic wind q_G and the surface roughness z_0 are specified, the friction velocity u_* and the cross isobar angle α_0 are determined by Eqs. 6.78 and 6.79.

The values of A and B can be found by analyzing wind profiles as described in the literature. Deacon (Ref. 10) critically examined the results of several meteorological studies and concluded that $A = 1.9$ and $B = 4.7$. Clarke and Hess (Ref. 11) examined a large body of observations made over the plains of Australia and concluded that $A = 1.1$ and $B = 4.3$*. The value $\kappa = 0.40$ was used in both studies.

* There exists widespread lack of uniformity on the definitions of A and B which here correspond to those used in Refs. 10 and 11.

6.4. THE CONVECTIVE ATMOSPHERIC BOUNDARY LAYER

At sunrise under clear skies the atmospheric boundary layer is normally in a state of static stability as a result of the outgoing nocturnal radiation flux. The increasing insolation, or solar radiative heat flux, subsequently warms the earth's surface which in turn heats the air in the surface layer. Strong vertical plumes or thermals may develop as this convective heating of the air by the surface intensifies. The convectively induced flows produce efficient mixing and give rise to the term "mixed layer" for the atmospheric boundary layer under these conditions. The ability to predict the height of the mixed layer is of great importance because this height determines the volume of air over which the pollutants emitted at the surface of the earth are dispersed. Unfortunately, the time-dependent behavior of the convective atmospheric boundary layer cannot be easily predicted. We briefly summarize below recent theoretical studies and some observations of the convective atmospheric boundary layer.

The potential temperature and turbulent heat flux profiles in the developing convective atmospheric boundary layer that is capped by a strong local inversion as depicted by Tennekes (Ref. 12) are shown in Fig. 6.8. We interpret this figure in detail to give insight into the mechanisms that are in operation under these conditions. Solar heating of the surface of the earth gives rise to a positive heat flux $q^T_{zo} = \rho c_p(\overline{\theta' w'})_0$ to the air in the surface layer. This heat flux represents the difference between the insolation and

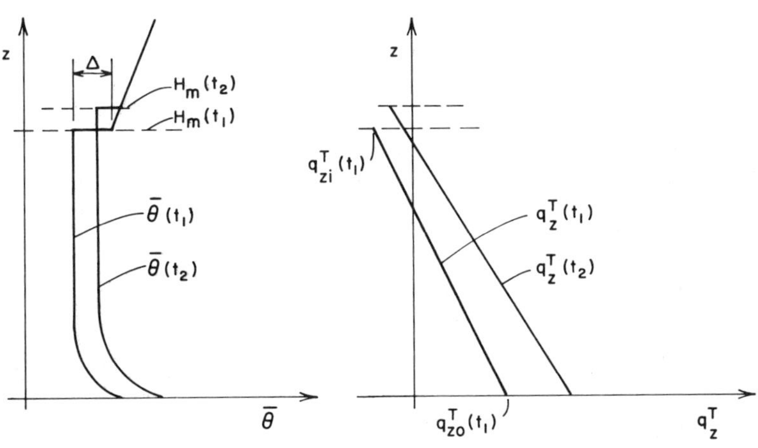

Figure 6.8. The convective boundary layer capped by an elevated inversion. The profiles of potential temperature and turbulent heat flux are displayed. [Adapted from H. Tennekes, *J. Atmos. Sci.*, *30*, 559 (1973). Reproduced with consent of the American Meteorological Society and H. Tennekes.]

the heat absorbed and reradiated by the surface. The potential temperature gradient is negative in the surface layer because of the positive turbulent heat flux, but it becomes small at higher elevations where turbulent mixing is efficient and the eddy diffusivity of heat is large. The turbulent heat flux decreases linearly with altitude because of the approximately uniform heating of the air within the mixed layer that it causes. The elevated inversion forms the upper boundary of the mixed layer at $z = H_m$ and it is shown as a discontinuous change in potential temperature in Fig. 6.8. In actuality this change in potential temperature will occur over a finite, but relatively small, height difference.

Turbulence within the mixed layer can cause mixing to occur across the inversion which marks its upper boundary through the action of penetrative convection. Bursts of turbulence from below penetrate through the inversion and envelope some of the air above. Air parcels captured by this process are quickly dispersed in the mixed layer whose height is thereby increased. The transport of air across the interface, which is called turbulent entrainment, brings warm air toward the surface, causing a locally negative heat flux that also warms the air within the mixed layer. The potential temperature gradient $d\theta/dz$ above the mixed layer is positive indicating that a weakly stable stratification is present above the strong local inversion.

At a later time the turbulent heat flux from the surface and the turbulent entrainment from the interface have supplied heat to the mixed layer, which causes H_m and θ to increase while Δ initially decreases as depicted in Fig. 6.8. The turbulent heat flux and its slope, which is the local heating rate, will also increase in the early part of the day as indicated.

The dynamics of the convective atmospheric boundary layer are discussed in the article by Tennekes (Ref. 12) who formulates the equations describing the evolution of H_m and Δ with time. These equations require a numerical solution by digital computer. In general the strength of the inversion Δ decreases during the morning hours and then increases while the height of the mixed layer H_m increases continuously, although not at a uniform rate. Subsidence in the atmosphere resulting from the large scale meteorological events and the presence of moisture may also influence the development of the mixed layer.

6.4.1. The Convective Boundary Layer under Weakly Stable Stratification

A special case that is subject to straightforward analysis occurs when a weakly stable stratification is present without a capping inversion. Turbulent entrainment is then absent and the heat flux into the mixed layer is solely the convective heating at the surface. The boundary layer dynamics can be

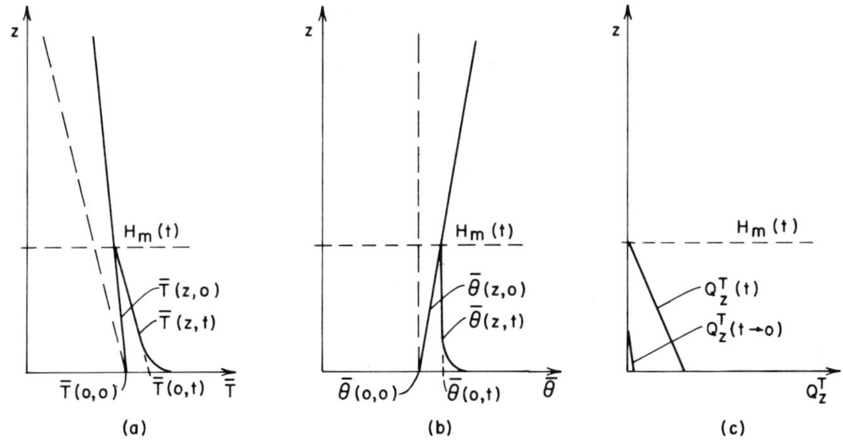

Figure 6.9. The convective boundary layer under a weakly stable stratification: (*a*) temperature versus altitude; (*b*) potential temperature versus altitude; (*c*) turbulent heat flux versus altitude.

completely described by a simple heat balance. We illustrate this case by the hypothetical temperature profile depicted in Fig. 6.9. The notation $\overline{T}(z, t)$ is used for time-averaged temperature to denote explicitly its dependence on both altitude and time which is measured from local sunrise. Accordingly the sunrise or initial temperature profile is described by

$$\overline{T}(z, 0) = \overline{T}(0, 0) - \lambda_0 z \tag{6.80}$$

where λ_0 is an average low level lapse rate at sunrise. At a later time the temperature profile within the mixed layer is

$$\overline{T}(z, t) = \overline{T}(0, t) - \Gamma z \qquad 0 < z \leq H_m(t) \tag{6.81}$$

because the lapse rate is the dry adiabatic value within this layer where turbulent mixing is high. An energy balance equates the increase of enthalpy of the air within the mixed layer at any moment to the total heat supplied by the turbulent heat flux at $z = 0$,

$$\rho c_p \int_0^{H_m(t)} [\overline{T}(z, t) - \overline{T}(z, 0)] dz = \int_0^t q_{z0}^T(t) dt \tag{6.82}$$

The turbulent heat flux is temporarily left in an unspecified time-dependent form. Inserting Eqs. 6.80 and 6.81 into Eq. 6.82, we find, by evaluating the integral on the left,

$$\rho c_p \left\{ [\overline{T}(0, t) - \overline{T}(0,0)] H_m - (\Gamma - \lambda_0) \frac{H_m^2}{2} \right\} = \int_0^t q_{z0}^T(t) dz \tag{6.83}$$

The surface temperatures in the square bracket are related,

$$(\Gamma - \lambda_0)H_m(t) = \overline{T}(0, t) - \overline{T}(0, 0)$$

and the height of the mixed layer is found from Eq. 6.83 as

$$H_m(t) = \left[\frac{2}{\rho c_p(\Gamma - \lambda_0)} \int_0^t q_{z0}^T(t)dt\right]^{1/2} \tag{6.84}$$

The ground level turbulent heat flux during the time interval τ_d between sunrise and sunset can be approximated by the sinusoidal function

$$q_{z0}^T(t) = q_{\max}^T \sin\frac{\pi t}{\tau_d} \qquad 0 \le t \le t_d \tag{6.85}$$

where $q_{\max}^T$ is the maximum turbulent heat flux near the time of local noon hour. By inserting Eq. 6.85 into Eq. 6.84 and evaluating the integral, we find the height of the mixed layer as

$$H_m(t) = \left[\frac{2q_{\max}^T \tau_d}{\pi\rho c_p(\Gamma - \lambda_0)}\left(1 - \cos\frac{\pi t}{\tau_d}\right)\right]^{1/2} \tag{6.86}$$

This equation agrees with a suitable specialization of a result given by Carson (Ref. 13).

In the early morning hours the heat flux is the linear function of time that is given by a Taylor expansion of the right side of Eq. 6.85

$$q_{z0}^T(t) = q_{\max}^T \frac{\pi t}{\tau_d} \qquad \text{for} \quad t \ll \tau_d \tag{6.87}$$

In this case the height of the mixed layer is given by

$$H_m(t) = \left[\frac{\pi q_{\max}^T}{\rho c_p \tau_d(\Gamma - \lambda_0)}\right]^{1/2} t \tag{6.88}$$

and increases linearly with time.

Turbulent entrainment can also be ignored if the strength of the inversion is very large, in which case the turbulence within the mixed layer in the early morning lacks the kinetic energy to penetrate its upper boundary. The height of the mixed layer is then constant until the turbulent heat supplied only at the surface weakens the inversion to the point that entrainment can occur. The time required to irradicate the initial strong inversion is readily calculated in terms of the initial values of the inversion strength Δ_0 and the height of the mixed layer H_{mo} (see Problem 6.8). The mechanism of turbulent entrainment is currently the subject of research with the goal of achieving a better knowledge of the factors that determine its rate.

6.4.2. The Diurnal Cycle of the Atmospheric Boundary Layer

The diurnal cycle of the atmospheric boundary layer has a strong influence on the transport and the chemical reactions that involve certain air pollutants. A recent mathematical model of this cycle is provided by Yamada and Mellor (Ref. 14) which we summarize in brief. This theory is based on a numerical solution to the set of partial differential equations which express the conservation of momentum, energy, and moisture. The turbulent fluxes of heat, momentum, and moisture are related to the gradients of the time-averaged quantities through eddy diffusivities that are complicated algebraic functions of the turbulence level, the velocity gradients, and the static stability as well as certain arbitrary constants. The theoretical model was compared with observations from an extensive experiment conducted at Wangara, Australia. The observed, time-dependent geostrophic wind was employed as an input for the model. The theory provides the detailed description of the boundary layer including the vertical profiles of such quantities as wind components, temperature, Reynolds stress, turbulent heat flux, and turbulent energy terms. One test of the model is its ability to predict the height of the atmospheric boundary layer as a function of time and this quantity is shown in Fig. 6.10.

The model predicts that the mixed layer grows to a maximum height during late afternoon when solar heating of the surface ceases. The turbulence persists within this region at a low level during the nighttime hours. A surface inversion layer is formed beneath the nocturnal mixed layer as a result of the outgoing nocturnal radiation. These two layers are dispersed at sunrise by the convective activity which re-establishes a new mixed layer. The various terms in the energy equation are calculated and buoyancy production balanced by viscous dissipation are found to dominate during daylight hours. Mechanical production and dissipation are the important terms in the turbulent energy equation at night. The two-day history of the calculated boundary layer height shown in Fig. 6.10 is in qualitative agreement with the observed height.

The above description of the diurnal cycle of the atmospheric boundary layer is also confirmed by the records of acoustic sounders. This is illustrated by Fig. 6.11 from the work of N. A. Shaw (Ref. 8) which was obtained near Melbourne, Australia. A nocturnal surface inversion layer is revealed at 0700 hr by the dark area that terminates at an altitude of about 150 m on the lower right corner of the record. The subsequent convective activity results in a mixed layer that reaches to 750 m at noontime and then decays during the late afternoon. The low-level nocturnal inversion then reforms during the early evening.

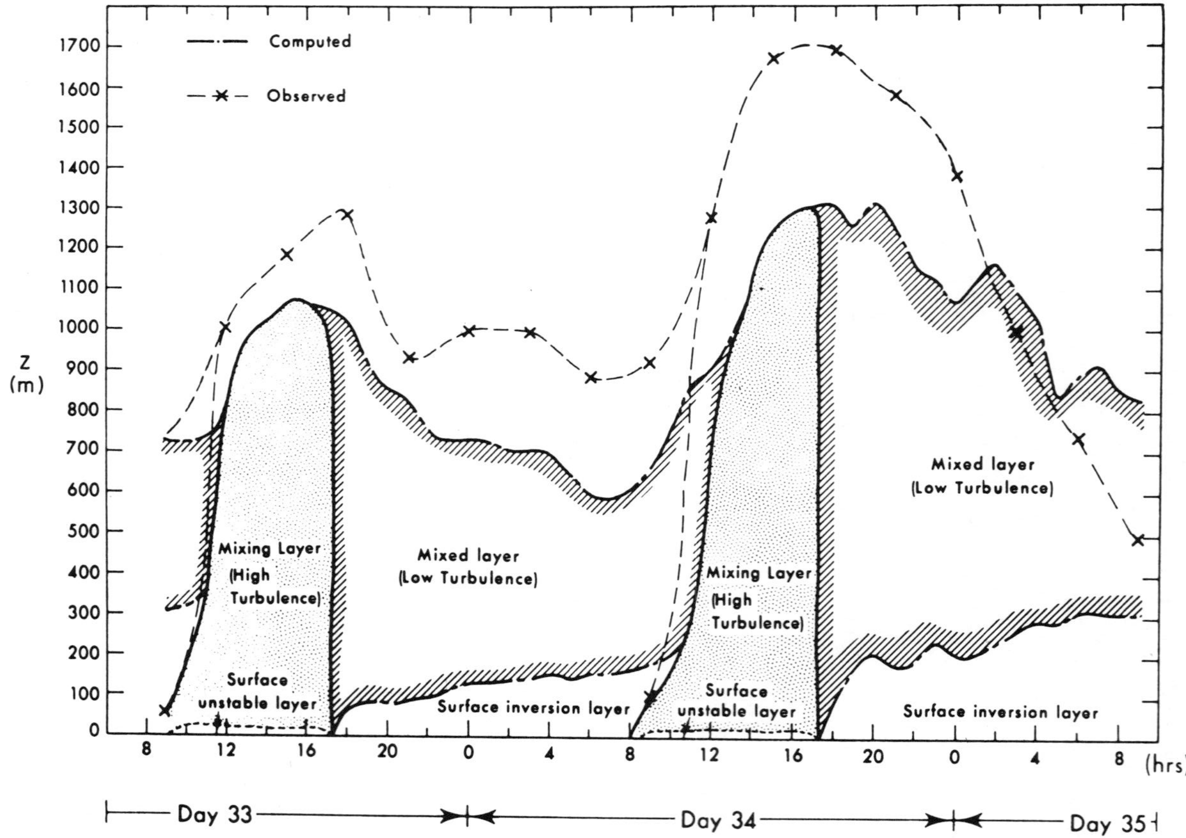

Figure 6.10. The diurnal variation of the atmospheric boundary layer. A comparison of the computation of Yamada and Mellor (Ref. 14) versus the observations made over the plains of Australia. [From T. Yamada and G. Mellor, J. *Atmos. Sci.*, **32**, 2326 (1975). Reproduced with consent of the American Meteorological Society and G. Mellor.]

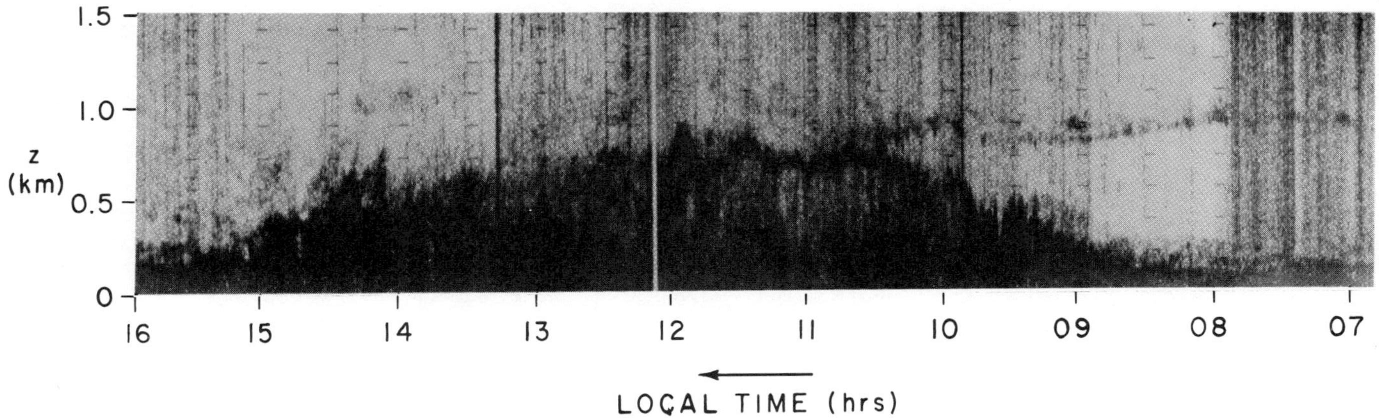

Figure 6.11. An acoustic sounding obtained near Melbourne, Australia, on September 26, 1970. The development and decay of the convective boundary layer during daytime hours is displayed. (From Ph.D. Thesis of N. A. Shaw, University of Melbourne, 1971. Reproduced with consent of N. A. Shaw.)

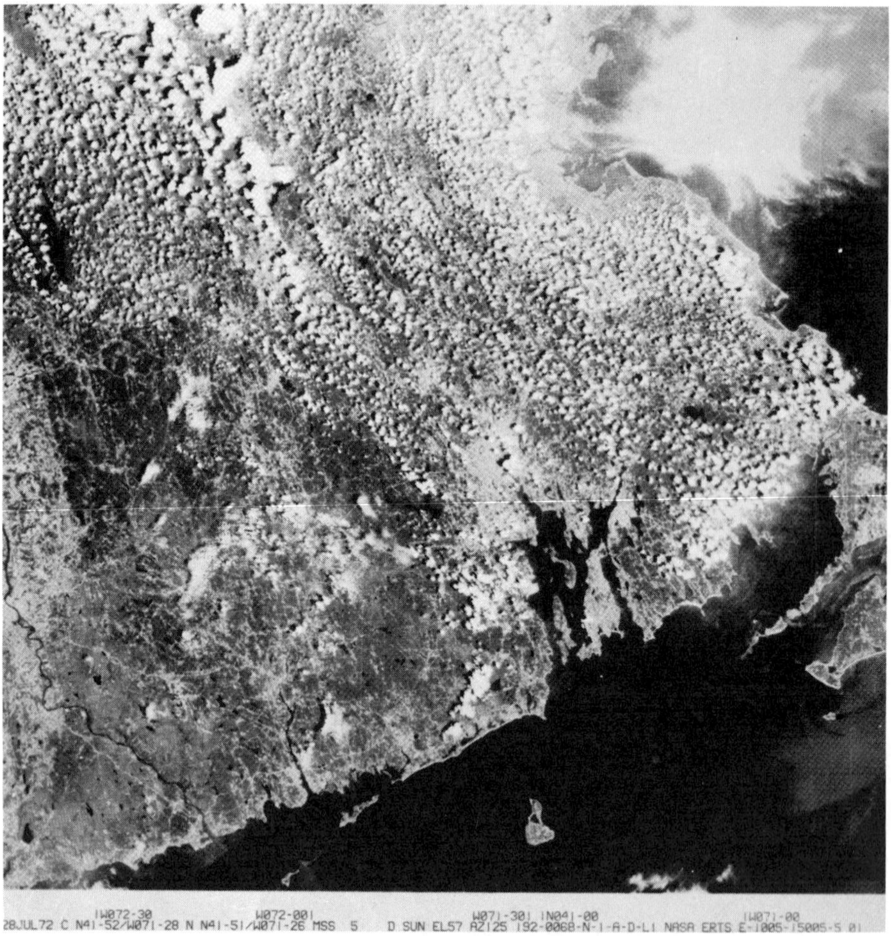

Figure 6.12. A photograph of southeastern New England taken from an Earth Resources Technical Satellite (ERTS) (NASA photo). This photograph was obtained at 10:01 EST on July 28, 1972 by the ERTS at an altitude of 915 km. The spectral sensitivity ranged from 0.6 to 0.7 μm for this photograph. The sun's elevation angle was 57° and its azimuth angle, measured clockwise from north, was 125°.

The low level cumulus clouds over southeastern Massachusetts result from convective activity and approximately mark the height of the mixed layer. The height of these clouds can be estimated from the width of their shadows and the sun's elevation angle. The absence of low level cumulus over eastern Connecticut and western Rhode Island suggests that the humidity was lower over these areas. The large cloud over Boston harbor on the upper right casts no discernible shadow because of the diffuseness of its edges. This cloud, which is associated with the synoptic scale activity, is probably at the same level as those small clouds southeast of Cuttyhunk Island on the right side of the lower border. The latter clouds do cast clear shadows on the ocean surface. The band of land along the Massachusetts shore, upper right, is evidence of a sea breeze whose presence at Boston is confirmed by radiosonde observations.

The convective activity in the atmospheric boundary layer that is induced by solar radiation frequency is manifest by the occurrence of low-level cumulus clouds. Vertically rising air parcels in the plumes or thermals associated with the convective boundary layer expand as their ambient pressure decreases. This expansion may be considered adiabatic until the relative humidity within the parcel becomes 100%. Further expansion then causes condensation to occur on the small particles that are present and clouds soon become evident. The strongly structured billows of the low-level cumulus are indicative of strong, vertically directed, convective velocity components. A striking photograph of low level cumulus clouds over south-eastern New England was taken by the ERTS satellite on July 28, 1972 and is shown in Fig. 6.12.

6.5. PRACTICAL CONCEPTS IN BOUNDARY LAYER METEOROLOGY

In this section we note several practical concepts relating to boundary layer meteorology.

6.5.1. Power Law Form of the Velocity Profile

From the discussion in Sections 6.3 and 6.4 it is apparent that the wind speed and direction in the atmospheric boundary layer are subject to a variety of influences. Often it is not possible to quantify these effects or to make a purely rational prediction of the vertical variation of wind speed. For this reason an empirical expression of the form

$$\frac{\bar{u}(z)}{\bar{u}_1(z_1)} = \left(\frac{z}{z_1}\right)^m \tag{6.89}$$

is used where $\bar{u}_1(z_1)$ is the observed mean wind speed at the height z_1, $\bar{u}(z)$ is the inferred wind speed at the height z, and m is the exponent that has been determined by experience. In a comprehensive recent survey Counihan (Ref. 15) recommends values of $m = 0.143$ for rural areas, 0.21 to 0.23 for suburban areas, and 0.28 for urban areas. A velocity profile of this type is best regarded as simply an empirical expression which is capable of representing a wide variety of conditions within the boundary layer. No indication of the turning of the wind vector with increasing altitude is suggested by Eq. 6.89. A relationship of this type is often used to represent the wind profile for purposes of calculating the dispersion of air pollutants.

6.5.2. The United States Air Pollution Potential Forecast Program

The National Weather Service informs regional meteorologists throughout the United States of impending conditions which are adverse to the dispersion of air pollutants. The details of this program are described in the report by E. Gross (Ref. 16) and in subsequent bulletins of the National Weather Service. We here summarize a few of the important concepts from these references. The following terms are defined:*

Mixing Height. The height in meters to which relatively vigorous mixing occurs. (This term clearly is synonymous with the height of the mixed layer when the boundary layer is in a convective condition. When the boundary layer is in a neutral or stable condition and the dispersive action of turbulence is less effective, the mixing height presumably corresponds to the height of the Ekman layer. When an elevated inversion is present the mixing height will correspond to the base of the inversion. In meteorological practice the mixing height is determined from radiosonde observations of temperature, humidity, and wind velocity profiles. For example the mixing height would clearly be defined if, at the same altitude, there occurred a sharp increase in temperature, a sharp change in relative humidity, and distinct changes in wind speed and direction. On numerous occasions the observed data may not permit a direct interpretation of mixing height, and its determination then is made by a subjective evaluation of the radiosonde data by the air pollution meteorologist.)

Transport Wind. The wind speed within the mixing layer determines in part the efficacy of the winds in the lower atmosphere in diluting and dispersing air pollutants. The transport wind (m/s) is defined by

$$q_T = \frac{1}{n} \sum_{i=1}^{n} q_{bi}(z_i) \tag{6.90}$$

where $q_{bi}(z_i)$ are the wind speeds as determined by radiosonde observations at n levels which lie within the mixing layer.

Ventilation Factor. Elementary theory for the dispersion of air pollutants from extended sources indicates that their concentrations vary inversely as the product of wind speed times mixing height. The ventilation factor is defined as the product $q_T H_m$(m^2/s), and the smallness of this quantity is one measure of adverse conditions for dispersal of air pollutants.

* Parenthetical comments in this subsection are intended to relate the material abstracted from Ref. 15 to the discussion in this text.

Air Pollution Potential. A measure of the conditions that are unfavorable to the dilution and dispersion of pollutants.

A set of quantitative criteria provides the basis on which the Weather Service evaluates the potential for high concentrations of air pollutants of a specified geographic area. These criteria are:

The wind speed at 1500 m above ground is <10 m/s.
- The rate of change of temperature with time at the 1500 m level is ≥ -5°C in 12 hr.
- The absolute vorticity η at $p = 500$ mb, corresponding to roughly 5200 m, is $\leq 1.0 \times 10^{-4}\ s^{-1}$. (The relative vorticity ζ is equal to $\eta - f$. Since $f \approx 1.0 \times 10^{-4}\ s^{-1}$ for the continental United States, this criterion is a test for negative relative vorticity which is an indication of subsidence in the atmosphere.)
- The change in absolute vorticity in the last 12 hr is $\leq 0.30 \times 10^{-6}\ s^{-1}$.
- The observed precipitation in 6 hr is less than 0.25 mm.
- The predicted relative humidity averaged from the surface to 500 mb is less than 80%. (The prediction in this case refers to the numerical forecast that is obtained using the so-called primitive equations.)
- The morning mixing height must be ≤ 500 m and the transport wind speed must be ≤ 4 m/s.
- The afternoon ventilation factor must be $\leq 6000\ m^2/s$ and the afternoon transport wind speed must be ≤ 4 m/s.

  The area affected must be as large as 4° latitude by 4° of longitude (about 400 × 400 km).
- The conditions are forecast to persist for 36 hr.

A numerical rating, termed the stagnation index, is formulated on the basis of the total or partial compliance of a region with the conditions enumerated above. The above criteria are clearly intended to test for the presence of a slow moving anticyclone combined with a small horizontal pressure gradient, the absence of precipitation, and the absence of conditions which will bring about a change of weather through the horizontal movement of the air mass over a region.

REFERENCES

1. R. W. Stewart, "Turbulence," 16 mm film, National Committee for Fluid Mechanics Films, distrubuted by Encyclopedia Brittanica Educational Corp., Chicago, Ill.

2. A. H. Shapiro, Editor, *Illustrated Experiments in Fluid Mechanics* (M.I.T. Press, Cambridge, Mass., 1972).
3. H. Tennekes and J. L. Lumley, *A First Course in Turbulence* (M.I.T. Press, Cambridge, Mass., 1972).
4. J. O. Hinze, *Turbulence* (McGraw-Hill Book Co., New York, 1975). Second Edition.
5. H. Schlichting, *Boundary Layer Theory* (McGraw-Hill Book Co., New York, 1960) Fourth Edition.
6. E. J. Plate, "Aerodynamic Characteristics of Atmospheric Boundary Layers," U.S. Atomic Energy Commission, TID-25465, 1971.
7. J. C. Wyngaard and O. R. Coté, "The Budgets of Turbulent Kinetic Energy and Temperature Variance in the Atmospheric Surface Layer," *J. Atmos. Sci.*, **28**, 190–201 (1971).
8. N. A. Shaw, "Acoustic Sounding of the Atmosphere," Ph.D. Thesis, University of Melbourne, Australia (1971).
9. H. Tennekes, "Similarity Laws and Scale Relations in Planetary Boundary Layers," in *Workshop on Micrometeorology*, D. A. Haugen, Editor (American Meteorological Society, 1973).
10. E. L. Deacon, "Geostrophic Drag Coefficients," *Boundary-Layer Meteorol.*, **5**, 321–340 (1973).
11. R. H. Clarke and G. D. Hess, "Geostrophic Departure and The Functions A and B of Rossby-Number Similarity Theory," *Boundary-Layer Meteorol.*, **7**, 267–287 (1974).
12. H. Tennekes, "A Model for the Dynamics of the Inversion Above a Convective Boundary Layer," *J. Atmos. Sci.*, **30**, 558–567 (1973).
13. D. J. Carson, "The Development of a Dry Inversion-capped Convectively Unstable Boundary Layer," *Quart. J. Roy. Meteorol. Soc.*, **99**, 450–467 (1973).
14. T. Yamada and G. Mellor, "A Simulation of the Wangara Atmospheric Boundary Layer Data," *J. Atmos. Sci.*, **32**, 2309–2329 (1975).
15. J. Counihan, "Adiabatic Atmospheric Boundary Layers: A Review and Analysis of Data From The Period 1880–1972," *Atmos. Environ.*, **9**, 871–905 (1975).
16. E. Gross, "The National Air Pollution Potential Forecast Program," National Meteorological Center, Washington, D.C. (1970).

PROBLEMS

6.1. Show that the angle ψ between the slope of the Ekman hodagraph and the chord $\mathbf{q}_b - \mathbf{q}_G$ is equal to 45° for all values of z. Note that for α_1 and α_2 as defined in the sketch below, $\tan\psi = \tan(\alpha_1 + \alpha_2) = 1$ implies

$$\tan\alpha_1 + \tan\alpha_2 = 1 - \tan\alpha_1 \tan\alpha_2$$

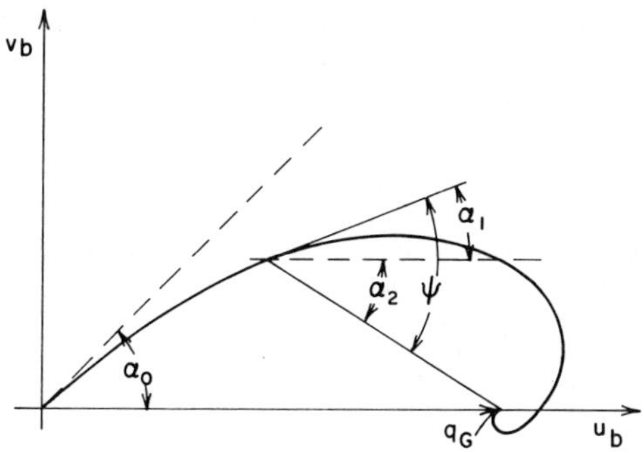

Also from the sketch it follows that

$$\tan \alpha_1 = \frac{\dfrac{dv_b}{dz}}{\dfrac{du_b}{dz}}$$

and

$$\tan \alpha_2 = \frac{v_b}{q_G - u_b}$$

The hodagraph is called an equiangular spiral because ψ = constant for all points on the curve.

6.2. What is the magnitude of the angle α_0 between the direction of the surface shear stress and the geostrophic wind for the Ekman spiral?

6.3. According to Eq. 6.65 the height of the boundary layer is given by

$$H_b = \pi \sqrt{\frac{2K_m}{f}}$$

Determine the value of $K_m(\mathrm{m}^2/\mathrm{s})$ at your latitude if $H_b = 500$ m. Compare this value of eddy kinematic viscosity with the kinematic viscosity ν of air at sea level conditions for the standard atmosphere using the information from Appendix II.

6.4. From the Ekman solution determine:

1. The value of $\tau_z^T(z)/\tau_z^T(0)$ where

$$\tau_z^T = [(\tau_{zx}^T)^2 + (\tau_{zy}^T)^2]^{1/2}$$

Assume ρ = constant for $0 < z \leq H_b$.

2. The value of u_*/q_G, where u_* is the friction velocity, as a function of f, H_b, and u_*.

3. The value of the production of the turbulent kinetic energy per unit mass P given by

$$P = -\overline{u'w'}\frac{du_b}{dz} - \overline{v'w'}\frac{dv_b}{dz}$$

Determine the production term at $z = 0$ and express this quantity in an appropriate dimensionless form.

6.5. Show that the average value of the transverse wind component of the atmospheric boundary layer that is described by the Ekman solution is given by

$$(v_b)_{\text{av}} = \frac{q_G}{\pi}\int_0^{\pi} e^{-az}\sin az\,d(az)$$

where the height of the ABL is defined by Eq. 6.65 and $(v_b)_{\text{av}}$ is defined by

$$(v_b)_{\text{av}} \equiv \frac{1}{H_b}\int_0^{H_b} v_b(z)\,dz$$

Evaluate the above integral and determine the numerical relationship between $(v_b)_{\text{av}}$ and q_G.

6.6. Discuss how the Buys–Ballot rule should be modified to apply at the surface of the earth as a result of the presence of the atmospheric boundary layer. Use the predictions of the Ekman solution to answer this question. The high altitude weather map supports the prediction of the Buys–Ballot rule quite well at the intermediate levels of the troposphere. Does the surface weather map support prediction of the Ekman solution? Explain why this is so.

6.7. The geostrophic wind is 15 m/s over an urban area where the surface aerodynamic roughness is 1.0 m. The cross isobar angle α_0 is 33.9° of arc. Use the similarity theory for the Ekman layer with $A = 1.9$, $B = 4.7$, and $\kappa = 0.40$ to determine, at your latitude:

1. The friction velocity u_* (m/s) and shear stress (N/m^2) if $T = 288$°K and $P = 101.3$ kPa.
2. The geostrophic drag coefficient u_*/q_G.
3. The height of the atmospheric boundary layer if the relation

$$\frac{H_b f}{u_*} = 0.20$$

which is derived from observations over a smooth terrain is applicable.

6.8. During early morning hours the height H_{mo} of an elevated inversion will remain nearly constant if its strength Δ_0 is adequately large. The profile of potential temperature is then depicted by the sketch below, $\theta =$ constant for $0 < z \leq H_{mo}$. The turbulent heat flux is given by Eq. 6.87 for $t \ll \tau$.

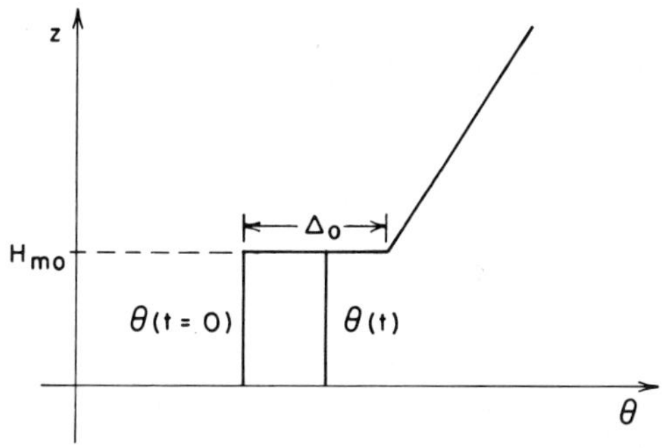

1. Show that the time required to reduce the strength of the inversion to zero is given by

$$t_m = \left[\frac{2\rho c_p \tau_d H_{mo} \Delta_0}{\pi q^T_{\max}}\right]^{1/2}$$

(This time is called the morning transient and is followed by the period in which the height of the layer increases. See discussion in Ref. 12.)

2. If $H_{mo} = 300$ m and $\Delta_0 = 5°\text{C}$ find the duration of the morning transient when $q^T_{\max} = 350$ W/m^2, $\rho = 1.20$ kg/m^3, and τ_d, the time interval between sunrise and sunset, is 14.5 hr. Recall that $c_p = \gamma R/(\gamma - 1)$.

6.9. An early morning radiosonde observation shows that the temperature profile has a constant lapse rate equal to 3.0°K/km. The noonday turbulent heat flux is 200 W/m^2 and the time between sunrise and sunset is 14.5 hr. Find the height of the mixed layer 5.5 hr after sunrise. The air density is 1.20 kg/m^3.

7

THE MOTION OF BUOYANT PLUMES

Steam generation plants associated with electric power stations or large factories release from tall stacks large volumes of combustion products at temperatures near 150°C ($\sim$300°F). Under these conditions the density of the released gases is about one half of the atmospheric density and an initial buoyancy force is present which is nearly equal to one half the weight of an equal volume of the ambient air. The effluent forms a buoyant plume whose initial motion is aided by the momentum of the products as they leave the stack. Under windless conditions, the plume rises vertically under the action of buoyancy as its diameter increases while the surrounding air is entrained by turbulent mixing. More often the plume is bent as a result of the wind that is usually present. In this chapter, we discuss the motion of the bent plume with the immediate goal of attempting to estimate the trajectory of its centerline. The extensive literature on this subject is summarized in recent survey articles by Briggs (Ref. 1), Fay (Ref. 2), and Briggs (Ref. 3).

The ground level concentrations of the species emitted by a large power plant are determined by the height of the stack h_s and the additional height of the plume z_p measured above the stack mouth that results from the vertical component of motion of the effluent through the atmosphere. The sum of these heights is designated as the effective stack height. The practical design problem is to loft the effluent to the effective stack height, $h_s + z_p$, that is required to maintain the ground level concentrations of pollutant components at acceptable values. This goal is achieved by the use of a stack of adequate height in combination with the favorable influence of the buoyancy of the effluent.

While the initial momentum of the plume usually does not add significantly to the final height reached by the effluent of a tall stack, the exit velocity at the stack mouth does constitute an important design constraint.

This is because the wind blowing over the stack creates a low pressure region on its downwind side which can entrain the stack gases if their exit velocity is too low. Dramatically higher concentrations of stack gases at the ground level are then evident near the stack. The entrainment of effluent by the stack's wake will be prevented if the exit velocity w_s bears a critical relationship to the wind speed $\bar{u}$,

$$w_s > R_v \bar{u} \tag{7.1}$$

The data of Fay et al. (Ref. 4) imply that the motion of the plume is independent of w_s if $R_v \geq 1.2$, whereas Briggs (Ref. 1), based on wind tunnel observations, quotes a value of 1.5 for this velocity ratio.

A stack effluent that has enough momentum to escape wake entrainment behaves, in the region near the mouth of the stack, like a jet of fluid issuing into a cross flow. The action of the wind and the mixing process is to cause the plume to bend over and to grow in diameter as it entrains the surrounding air and its initial vertical momentum is dissipated. Far from the stack the vertical motion of the plume results entirely from its buoyancy, while the horizontal motion is imparted entirely by the wind. The theoretical description given below applies to the motion of the plume in the far field region when these two effects and the continuous entrainment of the air by the stack effluent are the factors that control the trajectory of the plume.

Several quantities pertinent to the theory of plume rise are worth defining at the outset. The heat flux of a stack (W or Btu/hr) is essentially the flow rate of enthalpy measured with respect to atmospheric temperature and is given by

$$Q_s = C_{ps}(T_s - T_1)\rho_s w_s \pi b_s^2 \tag{7.2}$$

where b_s is the stack radius, C_{ps} is the specific heat at constant pressure of the stack gases, T_s is their absolute temperature, T_1 is the absolute temperature of the atmosphere, ρ_s is the density of the stack gases, and w_s is the stack gas vertical velocity; all quantities are evaluated at the stack exit height which is the origin of the coordinate system. The nomenclature used to describe plume rise is summarized in Fig. 7.1. A second important quantity is called the buoyancy flux and is given by

$$F_b = \frac{g(T_s - T_1)w_s b_s^2}{T_1} \tag{7.3}$$

Finally the buoyancy length scale is defined by

$$l_b = \frac{gQ_s}{\pi\rho C_p T_1 \bar{u}^3} \tag{7.4}$$

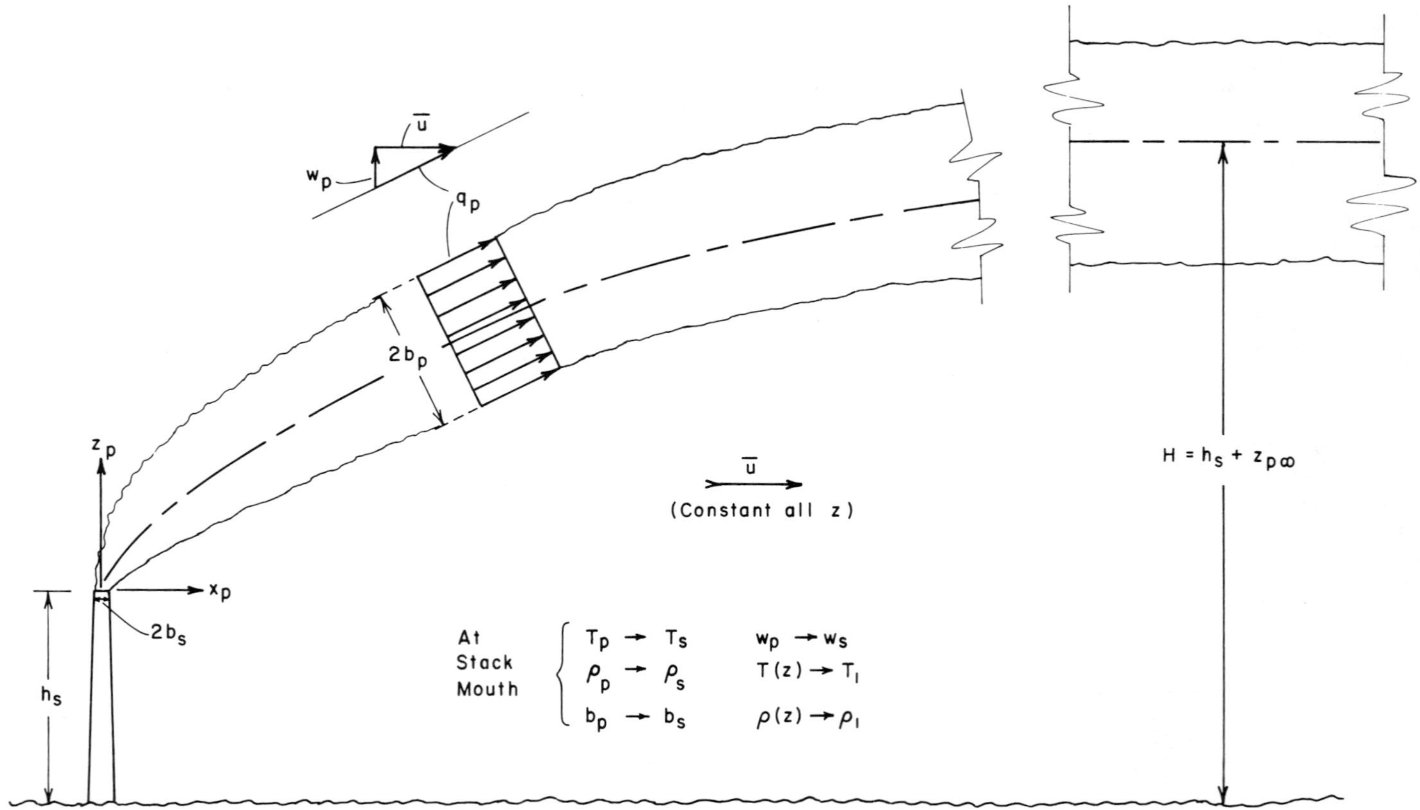

Figure 7.1. The nomenclature for the motion of a bent plume in a stable stratification.

where all quantities in the denominator represent the properties of the atmosphere evaluated at the mouth of the stack. Inserting Eq. 7.2 into Eq. 7.4 and taking the product ρC_p to be the same for the stack gases and the atmosphere, we find the expression that is normally used to calculate the buoyancy length as

$$l_b = \frac{F_b}{\bar{u}^3} \tag{7.5}$$

The buoyancy length scale can be interpreted as the radius of curvature of the plume trajectory at the stack exit. A high buoyancy flux or a low wind speed will result in a plume trajectory that is nearly a vertical line, while a nearly horizontal plume results under the opposite conditions.

One may argue that, in the case of the neutrally stable atmosphere, the buoyancy length scale is the fundamental quantity embodying all plume properties that influence its trajectory. An immediate consequence is that the height of the plume z_p at any downwind distance x_p must be related by a dimensionless functional relationship of the form

$$\frac{z_p}{l_b} = f\left(\frac{x_p}{l_b}\right) \tag{7.6}$$

A dimensional argument of this type is extremely powerful because the nature of the function on the right side of Eq. 7.6 can always be determined by recourse to the examination of the results of carefully conducted laboratory experiments or field observations. Early studies of plume rise have taken

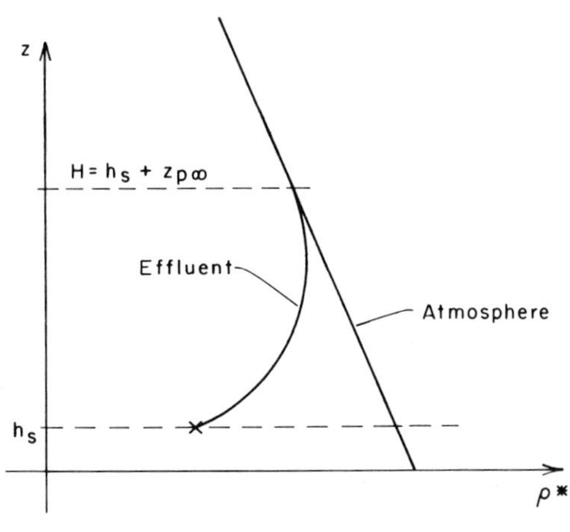

Figure 7.2. A schematic diagram of potential density versus altitude for the motion of a buoyant plume in a stable atmosphere of constant potential density gradient $d\rho^*/dz$. The mixing action causes the potential density of the plume to increase with altitude.

advantage of dimensional arguments of this type. In the following section we present the theory that will yield a specific algebraic relationship between plume rise and downwind distance.

In the discussion given in Section 1.6 it was noted that in the absence of mixing a buoyant effluent would rise at constant potential density to the level where this quantity is equal to the atmospheric potential density. The influence of mixing is clearly to cause the potential density of the effluent to increase during its ascent. For this reason the equilibrium height that is achieved is considerably lower than the level that would be predicted if mixing is ignored. This description is depicted schematically in Fig. 7.2.

7.1. THE ENTRAINMENT THEORY FOR THE MOTION OF A BENT PLUME

The motion of the bent plume was described on a theoretical basis by Slawson and Csanady (Ref. 5) as an extension of the earlier investigation by Morton, Taylor, and Turner (Ref. 6) of the motion of vertical plumes. The theory of Slawson and Csanady applies to the motion of the plume in the far field region where the following assumptions are reasonable:

1. Mixing or entrainment of the effluent gases with the atmosphere is the result of the vertical motion of the plume and can be quantitatively expressed by a universal entrainment coefficient.
2. The horizontal motion of the plume results solely from advection by the horizontal wind which is assumed to be independent of height.
3. Density may be considered to be constant in the equation for the conservation of mass, but the density differences, although small, must be retained where they express a buoyancy force. Accordingly the ratio of plume to atmospheric densities is approximated as unity and their difference divided by atmospheric density is small compared with unity. These assumptions essentially represent the Boussinesq approximations which are described in detail in a more general context by Seinfeld (Ref. 7) and the references that he cites.
4. The lower atmosphere has a constant or vanishing value of the Brunt–Väisällä frequency (see Section 1.6)

$$N = \left(- \frac{g}{\rho^*} \frac{d\rho^*}{dz} \right)^{1/2} \tag{7.7}$$

5. The flow both in the atmosphere and the plume is quasisteady and the pressure within the plume is equal to that of the local atmosphere. Time dependence of either the emission rate of the stack or the mean wind is excluded.

6. The velocity and other property profiles of the plume can be approximated by a top-hat distribution (see the sketch of the velocity profile shown in Fig. 7.1). While a Gaussian distribution is a more realistic representation of the property profiles, the top-hat distribution is equivalent if the mean property values are used.

The conservation of mass within the plume is expressed by

$$\bar{u} \frac{d}{dx_p} (\pi b_p^2 \rho_p) = 2\pi b_p \beta \rho w_p$$

In view of the Boussinesq approximations we may neglect the density variation in this equation and therefore the mass conservation is expressed by

$$\bar{u} \frac{d}{dx_p} (b_p^2) = 2 b_p \beta w_p \tag{7.8}$$

The derivative $\bar{u}(d/dx_p)$ is the particle derivative since the flow is quasi-steady and the time dependent part of the mean flow vanishes. Therefore, the left sides of the above equations represent, respectively, the time rate of change of mass and volume per unit length of plume. The entrainment coefficient β that results from vertical motion is defined by these equations and it is simply the rate of increase of mass in a unit length, per unit of lateral surface, per unit of vertical velocity of plume. Equation 7.8 contains the unknowns b_p and w_p and the entrainment coefficient which will be determined by appropriate experiments.

The vertical momentum equation for the bent plume is expressed by equating the rate of change of z directed momentum per unit plume length to plume vertical buoyancy force,

$$\bar{u} \frac{d}{dx_p} (\pi b_p^2 w_p \rho_p) = \pi b_p^2 (\rho - \rho_p) g \tag{7.9}$$

At this point we introduce the quantity f_b that is defined by

$$f_b \equiv \frac{b_p^2 (\rho - \rho_p) g \bar{u}}{\rho} \tag{7.10}$$

A direct solution for the motion of the bent plume is achieved if, following the treatment of Slawson and Csanady for the neutral atmosphere, we take f_b to be constant. If both air and the effluent gases are treated as perfect gases and molecular weight differences are ignored then

$$\frac{\rho - \rho_p}{\rho} = \frac{T_p - T}{T} \tag{7.11}$$

and we identify f_b as the far field buoyancy flux whose initial value is the quality F_b given by Eq. 7.3. Under these circumstances the equation for the conservation of momentum becomes

$$\frac{d}{dx_p}(b_p^2 w_p) = \frac{f_b}{\bar{u}^2} \tag{7.12}$$

which, together with Eq. 7.8, represents a system of ordinary simultaneous equations for b_p and w_p as a function of x_p. The height of the plume z_p can be related to downwind distance x_p by means of the kinematic relation,

$$\frac{w_p}{\bar{u}} = \frac{dz_p}{dx_p} \tag{7.13}$$

Equations 7.8, 7.12, and 7.13 contain the information necessary to determine the plume centerline trajectory.

We solve the system of equations by first integrating Eq. 7.12 and using the far field boundary condition, $b_p = 0$ at $x_p = 0$, we find

$$w_p = \frac{F_b x_p}{\bar{u}^2 b_p^2} \tag{7.14}$$

In the latter equation we have replaced the far field buoyancy flux f_b by its initial value F_b. The plume radius b_p as a function of downwind distance is found when we substitute Eq. 7.14 into Eq. 7.8 and integrate,

$$b_p = (\tfrac{3}{2} l_b \beta)^{1/3} x_p^{2/3} \tag{7.15}$$

We combine Eqs. 7.13, 7.14, and 7.15 to obtain

$$dz_p = (\tfrac{3}{2} l_b)^{1/3} \beta^{-2/3} x_p^{-1/3}\, dx_p$$

which can be integrated directly. With the boundary condition $z_p = 0$ at $x_p = 0$ the plume centerline trajectory in neutral stratification when $N = 0$ is given by

$$\frac{z_p}{l_b} = C_1 \left(\frac{x_p}{l_b}\right)^{2/3} \tag{7.16}$$

where

$$C_1 = \left(\frac{3}{2\beta^2}\right)^{1/3}$$

This theory for the plume rise in a neutral atmosphere predicts no limit to the extent of the vertical motion. This prediction is in agreement with what we found in the hypothetical case when a buoyant effluent rises adiabatically in a neutral or stable atmosphere with no mixing. Accordingly we will use

Eq. 7.16 to apply for the motion of a plume in the atmosphere under both neutral and unstable stratification. We momentarily defer the question of the appropriate value of β to use in these relations.

The effect of a stable stratification in the atmosphere was considered by Hewett et al. (Ref. 8), who added an energy equation that takes the final form

$$\frac{df_b}{dx_p} = -b_p w_p N^2 \tag{7.17}$$

According to this equation the buoyancy flux is not conserved during a stable stratification, $N > 0$. In this case the plume is limited in the extent of its vertical motion. Equation 7.17 is consistent with the formulation of Slawson and Csanady for the neutral case when $N = 0$ and the buoyancy flux is conserved. The equilibrium plume height that is given by the simultaneous solution to Eqs. 7.8, 7.12, 7.17, and the kinematic relation Eq. 7.13, which was given by Hewett et al. (Ref. 8), is here expressed as

$$\frac{z_{p\infty}}{l_b S^{2/3}} = C_2 \tag{7.18}$$

where

$$C_2 = 0.842\left(\frac{6}{\beta^2}\right)^{1/3}$$

and the dimensionless stability parameter S is given by $\bar{u}/l_b N$. The downwind distance $x_{p\infty}$ at which the plume reaches the equilibrium height can be estimated by the substitution of Eq. 7.18 into Eq. 7.16 which yields

$$\frac{x_{p\infty}}{Sl_b} = 2(0.842)^{3/2} = 1.55 \tag{7.19}$$

regardless of the value of the entrainment coefficient.

7.2. LABORATORY TESTS AND FIELD OBSERVATIONS OF PLUME RISE

An experimental test of the entrainment theory for the motion of a bent plume was conducted by Hewett, Fay, and Hoult (Ref. 8) under the favorable conditions that can be achieved in a wind tunnel which is explicitly designed for the simulation of flow in a stratified fluid. The tests were conducted in a thermally stratified, laminar flow of air using both heated air and helium to

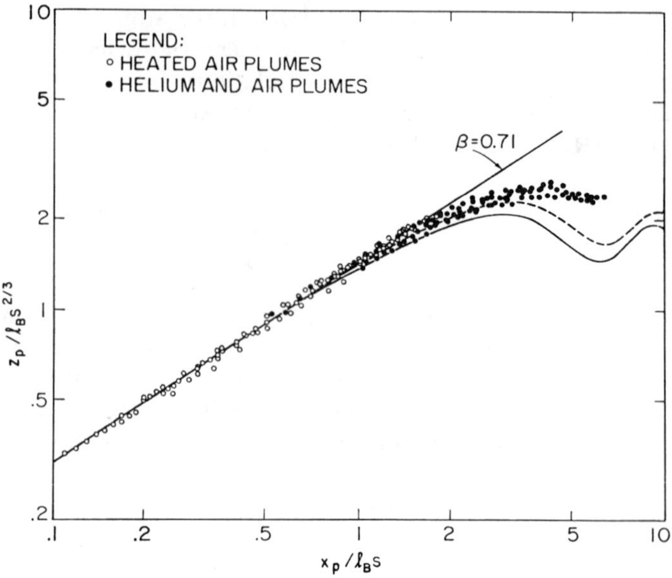

Figure 7.3. A comparison of entrainment theory with observation of plume rise in a wind tunnel. The straight line is Eq. 7.16 with $\beta = 0.71$. [From T. A. Hewett, J. A. Fay, and D. P. Hoult, *Atmos. Environ.*, **5**, 780 (1971). Reproduced with permission of the Pergamon Press and J. A. Fay.]

serve as the plume gases. The results of these experiments are depicted in Fig. 7.3 and they display excellent agreement with Eq. 7.16 for $0.1 < x_p/l_b S < 1.0$ when the entrainment coefficient $\beta = 0.71$. The results depicted in Fig. 7.3 also suggest that the downwind distance $x_{p\infty}$ corresponding to the maximum plume rise is somewhat larger than the value estimated by Eq. 7.19.

The practical use of entrainment theory to predict the plume rise from stacks requires extensive field observations to determine the appropriate value of β under atmospheric conditions. A study of this type was performed by Fay, Escudier, and Hoult (Ref. 4) and involved 137 sets of plume observations and nearly 1500 data points. The stack heights ranged from 21 to 183 m and the heat emission rates ranged from 0.4 to 111 MW. The plumes were divided into rising or leveled regions and separate values of the entrainment coefficient were derived for these two cases. Data for cases where the velocity ratio $w_s/\bar{u}$ was less than 1.2 were excluded in arriving at the recommended values of the entrainment coefficient. For the rising portion of the plume the statistical analysis of the field observations gave $\beta = 0.81$ and therefore $C_1 = 1.32$. For the level portion of the plume trajectory when the

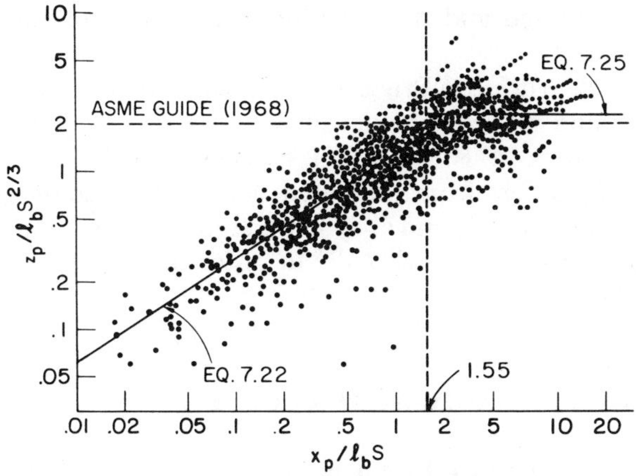

Figure 7.4. Observations of plume rise from tall stacks. The data of plume rise for $R_v > 1.2$ are plotted against downwind distance normalized with buoyancy length l_b and the stratification parameter S. [From J. A. Fay, M. Escudier, and D. P. Hoult, *J. Air Pollut. Control Assoc.*, **20**, 397 (1970). Reproduced with permission of the publisher and J. A. Fay.]

equilibrium height has been achieved, Fay et al. (Ref. 4) found $\beta = 0.55$ and thus $C_2 = 2.27$.

Figure 7.4 shows a comparison of the observed plume rise data and the predictions of Eq. 7.16 with $C_1 = 1.32$ and Eq. 7.18 with $C_2 = 2.27$. The scatter in the data shown in Fig. 7.4 is very large and is in sharp contrast to the results of wind tunnel tests depicted in Fig. 7.3. The scatter in the field observations of plume rise is to be expected for reasons such as the possible lack of quasisteadiness in the wind and source strength, the presence of atmospheric turbulence, as well as the greater difficulty in making field observations. This scatter should also serve as a warning of the reliability of plume rise predictions and as a reminder of the need for obtaining reliable input data for use in the plume rise equations.

The differences between the value of the entrainment coefficient observed for the rising plume in the wind tunnel, $\beta = 0.71$, and the corresponding field tests, $\beta = 0.81$, can be attributed to the influence of atmospheric turbulence which serves to enhance mixing. Fay et al. (Ref. 4) suggest the low values of $\beta = 0.55$ for the level plume may result from a different entrainment rate in the region where the motion of the plume tends to be oscillatory.

The transition of the plume motion to the equilibrium height in a stably stratified atmosphere is depicted for the wind tunnel data shown in Fig. 7.3 to occur at a value of $x_p/l_b S$ somewhat larger than the value of 1.55 that was estimated by the entrainment theory. The plume observations shown in Fig. 7.4 display sufficient scatter to obscure the downwind position at which equilibrium is achieved and the value predicted by entrainment theory is certainly adequate.*

A recent survey article by Briggs (Ref. 3) summarizes much of the more recent research and discusses values of the constants C_1 and C_2 found by various investigators. It is important to note that Briggs' definition of buoyancy flux F_b differs from that used herein by a factor of T_s/T_1 and, accordingly, the values of C_1 and C_2 which he recommends should be reduced by a multiplicative factor of $(T_1/T_s)^{1/3}$ in order to compare them with the values given above.

7.3. SUMMARY OF PRACTICAL FORMULAE

We summarize the practical formulae used to calculate plume rise by first repeating the basic definitions. The buoyancy flux is defined by

$$F_b = \frac{g(T_s - T_1)w_s b_s^2}{T_1} \tag{7.20}$$

and the buoyancy length scale by

$$l_b = \frac{F_b}{\bar{u}^3} \tag{7.21}$$

The stack heat flux, given by Eq. 7.2, is an auxiliary quantity of interest which does not enter explicitly into the calculation. A necessary condition is that $w_s > R_v \bar{u}$ where $R_v \approx 1.5$.

Under neutral or unstable stratification the plume rise is unlimited and is calculated from

$$\frac{z_p}{l_b} = 1.32\left(\frac{x_p}{l_b}\right)^{2/3} \tag{7.22}$$

where x_p and z_p are the centerline coordinates using the stack mouth as the origin of the coordinate system.

* The value of $x_p/l_b S = 1.55$ for transition is not numerically consistent with the values given for C_1 and C_2 because these quantities have been determined using different values of β for the rising and horizontal portions of the plume. The effect of this numerical inconsistency is to produce a discontinuity in the predicted motion of the plume between the end of the rising trajectory and the beginning of the horizontal trajectory. Thus Eqs. 7.16, 7.18, and 7.19 fail to intersect at a common point in Fig. 7.4. If we correct for the two different values of β that were used in determining $C_1 = 1.32$ and $C_2 = 2.27$, then we find $x_p/l_b S = 2.26$ (rather than 1.55). Both the wind tunnel data, Fig. 7.3, and the field observations on stack plumes, Fig. 7.4, are consistent with this improved estimate of the transition point. Clearly this improvement is of questionable significance in view of the large scatter in the observations displayed in Fig. 7.4.

When a stable stratification is present the plume rise occurs within the interval $0 < x_p < x_{p\infty}$ where

$$x_{p\infty} = 1.55\, Sl_b \tag{7.23}$$

and the stability parameter S is given by

$$S = \frac{\bar{u}}{l_b N} \tag{7.24}$$

Within this interval the trajectory of the plume is given by Eq. 7.22. For $x_p > x_{p\infty}$ the plume is advected by the wind at constant altitude above the stack mouth given by

$$z_{p\infty} = 2.27 l_b S^{2/3} \tag{7.25}$$

The so-called effective stack height H is given by adding $z_{p\infty}$ to the actual stack height h_s. Detailed discussions of plume rise theory and observations as well as an introduction to the extensive literature of this field are given in Refs. 1, 2, and 3 and the references quoted therein.

7.4. PLUME BIFURCATION

The splitting of a plume emitted by a tall stack, referred to as plume bifurcation, into two counter rotating line vortices is widely observed (Refs. 9 and 10) and is the basis for one analysis of plume rise (Ref. 11). The photograph of the bifurcated plume shown in Fig. 7.5 is provided by F. H. Fanaki and shows the formation of two distinct branches by the effluent of the 381 m stack located at Ontario, Canada. Usually the two separate plumes rise to the same height, but on occasions the vortices travel at two different altitudes.

Several factors contribute to the development of the bifurcation of a stack effluent. Near the mouth of the stack a moderate wind will develop shear stresses at the opposite sides of the plume thereby inducing concentrations of vorticity of opposite signs at these points. The bent plume is more buoyant at its core where the gases are the hottest and the density is the lowest. The central buoyancy also induces a motion within the plume which distorts its original circular boundary into a kidney-shaped cross section. On some occasions a full separation of the symmetric portions of the plume causes the two counter rotating vortices to be observed. Instrumented aircraft have been used to survey the concentration profiles within bifurcated plumes and these studies confirm the persistent nature of the two line vortices even after visual evidence of the plume is no longer present.

Figure 7.5. The bifurcated plume issuing from the 381 m stack of the International Nickle Company at Sudbury, Ontario. [From F. H. Fanaki, *Boundary Layer Meteorol.*, **9**, 480 (1975). Reprinted with permission of Reidel Publishing Co., Dordrecht, Holland and F. H. Fanaki.]

The exact circumstances that promote plume bifurcation are not fully understood and the literature is contradictory on the frequency of its occurrence. While some observers report it to be a rare event, Fanaki believes this effect to be often present. If this is correct then the empirically determined values of the entrainment coefficient that have been reported in the literature aleady account for the influence of bifurcation on the motion of the plume.

REFERENCES

1. G. A. Briggs, "Plume Rise," A.E.C. Critical Review Series, TID-25075, U.S. Atomic Energy Commission, 1969.
2. J. A. Fay, "Buoyant Plumes and Wakes," in *Annual Review of Fluid Mechanics* (Annual Reviews Inc., Palo Alto, Ca., 1973), Vol. 5, pp. 151–160.
3. G. A. Briggs, "Plume Rise Predictions," in *Lectures on Air Pollution and Environmental Impact Analysis*, D. A. Haugen, Editor (American Meteorological Society, Boston, 1975).
4. J. A. Fay, M. Escudier, and D. P. Hoult, "A Correlation of Field Observations of Plume Rise," *J. Air Pollut. Control Assoc.*, **20**, 391–397 (1970).

5. C. R. Slawson and G. T. Csanady, "On the Mean Path of Buoyant, Bent-over Chimney Plumes," *J. Fluid Mech.*, **28**, 311–322 (1967).

6. B. R. Morton, G. I. Taylor, and J. S. Turner, "Turbulent Gravitational Convection from Maintained and Instantaneous Sources," *Proc. Roy. Soc. (London), A*, **234** (1196), 1–23 (1956).

7. J. H. Seinfeld, *Air Pollution* (McGraw-Hill Book Co., Inc., New York, 1975).

8. T. A. Hewett, J. A. Fay, and D. P. Hoult, "Laboratory Experiments of Smokestack Plumes in a Stable Atmosphere," *Atmos. Environ.*, **5**, 767–789 (1971). Further details are given in J. A. Fay, M. Escudier, and D. P. Hoult, "A Correlation of Field Observations of Plume Rise," Fluid Mechanics Laboratory Publication No. 69-4, April 1969.

9. M. L. Khandekar and T. S. Murty, "A Note on Bifurcation of Buoyant Bent-over Chimney Plumes," *Atmos. Environ.*, **9**, 759–762 (1975).

10. F. H. Fanaki, "Experimental Observations of A Bifurcated Buoyant Plume," *Boundary-Layer Meteorol.*, **9**, 479–495 (1975).

11. J. Shwartz and M. P. Tulin, "Chimney Plumes in Neutral and Stable Surroundings," *Atmos. Environ.*, **6**, 19–35 (1972).

PROBLEMS

7.1. An old municipal power plant generates 110 MW of power when the fuel oil consumption is 80,400 lb/hr. The composition, by mass, of the fuel is 86% carbon, 11% hydrogen, 2.1% sulfur, 0.2% oxygen, and 0.7% water, nitrogen, and ash. The high heating value of the fuel is 18,300 Btu/lb.

1. What is the thermal efficiency of the plant?
2. What is the emission rate in gm/s of SO_2?

7.2. The stack height for the above plant is 331 ft, the stack inside diameter is 12 ft, the flue gas velocity is 58.5 ft/s, and the flue gas temperature is 300°F. The wind speed is 5 mph.

1. What heat flux is released by the stack if the atmospheric temperature at the stack height is 60°F, the molecular weight of the product is 29.3, and the specific heat at constant pressure of the flue gases is 0.28 Btu/lb °F?
2. What are the values of the buoyancy flux F_b and the buoyancy length l_b for this stack?
3. The temperature sounding of the lower atmosphere indicates that $\partial T/\partial z = 0$. What is the maximum height above ground (m), the so-called effective stack height, that the plume will achieve?
4. How far downwind (m) will the limiting height be reached?

7.3. A modern electrical generating unit produces 650 MW while burning coal at the rate of 442,000 lb/hr. The heating value of the coal is 13,500 Btu/lb. The products of combustion leave the stack, whose exit diameter is 19.5 ft, at a temperature of 255°F and an average velocity of 87 ft/s. The atmospheric temperature is 60°F.

1. What is the thermal efficiency of the unit?
2. What is the buoyancy flux F_b (ft^4/s^3)?
3. What is the stack heat flux Q_s (in W) if $\rho_s = 0.056$ lb/ft^3 and $c_{ps} = 0.28$ Btu/lb °F? Note that 1 W = 3.413 Btu/hr.

7.4. Calculate and compare the stack heat fluxes as a percentage of the heat input for the two generating units described in Problems 7.2 and 7.3.

7.5. Under other operating conditions the buoyancy flux of the unit in Problem 7.3 is 100,000 ft^4/s^3. If the wind speed is 5.0 mph, the lapse rate λ is zero, and the stack height is 350 ft where the atmospheric temperature is 60°F, find the following quantities:

1. The effective stack height (m).
2. The distance downwind that this height is first achieved (m).
3. If $d\theta/dz = 0$ and all other conditions remain unchanged, what will the height of the plume be over the center of a city located 3.25 km downwind?

7.6. One unit of the Paradise Steam Plant of the TVA generates 704 MW and has a stack which is 183 m high. The buoyancy flux is 985 m^4/s^3.

1. Find the plume rise when the wind speed $\bar{u} = 8.5$ m/s, $\Delta\theta/\Delta z$ is 17.3°K/km, and the ambient air temperature at the stack exit is 290°K. (The observed plume rise is 140 m.)
2. How far downwind from the stack is this height reached?

8

THE COMMUNITY AIR POLLUTANTS

Gaseous compounds and small particles that are present in the atmosphere in concentrations significantly higher than their normal natural level and which are potentially harmful to man or his environment are referred to as air pollutants. They may be of natural origin, as when they are created by a forest fire, but are more likely to be caused by the activities of man in relation to transportation, electric power generation, manufacturing, refuse disposal, and so on. Air pollutants that originate from the latter types of human activities are said to be anthropogenic and they may be divided into two categories according to the degree of hazard which they represent. Highly toxic substances such as asbestos fibers, mercury vapor, metallic beryllium or its compounds are designated as hazardous air pollutants. The human tolerance to these materials is relatively low, and they constitute a hazard to a small proportion of the total population mainly through occupational exposures. There exist numerous airborne compounds to which the human body has a somewhat higher level of tolerance but which are, nevertheless, damaging to human health, to other forms of animal life, to vegetation, and to some nonliving material objects. These substances are associated with common activities of man in the urban and industrial areas throughout the world and are referred to as the community air pollutants. Although the species that are classified as toxic substances or as community air pollutants are currently prescribed in the United States, we may expect these lists will change from time to time as more detailed information on composition and health effects of anthropogenic pollutants becomes available.

The community air pollutants are present to some extent in cities of moderate and large size throughout the world. Their presence is notorious in certain large cities where the combination of high emission rates and adverse meteorological conditions are coexistent. However, even rural areas in the United States are now known to be afflicted by the presence of certain of the community air pollutants which apparently have been carried by the winds from distant cities or industrialized areas. The anthropogenic community air pollutants are therefore geographically widespread in the industrialized nations, and they will constitute the principal focus of our discussion. Air pollutants are described as primary if they are directly emitted by a device or process. Secondary pollutants are those which are formed from primary pollutants by chemical reactions that occur in the atmosphere. Carbon monoxide, nitric oxide, and unburned hydrocarbons which are emitted by all types of automotive vehicles are examples of primary pollutants. Some of these compounds, referred to as precursors, are involved in atmospheric reactions in the presence of sunlight and produce a vast array of organic and inorganic species which are examples of secondary pollutants.

Carbon dioxide (CO_2) is emitted as a product of combustion, but it is not harmful to life and is not classified as a pollutant. Nevertheless, the increasing atmospheric concentration of CO_2 that results from the burning of fossil fuels is a source of concern because of its possible adverse influence on the climate of the earth.

The concentration of a pollutant is the amount, in units of mass or moles, of the species per unit volume. The density of a mixture of gases is the sum of the individual mass concentrations, c_i, and is given by

$$\rho = \sum_{i=1}^{n} c_i \tag{8.1}$$

The concentration of air pollutants is normally of such a level that the convenient units are milligrams per cubic meter (mg/m^3) or micrograms per cubic meter (μg/m^3). Because the density of air at ground level is approximately 1 kg/m^3, it is clear that the concentrations of the various pollutants are very small compared with the concentrations of the major species which control the properties of the mixture. For this reason the motion of the atmosphere is usually not influenced by the presence of a pollutant, which is then described as a passive contaminant. When the components of a mixture undergo chemical reactions with one another the molar concentration is the preferred quantity. This is the number of moles per unit volume which is found by dividing the mass concentration by the molecular weight of the species. Finally, the volume fraction f_v is also a frequently used measure of the amount of a pollutant that is present. For perfect gases the volume fraction is the same as the mole fraction and therefore,

$$f_v = \frac{c_i}{\rho} \frac{m_a}{m_i} \tag{8.2}$$

where m_a and m_i are the molecular weights of air and species in question and ρ and c_i are the corresponding densities at a specified pressure and temperature. These values are usually chosen to be 1 atm and 298.15°K for which air, whose molecular weight is taken to be 28.97, has a density of 1.184 kg/m^3. Thus f_v in parts per million (ppm) by volume is given by

$$f_v(\text{ppm}) = 0.02447 \frac{c_i(\mu\text{g/m}^3)}{m_i} \quad (8.3)$$

when c_i is the mass concentration specified in μg/m^3. The terms "parts per billion" (ppb) and "parts per trillion" (ppt) are also used to express volume fraction when convenient. In common usage and in the succeeding discussion the term "concentration" is used as defined above and also as a synonym for volume fraction.

8.1. THE COMMUNITY AIR POLLUTANTS AND THEIR SOURCES

The principal community air pollutants are as follows:

1. Carbon monoxide (CO) is a colorless, odorless gas which is present in the atmosphere at natural concentrations of 0.01 to 0.2 ppm. The worldwide production of CO by anthropogenic sources is estimated to be 270 million (metric) tons per year, but this rate is smaller than the natural production by a factor of approximately 13 (Ref. 1). Natural sources include the oxidation of naturally occurring methane, forest fires, volcanos, and electrical storms. Carbon monoxide is the least reactive of the gaseous community pollutants and decays at the estimated rate of 0.1% per hour under solar irradiation (Ref. 2). The residence time of CO in the atmosphere is therefore several months. The anthropogenic sources of CO are geographically concentrated on the global scale and cause high concentrations in urban or highly industrialized areas. Because of the low reactivity of carbon monoxide, its concentration in an air parcel, which is transported by the wind from an urban area, is decreased almost exclusively as a result of dilution through turbulent mixing.

2. Sulfur dioxide (SO_2) is a colorless gas which has a pungent odor and is detectable to the taste at concentrations of 0.3 to 1.0 ppm. The natural background of sulfur dioxide is low because of its high reactivity. Laboratory experiments have demonstrated the oxidation of SO_2 in air to form sulfur trioxide (SO_3) in the presence of sunlight or ultraviolet radiation in the 300 to 400 nm spectral region. This reaction is hastened by the presence of hydrocarbons and nitrogen dioxide. Sulfur trioxide, in turn, combines with atmospheric moisture to form droplets of dilute sulfuric acid at humidities

well below 100%. These droplets scatter light efficiently and contribute both to atmospheric haze and to the deterioration of visibility. Sulfuric acid droplets combine readily with other species to form sulfates which eventually are carried to the ground by rainfall. It is estimated that sulfates form 5 to 20% of the total particulate material that is present in the urban atmosphere.

3. Particulate material suspended in the atmosphere consists of a wide variety of chemical compounds and therefore spans a wide range of chemical reactivity. Two examples of particulate material that are produced by combustion are soot and fly ash. The former results from the incomplete combustion of fossil fuels and consists mainly of carbon with some combined hydrogen. Fly ash is made up of the fully oxidized products of combustion of the metallic components of fuels. Manufacturing processes may emit particulate matter of a wide variety of compositions according to the nature of materials being processed. Particulate material ranges in particle sizes from 10^{-3} μm or smaller to roughly 10 μm. Particles with a diameter of 100 μm have a settling velocity of 50 cm/s and are therefore rapidly removed by the action of gravity. Particles that are about 1 μm in diameter have the largest scattering power per unit cross section and they contribute efficiently to atmospheric haze. The human inner lung is prone to exposure to particles ranging from 0.1 to 1.0 μm in approximate diameter. Particles in this size range constitute respirable dust. Because both ash-producing and sulfurous ingredients are found in fossil fuels, notably coal, particulate materials and sulfur dioxide are frequently found to be coexisting products of combustion.

Particulate lead compounds, which are released into the atmosphere by automotive vehicles burning leaded gasoline, are the subject of specific concern and regulation. Tetraethyl lead [$Pb(C_2H_5)_4$] for many years has been added to gasoline to suppress detonation and to promote a slower, more controlled combustion that results in the efficient use of the fuel. The average lead content of regular gasoline sold in the United States during the mid 1970's was 1.3 grams per gallon. Premium grade gasolines often contained nearly twice as much lead. About 70% to 80% of the lead in gasoline is released to the atmosphere, while the remaining fraction is retained in the engine and its lubrication system. The lead released at the tailpipe of the automobile is mostly combined in the halogenated compounds PbClBr and $NH_4Cl \cdot 2PbClBr$, which are converted to lead oxide and carbonates in the atmosphere. About 50% of the particulate material in the atmosphere that bears lead-containing compounds is contained in the size range below 5 μm in diameter. Some airborne lead originates from smelters located in sparsely populated areas that refine metallic lead and copper.

A mixture of finely divided particulate material in atmospheric air is an example of an aerosol. A brief introduction to the physical properties of aerosols is given in Section 8.5 below, and the references cited therein provide a more detailed coverage.

4. Hydrocarbons of a wide range of compounds are emitted by internal combustion engines that do not achieve complete combustion, by refineries, and by fuel storage and handling facilities. Methane (CH_4) is present in the natural atmosphere at concentrations of about 1.5 ppm as a result of the decay of vegetation. The atmosphere may also contain more complex hydrocarbons called terpines that are emitted by trees and shrubs. The anthropogenic hydrocarbons are released against this natural background and their properties vary widely according to their chemical structure. A brief summary of nomenclature and structure of a few important hydrocarbon compounds is given in Table 8.1. The reactivity of hydrocarbons varies according to their chemical structure. The single bonded hydrocarbons of the paraffin family, such as methane, are quite inert. Usually the nonmethane hydrocarbon (NMHC) components are of interest in the context of air pollution. Compounds of the alkene group, also called olefins, that contain doubly-bonded carbon atoms are, in contrast to the paraffins, highly reactive. The reactive components of a hydrocarbon mixture (RHC) are those components that exceed a certain rating on a scale of reactivity such as is described below. An analysis of the composition by classes of automotive exhaust gases is given in Table 8.6 below.

5. Nitric oxide (NO) and nitrogen dioxide (NO_2), collectively referred to as NOX, are gaseous species that are formed in combustion processes wherein atmospheric nitrogen and oxygen are subjected to elevated temperatures. These species are present in the pollution-free air at concentrations of 0.2 to 2 ppb for NO and 0.5 to 4 ppb for NO_2; their residence times in the atmosphere are estimated at 4 days and 3 days, respectively (Ref. 3). They are key components in the formation of photochemical smog reactions as discussed below.

6. Photochemical oxidants is the term that was originally applied to the class of compounds that convert potassium iodide in solution into molecular iodine. Ozone (O_3), nitrogen dioxide (NO_2), and peroxyacetyl nitrate (PAN) all satisfy this definition which has more recently been narrowed to include only O_3 and PAN. Ozone is normally present in the troposphere at concentrations of about 10 to 20 ppb as a result of diffusion from the stratosphere where it is formed by the natural photochemical processes discussed earlier. Both O_3 and PAN are secondary pollutants formed by photochemical reactions involving precursor species such as NO, NO_2, and reactive hydrocarbons (RHC).

The estimated total emissions of some of the community air pollutants released in the United States during the year 1968 are listed in Table 8.2 (Ref. 4).

Table 8.1. The Chemical Structure of Selected Hydrocarbons of Several Important Classes

Class				
Alkanes (Paraffins)	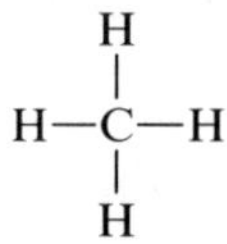 Methane (CH_4)	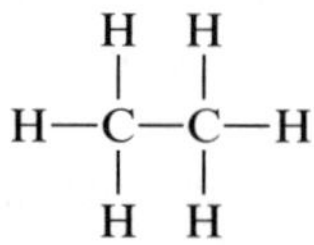 Ethane (C_2H_6)	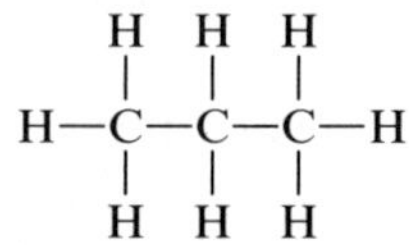 Propane (C_3H_8)	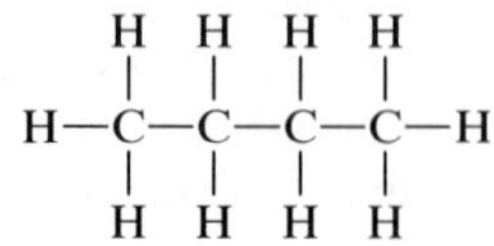 *n*-Butane (C_4H_{10})
Alkenes (Olefins)		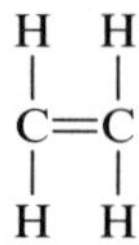 Ethylene (C_2H_4)	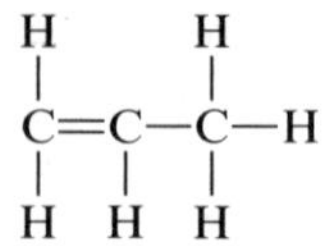 Propene (C_3H_6)	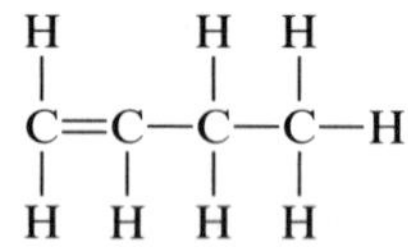 1-Butene (C_4H_8)
Alkynes	H—C≡C—H Acetylene (C_2H_2)			
Arenes (Aromatics)	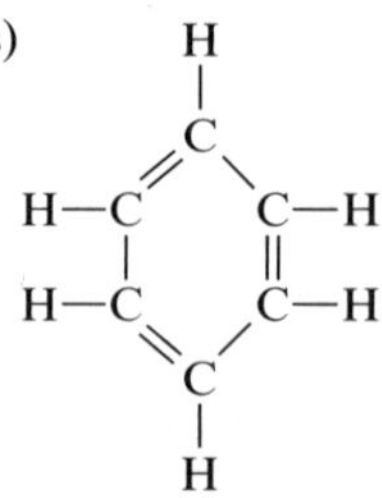 Benzene (C_6H_6)	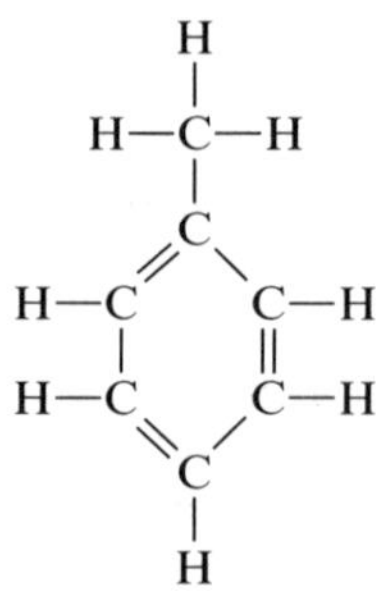 Toluene ($C_6H_5CH_3$)		
Terpines	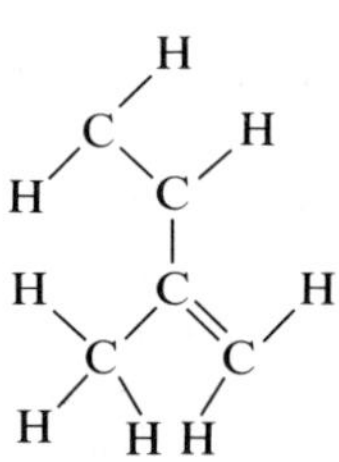 Isoprene (C_5H_8)			

Table 8.2. Estimated Emissions of Pollutants by Source Categories in the United States during 1968. Figures Represent 10^{+6} Metric Tons per Year (Ref. 4).

	Particulates	SO_2, SO_3	NOX	CO	HC
1. Stationary fuel combustion	8.1	22.1	9.1	1.7	0.6
2. Mobile fuel combustion	1.1	0.7	7.3	57.9	15.1
3. Incineration	0.9	0.1	0.5	7.1	1.5
4. Industrial processing	6.8	6.6	0.2	8.8	4.2
5. Solvent evaporation	—	—	—	—	3.9
Total	16.9	29.5	17.1	75.5	25.3

This information serves to emphasize the importance of controlled combustion processes as a source of various pollutants. The first three combustion-related source categories accounted for 60% of all particulates, 78% of all SO_2 and SO_3, 99% of the NOX, 88% of the CO, and 68% of the hydrocarbons (HC). The production of several of the gaseous pollutants formed in combustion processes bears a strong similarity in a variety of devices and is worthy of specific attention.

The gross composition of the exhaust products of an internal combustion engine as a function of air/fuel ratio is presented in Fig. 8.1 from Agnew (Ref. 5). The temperature of a hydrocarbon-air flame, which is not plotted in this figure, has a maximum near the stoichiometric fuel/air ratio. Nitric oxide is formed as a result of exposure of atmospheric N_2 and O_2 to the elevated temperature, and the amount produced is a maximum at the stoichiometric ratio and decreases markedly both at rich mixture ratios, where CO production is high because of the inadequate supply of oxygen, and also at lean mixture ratios where CO was not discernible. Figure 8.1 shows that significant unburned hydrocarbons are present not only at the stoichiometric ratio but also at the lean mixtures where excess oxygen is present. This incomplete combustion occurring at lean mixtures results mostly from quenching of the burning process in the immediate vicinity of the relatively cool walls of the combustion chamber.

Attempts to reduce the air pollutants emitted by automotive sources can be interpreted in terms of the information contained in Fig. 8.1 together with the knowledge that smooth combustion results only within a narrow range of fuel-air ratios close to the stoichiometric mixture. Earliest attempts to reduce pollutants were based on the use of a rich mixture to lower NO production. The CO and HC were to be controlled by adding oxygen to burn these components after they had left the combustion chamber. This

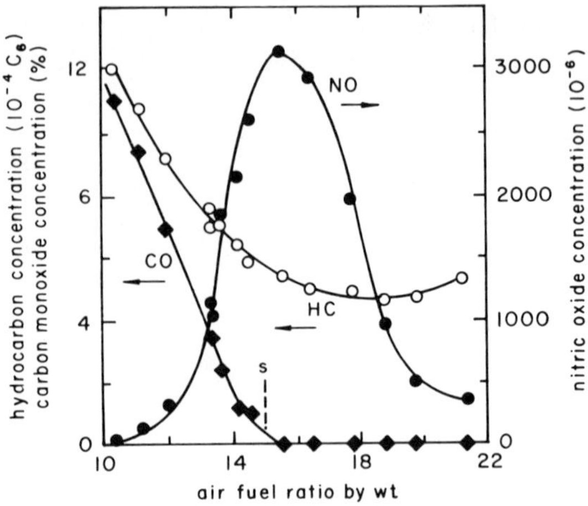

Figure 8.1. The relationship between the composition of engine exhaust products and air/fuel (mass) ratio. The stoichiometric ratio (s) is approximately 15. [From W. G. Agnew, *Proc. Roy. Soc.*, London, **A307**, 166 (1968). Reproduced with permission of The Royal Society and W. G. Agnew.]

method of controlling the emission of pollutants causes high fuel consumption and has been supplanted by alternative strategies. One method now used is to provide a small region of intense combustion which provides the ignition source for the main chamber where a relatively lean mixture ratio is used. The net effect is to allow the overall combustion process to occur at lean mixtures where the production of NO, HC, and CO is minimized.

8.2. HEALTH EFFECTS OF POLLUTANTS

The reaction of individuals to a given dosage of an air pollutant is highly variable and depends on such factors as age, existing state of health, medical history, smoking habits, occupational exposures, and hereditary factors. Thus physiological reactions to air pollutants must be stated in statistical terms for a population of specified composition. It is, of course, ethically unacceptable to intentionally subject humans in controlled laboratory experiments to doses of air pollutants which might prove to be detrimental to their health. For these two reasons our knowledge of the physiological effects of pollutants on the human system is inadequate and primarily is based on observations of human populations in their usual environments, while engaged in normal activities, and subjected to the varying concentrations of the various air pollutants that are produced in their surroundings. Systematic observations of this type are called epidemiological studies, and they are difficult to perform in a conclusive manner because not all pertinent factors are observable and because of the difficulty in selecting both the sample and the control populations. This situation is in sharp contrast to laboratory experiments in which a particular set of desired conditions can be repeatedly reconstructed to permit ever more careful measurements and more detailed analysis.

Numerous epidemiological studies have been discussed in two review articles. Anderson (Ref. 6) concludes that, despite the limitations of the published epidemiological data, a health hazard is associated with air pollution which is similar to the hazard of smoking in its biological effect. Higgins and McCarroll (Ref. 7) concluded that persons with pre-existing cardiac and respiratory ailments are more susceptible to the effects of air pollution than are persons in good health. They also concluded that a lower limit for any single pollutant which is considered safe for all persons cannot be specified, and adverse health effects are probably associated with the cooperative effects of several pollutants acting in concert.

The following discussion of physiological and health effects of the specific air pollutants is a summary of the information contained in Refs. 8 to 12 which contain the results of numerous epidemiological studies of the various community pollutants.

8.2.1. Carbon Monoxide

Carbon monoxide is a hazard to humans because of its strong affinity for the hemoglobin in the bloodstream. Normally the hemoglobin combines in the lungs with oxygen which is conveyed throughout the body. Even small concentrations of CO, which combines with the hemoglobin to form carboxyhemoglobin (COHb), interferes with this process and can cause physiological effects. Since CO is colorless and odorless the exposure to its effects cannot be easily detected. The severity of the physiological effects is determined by the concentration of CO and the duration of the exposure as is summarized in Fig. 8.2 from Ref. 13. Thus, for example, a 3 hr exposure to a level of 50 ppm of CO by a person engaged in heavy work results in about a 5% content of COHb in the bloodstream, which can cause effects such as the loss of visual acuity and the impairment of time interval discrimination. The more mild adverse effects of CO are completely alleviated when the subject is allowed to breathe pure air and the level of COHb in the bloodstream is reduced.

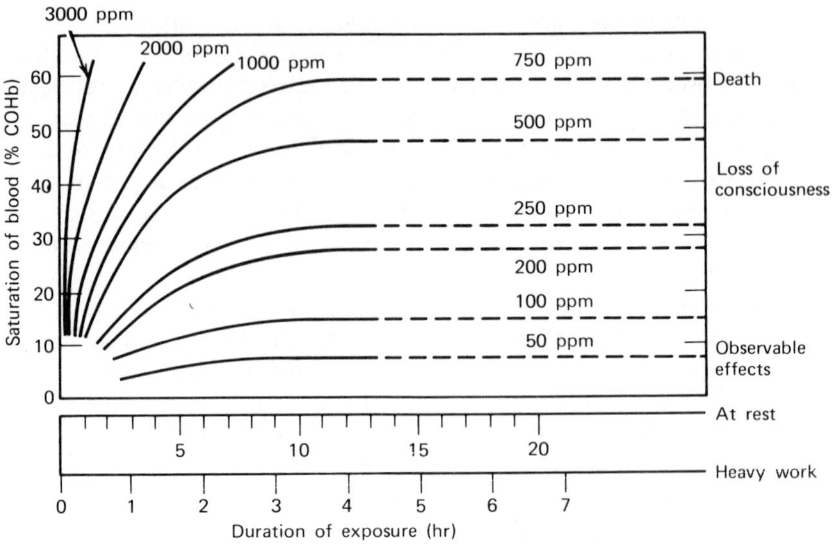

Figure 8.2. The physiological effects of exposure to carbon monoxide. The graph shows carboxyhemoglobin as a percentage of saturation versus duration of exposure for various concentrations (ppm) of CO. (Reprinted with permission from *Environ. Sci. and Technol.*, **5**, 213 (1971). Copyright by the American Chemical Society.)

8.2.2. Sulfur Oxides in Combination with Particulate Material

The oxides of sulfur, SO_2 with small amounts of SO_3, are invariably found with particulate material in the products of combustion of fossil fuels. The widespread use of coal during the last century resulted in this mixture being a common combination of pollutants found in large cities throughout the world. For this reason there exists an extensive backlog of experience with these materials. It is generally recognized that gaseous pollutants in the presence of respirable particles are synergistic in their impact on the respiratory system. For example, concentrations of SO_2 equal to 0.037 to 0.092 ppm with particulate concentrations of 185 $\mu g/m^3$ represent the levels that cause increased frequency of respiratory symptoms and lung disease. Data from air pollution episodes in England (Ref. 9) indicate that a 24 hr average of 0.25 ppm of SO_2 with 750 $\mu g/m^3$ of smoke particles is accompanied by an increase in the daily death rate. A sharp increase in death due to bronchitis is observed, and persons in the older age categories contribute most heavily to the mortality statistics.

8.2.3. Hydrocarbons

Direct health effects on humans by the anthropogenic hydrocarbons found in the atmosphere have not been demonstrated although injury to plants is known to occur. Aromatic compounds are irritating to the mucous membranes at concentrations greater than 25 ppm. The more offensive action of hydrocarbons is to serve as precursors for the photochemical smog reactions.

8.2.4. Oxides of Nitrogen

Little data exist on the health effects of NO and NO_2 because exposure to these compounds is usually accompanied by the concurrent exposure to other pollutants. The airway resistance of the lungs is known to increase when moderate exposures to NOX occur. Exposures to concentrations of NO_2 far higher than those observed in the atmosphere have caused death. The principal epidemiological study of the effects of NO_2 was conducted on the population of Chattanooga, Tennessee (Ref. 11), where manufacturing operations produced an atmosphere contaminated primarily by this compound. A conclusion of this study was that illness rates are higher than normal if the NO_2 concentration exceeds 0.06 ppm. This study has been criticized for several reasons, but it remains as the main source of data on the health effects of NO_2.

8.2.5. Photochemical Oxidants

Eye irritation is associated with short-term exposure to oxidants in the concentration range of 0.1 ppm and above. Impairment has been noted in the performance of athletes who were exposed for 1 hr to concentrations of 0.03 to 0.30 ppm. Somewhat longer exposures to ozone, 8 hr at the 0.3 ppm level, represent the threshold for nasal and throat irritation. Increasing concentrations impair the pulmonary function and short-term exposure to 9.0 ppm produces severe illness. Photochemical oxidants, O_3 and PAN, are often accompanied by other species such as nitrous (HNO_2) and nitric (HNO_3) acids, oxygenated hydrocarbons such as the aldehydes, and an array of related compounds that irritate the eyes and also the respiratory system.

8.2.6. Particulate Lead Compounds

Airborne lead compounds constitute a health hazard, especially to children in the 1 to 5 years age levels who are known to absorb and retain this element more readily than adults. The control of the emissions of lead is intended to keep its concentration from exceeding 30 μgm per deciliter in the blood of children. Concentrations higher than this level are associated with an impairment of cell functions that are regarded as adverse to the health of chronically exposed children. Excess exposure to lead has adverse influence on the blood-forming, nervous, and kidney systems, and on other body functional systems as well. Lead may also enter the body by ingestion of contaminated food and water.

8.3. THE AMBIENT AIR QUALITY STANDARDS

As a result of legislative action by the United States Congress, a set of goals for allowable air pollution concentrations have been established (Ref. 14). These ambient air quality standards, summarized in Table 8.3, are based on the health effects listed in Refs. 8 to 12 and their violation necessitates ameliorative action on the part of state or municipal governments. Many localities in the United States have taken major steps toward compliance with the ambient air quality standards for sulfur dioxide and particulates. Pollutants that result from automobiles are generally proving more difficult to control.

The ambient air quality standards have been criticized as being excessively stringent and based on inadequate information on health hazards. As a result of this criticism, a study was undertaken by the National Academy of Science and the National Academy of Engineering (Ref. 15) which concluded that the available evidence supported the recommended standards and "... found no substantial basis for changing the standards" even though information on health effects was inadequate. The panels involved in the study considered the health effects for SO_2 and CO to be best understood while those for NO_2 and photochemical oxides to be poorly understood. Epidemiological studies combined with research on animals were recommended to clarify the unknown health effects. The panels additionally stated that a particular segment of the population—a group of 20% of the total which consists of those who suffer from respiratory ailments and cardiovascular disease, and persons of normal health who are very young or very old—are particularly sensitive to the adverse effects of air pollution. The study concluded that this group and the general population were both considered to be adequately protected, to the extent to which the risks can be evaluated, by the ambient air quality standards promulgated mainly in 1971.

8.4. PHOTOCHEMICAL SMOG REACTIONS

The injection of nitric oxide, carbon monoxide, and hydrocarbons into the atmosphere by automobiles promotes, in the presence of sunlight, a series of reactions that form photochemical smog. Sulfur is present in gasoline in concentrations ranging from 0.1 to 0.3% and results in the formation of SO_2, and it may also participate in the reaction sequence. Particles are formed that contribute to atmospheric haze and the loss of visibility, and the concentrations of both O_3 and NO_2 may rise to levels that produce physiological reactions, damage to vegetation, and the deterioration of rubber products. The occurrence of photochemical smog was first a serious problem in areas such as Los Angeles where both climatic and geographic considerations cooperate perversely. More recently, the development of photochemical smog has afflicted large cities throughout the world as their populations of automobiles have increased.

Professor A. J. Haagen-Smit (Ref. 16) first identified the eye irritation, crop damage, and deterioration of rubber with the release of hydrocarbons and oxides of nitrogen in the atmosphere and their subsequent irradiation by the solar ultraviolet spectral region. He employed a plexiglass chamber to irradiate with sunlight the vapors of gasoline or specific pure hydrocarbons mixed with the oxides of nitrogen. Persons who were known to be sensitive to eye irritation induced by smog found similar reactions produced in Haagen-Smit's chamber. He also noted that aerosols, that is, a dispersion of fine particles in a gas, were also formed in the chamber when mixtures of NO_2 and hydrocarbons were irradiated by sunlight. This effect was pronounced in hydrocarbon compounds containing a double bond in a ring such as cyclohexene and was associated with the opening of the ring, with the formation of nonvolatile oxidation products that constituted the aerosol particle. This observation agreed with the known fact that smog droplets possess an oily nature. The use of a chamber with transparent walls, now referred to as a smog chamber, to test the photochemistry of mixtures of hydrocarbons and oxides of nitrogen has since been refined and represents an important experimental technique for the laboratory study of the smog type reactions that occur in the polluted atmosphere.

This histories of the concentrations of CO, NO, NO_2, and O_3, shown in Fig. 8.3 for the city of Los Angeles, display a typical development of the photochemical smog reactions. These curves show that CO and NO, both

Table 8.3. Ambient Air Quality Standards

Pollutants	National Primary Standard	National Secondary Standard
Sulfur dioxide	1. Annual arithmetic mean of 80 μgm/m^3 (0.03 ppm).	1. 1300 μg/m^3 3 hr concentration, not to be exceeded more than once per year.
	2. A maximum 24 hr concentration of 365 μgm/m^3 (0.14 ppm), not to be exceeded more than once per year.	
Particulates	1. Annual geometric mean of 75 μgm/m^3. 2. A maximum 24 hr concentration of 260 μgm/m^3, not to be exceeded more than once per year.	1. Annual geometric mean of 60 μgm/m^3. 2. A maximum 24 hr concentration of 150 μgm/m^3, not to be exceeded more than once per year.
Carbon monoxide	1. Maximum 8 hr concentration, not to be exceeded more than once per year of 10 mgm/m^3 (9 ppm).	1. Same as primary standard.
	2. Maximum 1 hr concentration, not to be exceeded more than once per year, of 40 mgm/m^3 (35 ppm).	2. Same as primary standard.
Ozone	1. Maximum 1 hr concentration, not to be exceeded more than 3 days during each 3 year period ending each calendar year, of 235 μgm/m^3 (0.12 ppm).	1. Same as primary standard.

Hydrocarbons (corrected for naturally occurring methane)	1. Maximum 3 hr concentration (6 to 9 a.m.), not to be exceeded more than once per year, of 160 μgm/m^3 (0.24 ppm).	1. Same as primary standard.
Nitrogen dioxide	1. Annual arithmetic mean of 100 μgm/m^3 (0.05 ppm).	1. Same as primary standard.
Lead	1. 1.5 μgm/m^3 averaged over the 3 months in each calendar quarter.	1. Same as primary standard.

Notes: 1. National standard promulgated on April 28, 1971 and described in the Federal Register, Vol. 36, No. 84, Part II, April 30, 1971. Revisions were announced in the Federal Register, Vol. 43, No. 194, Oct. 5, 1978, and in Vol. 44, No. 28, Part V, Feb. 8, 1979.

2. The phrase "24 hr concentration" means the integrated average concentration measurement in which the sampling time is 24 hr, and so on.

3. Particulate concentration is measured by a high volume sampler with a filter capable of collecting particles of size D, where $0.1 < D < 100$ μm. A sampling time of 24 hr is recommended. The geometric mean, $\overline{C}$, is defined by

$$\overline{C} = (C_1 \times C_2 \times \cdots \times C_n)^{1/n}$$

The minimum frequency of sampling is one 24 hr sample every 6 days. The number of samplers in a region is specified by demographic considerations.

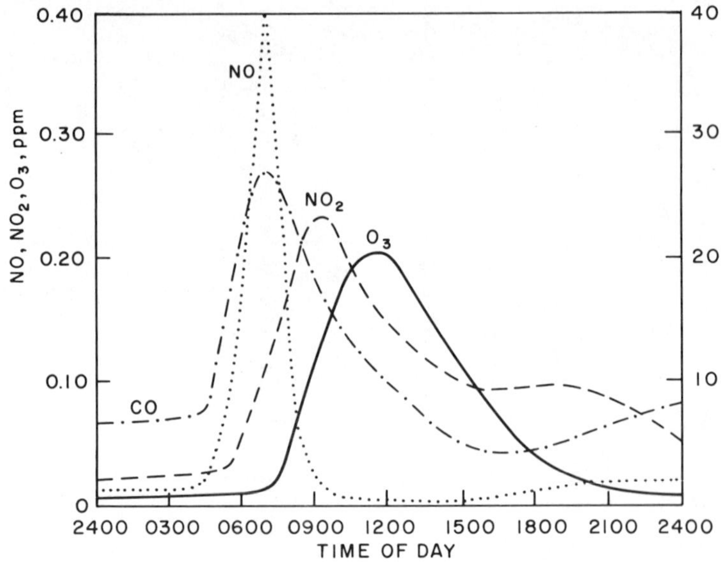

Figure 8.3. The hourly average concentrations of some primary and secondary pollutants in Los Angeles on July 19, 1965. (Reprinted from Ref. 12, "Air Quality Criteria for Photochemical Oxidants," U.S.E.P.A., April 1978.)

primary pollutants, reach their maximum levels during the morning hours of high traffic activity. The accumulation of these pollutants may be aided by poor atmospheric dispersion resulting from a nocturnal radiation inversion. The NO_2 concentration increases after NO has reached its maximum. Since the NO/NO_2 ratio in automobile exhaust products is about 5 to 10, the high NO_2 level that is shown at 0900 hr in Fig. 8.3 cannot be a result of primary emissions because this peak reading is both too high and too late. Finally, O_3 reaches a peak at about noontime, lagging the NO_2 maximum by several hours and becoming significant only after the NO concentration has decreased to low levels. The NO_2 concentration typically tends to build up again in the evening although this effect is not pronounced in Fig. 8.3. The explanation for this sequence of events is discussed in the following sections.

8.4.1. The NO—NO_2—O_3 Photochemical Reaction Sequence

We gain a limited insight into the above description of atmospheric pollutant concentrations by considering the reactions that ensue when a mixture of NO and NO_2 is subjected to ultraviolet radiation. The first reaction to occur is the photodissociation of NO_2 by photons in the 300 to 400 nm ultraviolet spectral region,

$$NO_2 + h\nu \xrightarrow{K_1} NO + O \qquad \text{(R-1)}$$

The atomic oxygen produced by reaction R-1 is a highly reactive species which will combine through collisions with the molecular oxygen that is a major component of the atmosphere to produce O_3,

$$O_2 + O + M \xrightarrow{K_2} O_3 + M \qquad \text{(R-2)}$$

The third body M is necessary to carry off the excess energy and may be any air molecule. Ozone reacts very rapidly through collisions with the NO originally present or formed by the first reaction to yield NO_2 and O_2,

$$O_3 + NO \xrightarrow{K_3} NO_2 + O_2 \qquad \text{(R-3)}$$

Thus NO and O_3 are formed by R-1 and R-2 and are consumed by R-3.

The rates associated with each of the above reactions taking place individually are designated by K_1, K_2, and K_3, respectively. They are defined by

$$\frac{d[O]}{dt} = K_1[NO_2] \qquad (8.4)$$

for the production of O by R-1,

$$\frac{d}{dt}[O_3] = K_2[O][O_2][M] \qquad (8.5)$$

for the production of O_3 by R-2, and

$$\frac{d[NO_2]}{dt} = K_3[NO][O_3] \qquad (8.6)$$

for the production of NO_2 by R-3. The value of K_1 depends upon the intensity of the ultraviolet radiation and therefore upon the altitude of the sun and the atmospheric ultraviolet absorption. The square bracket in the above equations refers to the concentration of the indicated species. Normally in the study of reaction kinetics the concentration units are expressed in moles per unit volume. In studies of air pollution chemistry the concentration of pollutants is conventionally specified in units of ppm by volume referred to air at 1 atm of pressure and a temperature of 25°C. In Eq. 8.4 for the formation of atomic oxygen the units of K_1 are min^{-1} and, since the rate of formation of O depends only on the concentration of NO_2 for fixed value of K_1, the reaction is said to be unimolecular. Equation 8.6 illustrates a bimolecular reaction where the rate depends upon the concentrations of two species undergoing a reaction as a result of thermal collisions. In this case the production of NO_2 varies directly as the concentration of each substance, and K_3 will have units typically expressed as ppm^{-1} min^{-1}. Finally, Eq. 8.5 expresses a termolecular reaction whose rate constant has units of ppm^{-2} min^{-1}. In the latter case the values of O_2 and M may be treated as constants which can be incorporated into a modified K_2 that is termed a pseudo-monomolecular reaction rate.

The net rate of formation of the atomic oxygen is the difference between the production by R-1 and the consumption by R-2 and is expressed by

$$\frac{d[\mathrm{O}]}{dt} = K_1[\mathrm{NO_2}] - K_2[\mathrm{O}][\mathrm{O_2}][\mathrm{M}] \tag{8.7}$$

In a similar fashion the net production rates of NO_2 and O_3 are given by

$$\frac{d[\mathrm{NO_2}]}{dt} = K_3[\mathrm{NO}][\mathrm{O_3}] - K_1[\mathrm{NO_2}] \tag{8.8}$$

$$\frac{d[\mathrm{O_3}]}{dt} = -K_3[\mathrm{NO}][\mathrm{O_3}] + K_2[\mathrm{O}][\mathrm{O_2}][\mathrm{M}] \tag{8.9}$$

Conservation of atomic species is clearly applicable at each moment and therefore, for example, the sum of the nitrogen atoms present is equal to their initial value,

$$[\mathrm{NO}]_0 + [\mathrm{NO_2}]_0 = [\mathrm{NO}] + [\mathrm{NO_2}] \tag{8.10}$$

where the subscript zero refers to the initial concentration. Differentiation of Eq. 8.10 gives

$$\frac{d[\mathrm{NO}]}{dt} = -\frac{d[\mathrm{NO_2}]}{dt} \tag{8.11}$$

Corresponding relations also hold for oxygen.

In view of Eq. 8.10 the quantity [NO] in Eqs. 8.8 and 8.9 can be eliminated in favor of $[NO]_0 + [NO_2]_0 - [NO_2]$. Equations 8.7, 8.8, and 8.9 then constitute three simultaneous ordinary differential equations in [O], $[NO_2]$, and $[O_3]$. These equations with the initial values $[NO]_0$ and $[NO_2]_0$ and $[O_3] = [O] = 0$, together with the appropriate values of K_2, K_3, and the time-dependent K_1, allow the time-dependent evolution of the species concentration to be calculated by an appropriate numerical means.

Detailed calculations of this type, which also contain additional reactions that are not of consequence to this discussion, have been performed by Demerjian et al. (Ref. 17), and they show that a steady state value of [O] about 10^{-7} times the values of [NO], [NO_2], and [O_3] is instantly established. Furthermore, the steady state condition in which

$$\frac{d}{dt}[NO_2] = \frac{d}{dt}[O_3] = \frac{d}{dt}[NO] = O \tag{8.12}$$

is achieved within a time period of about one minute. For many practical purposes the concentration of [O_3] is found directly by setting the left side of Eq. 8.8 equal to zero,

$$[O_3] = \frac{K_1}{K_3}\frac{[NO_2]}{[NO]} \tag{8.13}$$

The condition expressed by Eq. 8.13 is widely referred to as the photostationary state and was discussed in detail by Leighton (Ref. 18). Recent estimates of the rate constants (Ref. 17) are $K_1 = 0.48 \text{ min}^{-1}$ (corresponding to a solar zenith angle of 40°) and $K_3 = 23 \text{ ppm}^{-1} \text{ min}^{-1}$; these values give $K_1/K_3 = 0.021$ ppm. The photostationary state also applies to Eq. 8.7 which simplifies to an algebraic expression and yields the [O] in terms of K_1, K_2, and the concentration of the other species.

We note that for an initial mixture of pure NO and NO_2 the instantaneous value of [NO] is given by

$$[NO] = [NO]_0 + [O_3] \tag{8.14}$$

The quantities [NO] and [NO_2] can then be eliminated from Eq. 8.13 using Eqs. 8.10 and 8.14 to yield the algebraic expression

$$[O_3] = \frac{K_1}{K_3}\frac{[NO_2]_0 - [O_3]}{[NO]_0 + [O_3]} \tag{8.15}$$

The latter equation is readily solved for the ozone concentration in terms of the initial concentrations of NO and NO_2 and the time dependent K_1/K_3 rate ratio,

$$[O_3] = \frac{1}{2}\left([NO]_0 + \frac{K_1}{K_3}\right)\left(1 + \frac{4\dfrac{K_1}{K_3}[NO_2]_0}{\left([NO]_0 + \dfrac{K_1}{K_3}\right)^2}\right)^{1/2} - \frac{[NO]_0 + \dfrac{K_1}{K_3}}{2} \tag{8.16}$$

If the inequalities

$$[NO]_0 \gg \frac{K_1}{K_3} \qquad \text{and } ([NO]_0)^2 \gg 4\frac{K_1}{K_3}[NO_2]_0$$

are satisfied, Eq. 8.16 reduces to the result that the photostationary equilibrium based on the initial values, $[NO]_0$ and $[NO_2]_0$, is applicable,

$$[O_3] = \frac{K_1}{K_3}\frac{[NO_2]_0}{[NO]_0} \tag{8.17}$$

We can attempt to analyze the observations made in the Los Angeles area that are depicted in Fig. 8.3 by means of Eq. 8.17. At 0600 hr the values $[NO]_0 = 0.4$ ppm and $[NO_2]_0 \approx 0.05$ ppm apply, and with $K_1/K_3 = 0.021$ ppm for a 40° solar zenith angle, the above inequalities are fulfilled and Eq. 8.17 should give a reasonable estimate of the noontime value of $[O_3]$. An ozone concentration of 2.6×10^{-3} ppm is predicted while the observed value is about 100 times larger. It is clear that a model that is based solely on the occurrence of reactions R-1, R-2, and R-3 fails to describe the reaction sequence controlling the concentrations shown. The resolution of this paradox is discussed in the next section. Figure 8.3 also shows that the sum $[NO] + [NO_2]$ is not constant but decreases with time, and it suggests that there is a conversion of the oxides of nitrogen into other species.

The production of O_3 by the photochemical reaction R-1 ceases when the radiation source is extinguished and $K_1 = 0$. Atomic oxygen is depleted and only reaction R-3, which in this context is sometimes referred to as the dark reaction, controls the rate of depletion of O_3 that is given by

$$\frac{d[O_3]}{dt} = -K_3[O_3][NO] \tag{8.18}$$

Fresh injection of NO into the atmosphere by afternoon commuter traffic will therefore cause a depletion of O_3 with an attendant increase in NO_2.

8.4.2. The NO—NO_2 Conversion

The high concentration of O_3 at noontime that is displayed in Fig. 8.3 is explained by the conversion of the NO, which is released by automotive sources during the early morning, into NO_2 by reactions that occur in the atmosphere. It is this conversion that depletes the early morning peak NO concentration, that causes an NO_2 maximum at 0900 hr, and that allows the ultraviolet intensity, which increases with the higher solar elevation angle, to form high concentrations of O_3 at midday. Experiments conducted in smog chambers, similar to the device first used by Haagen-Smit, show that mixtures of NO and a wide variety of hydrocarbons will form NO_2 when they are irradiated with ultraviolet energy. Even a mixture of NO, CO, and water vapor will produce NO_2 when irradiated with ultraviolet light.

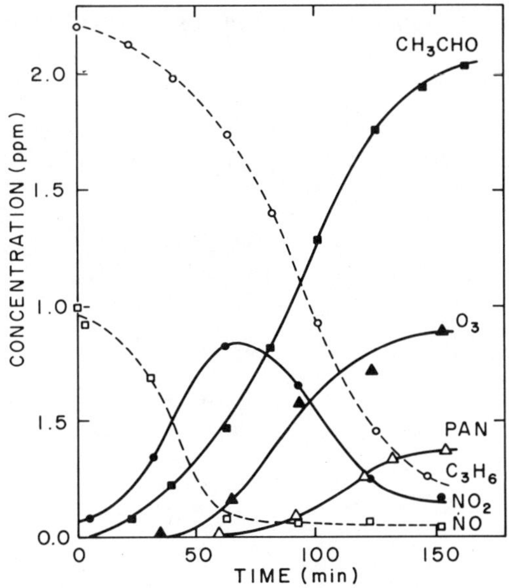

Figure 8.4. Smog chamber tests of C_3H_6, NO, NO_2, and air. The concentration histories are given for the reactants and some of the products. (Reprinted from Bernard Weinstock, Editor, *Photochemical Smog and Ozone Reactions*, Advances in Chemistry Series No. 113. Copyright 1972 by the American Chemical Society. Reprinted with permission of the copyright owner and H. Niki.)

In Fig. 8.4 the experimental results of smog chamber tests are shown for an initial mixture of 2.23 ppm of propene, C_3H_6, with 1.0 ppm of NO and 0.05 ppm of NO_2 in pure air at 50% relative humidity. The propene is converted into formaldehyde (HCHO) and acetaldehyde (CH_3CHO) although only the latter is shown in the figure. Here again we find the trend displayed in Fig. 8.3 that the peak O_3 level does not occur until after NO is substantially reduced. Part of the loss of NO_2 is accounted for by the formation of PAN and the remaining loss is presumed to be due to the formation of nitric acid which was not measured. In this case two primary pollutants, which individually are rather harmless, interact synergistically to form species—NO_2, HCHO, CH_3CHO, and PAN—that may have a serious effect on both animal and plant life. The structural arrangement of these products of smog reactions is given in Table 8.4.

In order to classify and quantify the propensity of various hydrocarbons to participate in smog reactions, several different reactivity scales have been proposed. One of these is based on the O_3 yield in the presence of ultraviolet radiation. A recent reactivity scale proposed by Pitts and colleagues (Ref. 20)

Table 8.4. Compounds Formed by Photochemical Smog Reactions

Aldehydes (Oxygenated Hydrocarbons)	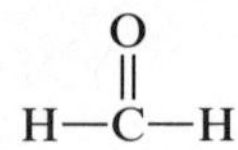 Formaldehyde ($HCHO$)	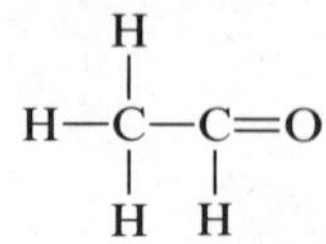 Acetaldehyde (CH_3CHO)
	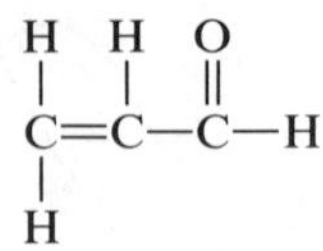 Acrolein (C_2H_3CHO)	
An Organic Nitrate	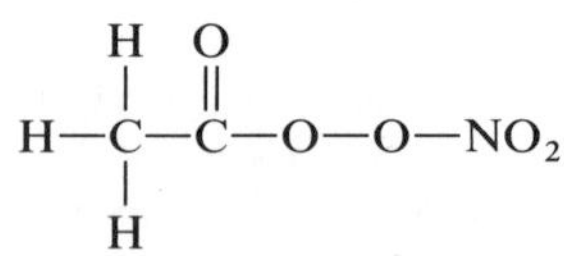 Peroxyacetyl nitrate (PAN) ($CH_3COO_2NO_2$)	

is based on the rate of reaction of the hydrocarbons with the hydroxyl radical OH which is regarded as the primary chain carrying species in the photochemical reaction sequence. A comparison of these two reactivity scales for selected hydrocarbons is presented in Table 8.5 based on information given by Pitts et al. The high reactivity of olefins and of propene in particular is demonstrated by both reactivity scales. The naturally emitted terpinic hydrocarbons possess the highest rating on the OH reactivity scale.

The reactivity ratings of hydrocarbons given above are confirmed in a qualitative way by the information presented in Table 8.6 compiled from data given in Ref. 10. Here the composition of diluted automotive exhaust products is given by hydrocarbon classes before and after the mixture has been irradiated by ultraviolet light. The major compound in each class is also given. The table shows that the alkene fraction of the mixture is reduced from 16% to 5% as a result of ultraviolet irradiation. All other

Table 8.5. A Comparison of Two Reactivity Ratings of Hydrocarbon Compounds (based on information given in Ref. 20)

Hydrocarbon Classes	Compound	O_3 Rating[a]	OH Reactivity Rating[b]
Alkanes	Methane	I	I (1)
(Paraffins)	Ethane	I	II (35)
	Propane	II	III (270)
	Butane	II	III (375)
Alkenes	Ethene	III	III (985)
(Olefins)	Propene	III	IV (3145)
	1-Butene	III	IV (4440)
Arenes	Benzene	I	III (180)
(Aromatics)	Toluene	III	III (750)
Alkynes	Acetylene	I	II (21)
Terpines	Various	III	IV (7000) to V (40,000)
	Carbon monoxide		II (18)

[a] Ratings defined: Class I—low O_3 yield; Class II—intermediate O_3 yield during first day; Class III—high O_3 yield within a few hours.
[b] Numbers in parentheses represent the reaction rate with the hydroxyl radical relative to methane. The roman numerals in this column represent classification based on a half-life in the presence of 10^{+7} OH radicals/cm^3. The half-lives are: I—more than 10 days; II—1 to 10 days; III—0.1 to 1.0 days; IV—0.01 to 0.1 days; V—<0.01 days.

categories increase by small amounts. Atmospheric tests also confirm that the alkene fraction decreases more rapidly with time after injection into the atmosphere than do the less reactive hydrocarbon components.

Because the hydrocarbon mixture emitted by automotive sources contains compounds possessing a wide range of reactivities, the prediction of the history of the O_3—NO—NO_2—RHC concentrations in the atmosphere is difficult and in most cases relies on a simplified version of the chemical reaction sequence. Simplistic approaches to the problem are forced by the large number of hydrocarbon species of widely varying reactivity, by the lack of adequate data on the actual reaction mechanisms and rates, by lack of detailed data on emission rates, and by the difficulty in incorporating the reaction sequence of moderate complexity into a model of atmospheric dispersion.

Table 8.6. Classes of Hydrocarbons in Diluted Automotive Exhaust Gases before and after Irradiation by Ultraviolet Light (based on data presented in Ref. 10.)

Hydrocarbon Classes	Nonirradiated	Irradiated
Alkanes (Paraffins)	23% (38% butane)	27% (38% butane)
Alkenes (Olefins)	16% (42% propene)	5% (50% propene)
Alkynes	29% (96% acetylene)	30% (98% acetylene)
Arenes (Aromatics)	7% (58% toluene)	10% (60% toluene)
Other Hydrocarbons	25%	28%

Note: Parenthetic insertions represent the major compounds in each class and the percentage of the total class population represented. Thus, propene represents 42% of the olefins prior to irradiation.

The conversion of NO to NO_2 through the hydrocarbon reaction sequence is sufficiently slow so that it only imperceptibly displaces the photostationary equilibrium expressed by Eq. 8.14. Therefore smog chamber tests generally show compliance with photostationary equilibrium even in the simultaneous presence of hydrocarbon reactions. The situation is not as clear in the atmosphere. Calvert (Ref. 21) concludes that, within somewhat large margins of error, photostationary equilibrium does apply for specified elevated air parcels in the Los Angeles basin between 0900 and 1430 hr local time. The composition of the pollutants in these air parcels, whose nominal altitudes varied from 60 to 250 m above ground level, was monitored with the aid of helicopters bearing instrumentation capable of monitoring the concentrations of NO, NO_2, O_3, CO, and hydrocarbons. On the other hand, Taylor and Eschenroeder (Ref. 22) find that the ratio $[O_3][NO]/[NO_2]$ is dramatically higher than K_1/K_3 near ground level in the Los Angeles area. They attribute this discrepancy to spatial nonuniformity of the concentrations of the various species. It should be noted that ground level measurements of ozone may be strongly influenced by the fresh injection of NO and NO_2 and the destruction of O_3 as a result of its contact with ground surfaces.

8.4.3. The Chemical-Meteorological Interaction of Photochemical Oxidant Pollutants

The fair-weather diurnal cycle of the atmospheric boundary layer exerts a major influence on the $NO—NO_2—O_3—RHC$ reaction cycle. Under clear skies by daylight O_3 will often reach its peak concentration during the early afternoon. The decrease of the K_1 reaction rate as a result of the setting sun allows O_3 and NO to consume each other, with the production of NO_2, until either species is depleted. With efficient conversion of NO to NO_2 by hydrocarbons, for example, we might expect mainly O_3 and NO_2 to remain at the time of sunset. At the upper levels of the mixed layer this excess of O_3 becomes available to participate in the long distance transport. Near the surface of the ground the low-level nocturnal radiation inversion stabilizes the layer adjacent to the surface and creates a surface layer with poor dispersive properties. Ozone at ground level can be depleted by either fresh emissions of NO from automobiles or by surface depletion reactions. Under clear nighttime skies surface-level ozone concentration reaches its minimum values during nighttime hours. Soon after sunrise there is the nearly simultaneous fresh injection of NO and HC by early morning traffic, the breakup of the nocturnal inversion, and the development of a significant intensity of ultraviolet radiation to promote the photochemical reactions. The "excess" ozone above the nocturnal surface layer can be brought to the surface by the breakup of the nocturnal inversion. This effect can increase surface level readings of O_3 before the local NO to NO_2 conversion has become significant. During the course of the day the latter reactions become important, and in the presence of sunlight, bring about the development of the high values of ozone concentration that are observed at midday.

In the eastern United States it has been found that the daily maximum of O_3 may occur during nighttime hours. This will happen, for example, under cloudy skies when a nocturnal inversion is absent. In this case there is relatively good mixing with air from aloft that may contain O_3 transported from upwind sources. Thus the $NO—NO_2—O_3—RHC$ reaction sequence interacts strongly with the state of development of the atmospheric boundary layer. The details of this interaction, a more thorough understanding of the $NO—NO_2$ conversion, the nature of the seasonal variation of these reactions, and the alternate methods of ameliorating the photochemical smog reactions are areas of current study.

A recent discovery has been the long-distance transport of ozone. It is now well documented that ozone formed near a large city can be advected by atmospheric circulation for hundreds of kilometers. In many areas in the eastern United States there exists a high background of ozone which presumably originates in the industrial and urban regions to the west. Rural areas frequently show higher long-term average ozone concentrations and more frequent violations of the ambient air quality standard for ozone than do urban areas in the same region. Presumably the high injection of NO by urban automotive traffic contributes to the reduction of O_3 in the cities, while long distance transport produces high O_3 levels in the country. Ozone of stratospheric origin (Ref. 23) is injected into the troposphere and contributes to some extent to the high concentrations observed near the surface. While the interplay of these effects is not fully understood, it is clear that attempts to bring the concentrations of O_3 down to the value specified in the U.S. National Ambient Air Quality Standards will probably require regional, national, and possibly international cooperation. These topics were discussed in detail in the papers presented at the international conference on photochemical oxidant pollution (Ref. 24) and in a subsequent analysis (Ref. 25).

8.5. THE ATMOSPHERIC AEROSOL

Finely divided particles, either in the liquid or solid state, are present in both the polluted and the unpolluted atmosphere. The mixture of these particles and atmospheric air constitute an aerosol. The aerosol of the polluted atmosphere may be comprised of either primary particles, which are directly injected into the atmosphere by a pollution source, or secondary particles which are produced in the atmosphere by gas-to-particle conversion processes from primary or secondary gaseous pollutants. In this section we introduce some basic concepts and properties relating to the atmospheric aerosol. More detailed discussion of the properties of aerosols are given by Davies (Ref. 26) and by Friedlander (Ref. 27).

8.5.1. Particle Size Distribution

The particles in a small sample of an aerosol are invariably found to possess a wide range of sizes. An upper limit to these sizes will be established by sedimentation—the falling out of particles as a result of gravitational settling (see Problems 8.4 and 8.5). The sizes of the particles in an aerosol can be measured by means of a cascade impactor, shown schematically in Fig. 8.5, which operates on the principle of inertial separation. In this device the largest particles are captured on the first encountered impaction plate because their inertia prevents them from following the curved streamlines of the gas flow through the passage between the jet and the plate. The smaller jet diameter of the next stage produces a higher jet velocity and a

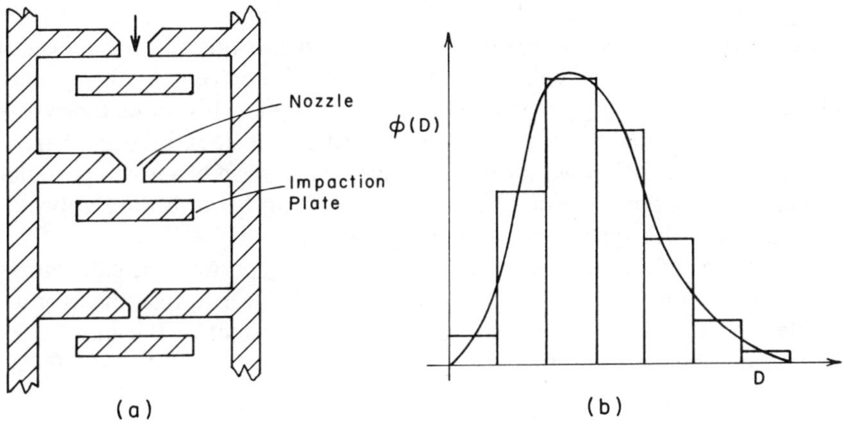

Figure 8.5. (*a*) A schematic diagram of a cascade impactor and (*b*) the size histogram approximation to the size distribution function.

sharper curvature of the streamlines and results in the capture of somewhat smaller particles. A large number of stages will then provide a separation into a corresponding number of size ranges from which a histogram of particle sizes can be prepared. The size histogram is an approximation to the curve, which is often smooth with but a single maximum, that is the particle size distribution function $\phi(D)$. When, as is the usual case, there are many particle sizes present, the particulate phase of the aerosol is said to be polydisperse.

The function $\phi(D)$ is defined such that $\phi(D)dD$ is the fraction of all particles that lie within the size range $D \pm dD/2$. The product $C_n\phi(D)$, where C_n is the total number of all sizes per unit volume, is the actual number per unit volume of particles in the same size range. An equivalent definition is that $\phi(D)$ is the function whose integral over a given size interval, D_1 to D_2, is the probability of occurrence of a particle within this size range. Thus

$$\int_{D_1}^{D_2} \phi(D)dD \equiv P[D_1 < D < D_2] \tag{8.19}$$

where $P[D_1 < D < D_2]$ designates the probability of finding a particle within the range D_1 to D_2. From this definition it follows that

$$\int_{0}^{D_\infty} \phi(D)dD = 1.00 \tag{8.20}$$

since all particles lie within the size interval 0 to D_∞. The atmospheric aerosol has a very large component of particles of very small sizes and is usually represented by a graph of $\log[\phi(D)]$ versus $\log D$. Friedlander (Ref. 27) has examined a theoretical model which is based on the production of intermediate particles by collisions and coalescence of very small particles and depletion of large particles by sedimentation. This model predicts that $\phi(D)$ varies as D^{-n} where $n \approx 4$ for intermediate sizes. Figure 8.6 from Whitby et al. (Ref. 28) shows data on the size distribution of aerosol particles sampled in Minneapolis, Fort Collins, and Los Angeles; this data approximately displays this type of power law dependence.

A knowledge of the size distribution function $\phi(D)$ permits the calculation of some of the important properties of an aerosol. For example, the average particle diameter D_{av} is found by summing the product of the number of each size times the size itself,

$$D_{av} = \int_0^{D_\infty} \phi(D) D dD \tag{8.21}$$

The total surface area per unit volume of the particles within an aerosol is given by

$$S_t = \pi C_n \int_0^{D_\infty} \phi(D) D^2 dD \tag{8.22}$$

The total volume of the particulate phase per unit volume of space is given by

$$V_t = \frac{\pi}{6} C_n \int_0^{D_\infty} \phi(D) D^3 dD \tag{8.23}$$

The above three integrals represent the first, second, and third moments of the particle size distribution function. The interesting physical significances of higher moments of the particle size distribution function are discussed by Friedlander (Ref. 27).

The chemical composition of the particles of the aerosol of a polluted atmosphere is quite complex, but can be determined by special methods of chemical analysis in spite of the smallness of the sample of material that is usually available. The particles of the polluted aerosol may possess a compositional variation that is size dependent. For example, the particles formed by burning leaded gasoline in automobiles often comprise a much smaller size range than the particles contained in fly ash which have large fractions of sodium, vanadium, and iron, and very little lead. An advantage of the cascade impactor is that it provides size-graded samples that can be analyzed by modern microanalytical techniques. This type of information, in principle, can be used to define a size-composition probability distribution function. The chemical composition of the polluted aerosol, even without a knowledge of its size gradation, can provide valuable information on the source material of the pollution by methods that are described below.

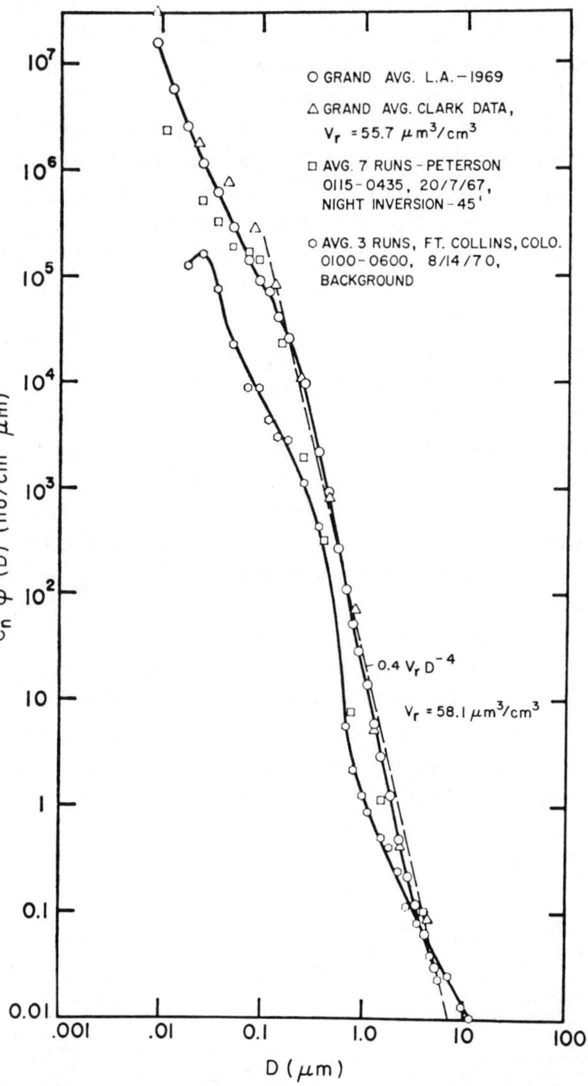

Figure 8.6. The size distribution of several atmospheric aerosols. The Clark and Peterson data refer to the aerosol in Minneapolis during 1969. The remaining data are from Los Angeles and Fort Collins. (From Whitby et al., "The Aerosol Size Distribution of Los Angeles Smog," in G. M. Hidy, Editor, *Aerosols and Atmospheric Chemistry*, Academic Press, 1972. Reprinted with permission of Academic Press and K. T. Whitby.)

8.5.2. The Optical Properties of Aerosols

The presence of the atmospheric aerosol is often made readily apparent by the impairment of visibility caused by the scattering of light by the particulate phase. The scattering power of spherical particles can be calculated from the solution of Gustave Mie to the Maxwell equations (Refs. 29 and 30). This solution addresses the problem of the scattering by a single homogeneous sphere of diameter D and refractive index m of a planar monochromatic electromagnetic wave of specified state of polarization (see Fig. 8.7). An observer, or his instrument, faces the particle at the angle γ_s measured from the forward direction. The plane of the observer is defined by the line from the source to the center of the particle and by the line from the latter to the observer. The refractive index of the particle is in general complex, $m = n_1 - in_2$. The complex portion of the refractive index n_2 is responsible for the absorption and degradation of radiant energy into heat and is usually a sensitive function of wavelength. For example, $n_2 = 0.43$ for water when $\lambda = 15$ μm in the infrared while at the visible wavelength of 0.6 μm, $n_2 \approx 10^{-7}$ and is commonly considered to be zero. A particle that absorbs no radiation, $n_2 = 0$, is said to consist of a dielectric material.

The beam of light incident upon a droplet may be attenuated either by absorption, if present, or simply by deflection from rectilinear propagation. The latter process is called scattering and, if $D \gg \lambda$, then the scattering can be described in terms of the wave effects known as refraction, reflection, and diffraction. In this limit the optically black particle ($n_2 \to \infty$) will absorb all radiation incident upon its geometric cross section; the dielectric particle will redirect an equivalent amount by refraction and reflection. In addition both particles will deflect by diffraction an amount of light passing

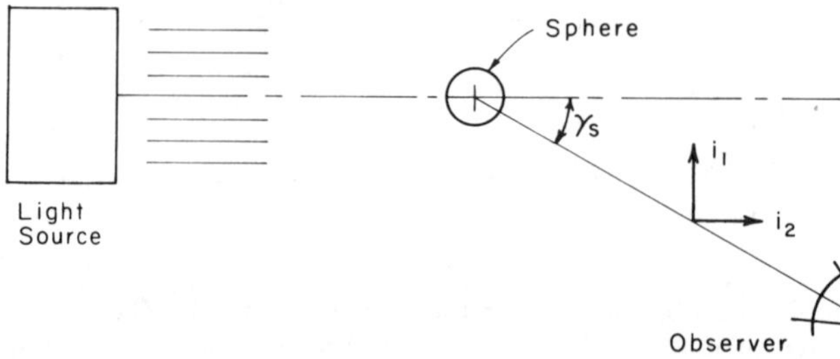

Figure 8.7. The scattering of light by a sphere. The particle of diameter D and refractive index m is illuminated by monochromatic light of wavelength λ. The light scattered at the angle λ_s from the forward direction is made up of two components that are polarized perpendicular and parallel to the plane of the observer.

nearby that is exactly equivalent to that which is incident upon them. Thus both particles will remove an amount of radiation from the incident beam of parallel light equal to twice that which is intercepted by the geometric cross section.

The solution of G. Mie gives quantities such as the intensity and polarization of light scattered at an arbitrary angle. The amount of light removed from the beam both by absorption and by scattering is, by definition, the amount that would fall upon an area equal to the extinction cross section, $Q_{ext}(\pi/4)D^2$. Thus the efficiency factor for extinction Q_{ext} is the number which multiplied by the cross section of the sphere will give the extinction cross section. For both the dielectric particle and the perfect absorber $Q_{ext} \to 2.00$ for $D/\lambda \to \infty$, in view of the discussion in the above paragraph. For particles of arbitrary size and refractive index the efficiency factor for extinction is given by Mie theory in the form

$$Q_{ext}(x_m, m) = \frac{2}{x_m^2} \sum_{n=1}^{\infty} (2n + 1) Re(a_n + b_n) \tag{8.24}$$

where $Re(a_n + b_n)$ designates the real part of the Mie scattering functions a_n and b_n that are somewhat complicated functions of $x_m = \pi D/\lambda$ and m. The number of terms required for convergence of the series represented by Eq. 8.24 increases with increasing x_m; a detailed understanding of the character of Q_{ext} versus x_m has been achieved only after digital computers became available to facilitate this calculation. A discussion of the programming of the quantities a_n and b_n and the calculation of related quantities is given by Maguire (Ref. 31). A graph of Q_{ext} versus x_m is shown in Fig. 8.8

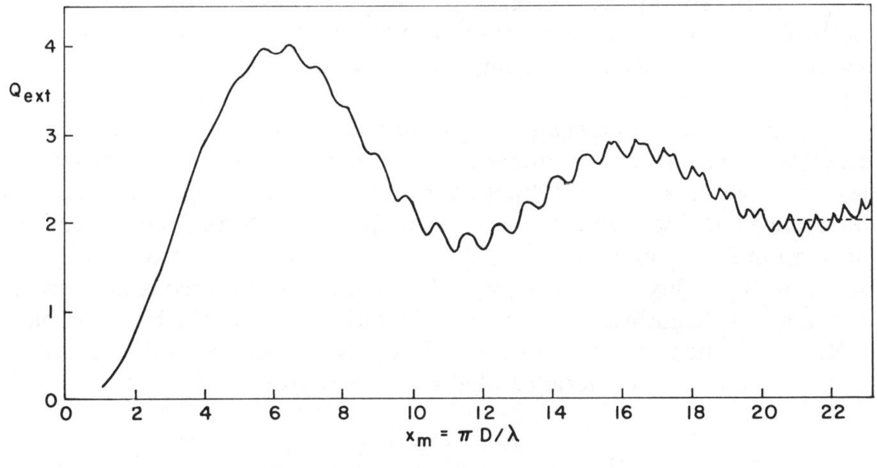

Figure 8.8. The efficiency factor for extinction versus $x_m = \pi D/\lambda$ for $m = 1.333$.

for $m = 1.333 - i0$ which corresponds closely to liquid water surrounded by a gas or gas–vapor mixture. The large scale oscillatory behavior of Q_{ext} versus x_m results from interference effects between the centrally transmitted and the diffracted rays.

When $D \ll \lambda$, the Rayleigh scattering limit, the particle behaves as an electric dipole and one term of Eq. 8.24 is sufficient to give Q_{ext}. In this case for m real ($n_2 = 0$), Eq. 8.24 reduces to

$$Q_{ext} = \frac{8}{3}\left(\frac{\pi D}{\lambda}\right)^4 \left(\frac{m^2 - 1}{m^2 + 2}\right)^2 \qquad (8.25)$$

The fourth power dependence of Q_{ext} on $\pi D/\lambda$ is the conspicuous result from Rayleigh scattering which implies that the shorter wavelengths of polychromatic solar light are scattered with greatest effectiveness. Therefore particles, such as atmospheric molecules, scatter the violet, blue, or green components of white light far more effectively than the longer wavelength yellow, orange, or red components. Thus the clear sky is seen as blue as a result of molecular scattering. Clouds appear white because they are composed of droplets which are large compared with the wavelength of visible light and which scatter all these wavelengths with about equal effectiveness.

Mie theory also gives the amount of radiation scattered by the particle. Thus the scattering cross section is defined by $Q_{sca}(\pi/4)D^2$, where the efficiency factor for scattering Q_{sca} is also given by an appropriate infinite series (Refs. 29 and 30). The absorption cross section $Q_{abs}(\pi/4)D^2$ determines the amount of radiation converted to heat, and the efficiency factor for absorption is simply the difference $Q_{ext} - Q_{sca}$. Thus for dielectric particles for which $n_2 = 0$ and absorption is absent $Q_{abs} = 0$, $Q_{ext} = Q_{sca}$, and scattering is the only mechanism that is present to attenuate the incident beam.

The transmission of a beam of light through an aerosol is of considerable practical importance. The apparatus required to perform a transmission test is shown in Fig. 8.9. A light source creates a beam of nearly parallel monochromatic light which is incident upon an aerosol made up of particles uniform in size (a monodispersion) that are uniformly distributed in space. A certain light flux F_0 (W) is emitted by the source and the irradiance, or flux per unit beam area dF/dA, is uniform throughout the beam on each plane perpendicular to the optic axis. The loss of light flux as a result of the combined effects of scattering and absorption is given by

$$dF = -\frac{F}{A_b}\left(Q_{ext}\frac{\pi}{4}D^2\right)(C_n A_b dx) \qquad (8.26)$$

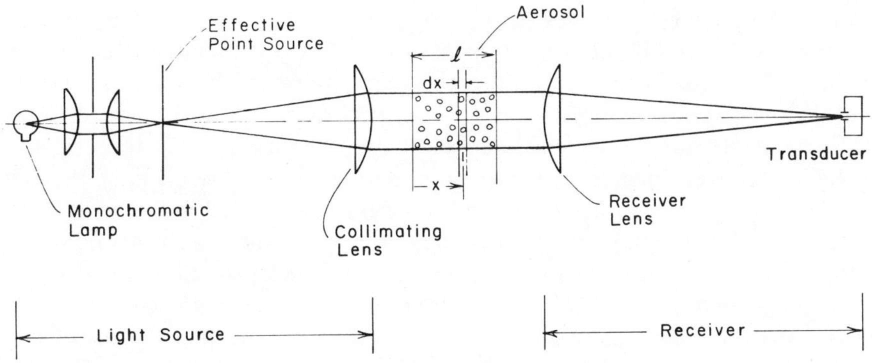

Figure 8.9. The apparatus to measure the transmission of light through an aerosol. The collimating lens forms a beam of nearly parallel light.

since $C_n A_b dx$ represents the total number of particles in the beam length dx (Fig. 8.9) and $Q_{ext}(\pi/4)D^2$ is the extinction cross section per particle. Integrating Eq. 8.26 between the limits of $x = 0$ where $F = F_0$ and $x = l$ where $F = F_1$, we obtain the transmission equation for a monodispersion

$$\frac{F_1}{F_0} = \exp\left(-\frac{\pi}{4} Q_{ext} D^2 C_n l\right) \tag{8.27}$$

where Q_{ext} is the size dependent quantity given by Eq. 8.24.

Normally a polydispersion is present and it is necessary to sum over all sizes in proportion to their relative probability of occurrence. In this case the transmission equation takes the form

$$\frac{F_1}{F_0} = \exp\left[-\frac{\pi}{4} C_n l \int_0^{D_\infty} Q_{ext}(D, m)\phi(D) D^2 dD\right] \tag{8.28}$$

The term in the exponent of Eq. 8.28 is the optical depth for extinction, $b_{ext} l$, where extinction coefficient b_{ext} is given by

$$b_{ext} = \frac{\pi}{4} C_n \int_0^{D_\infty} Q_{ext}(D, m)\phi(D) D^2 dD \tag{8.29}$$

Thus, the transmission law becomes

$$\frac{F_1}{F_0} = \exp(-b_{ext} l) \tag{8.30}$$

Just as the efficiency factor for extinction is the sum of the scattering and absorption efficiency factors, the extinction coefficient can be expressed as the sum of the scattering coefficient b_{sca} plus the absorption coefficient b_{abs}.

Again for dielectric particles $n_2 = 0$, $b_{abs} = 0$, and $b_{ext} = b_{sca}$ and no distinction need be made between these quantities. The optical apparatus shown in Fig. 8.9 measures the total amount of light removed by both scattering and absorption and therefore, in general, provides a measurement of b_{ext}. An optical instrument, called a nephelometer, which measures the amount of light scattered in a certain direction, detects the scattered light that is closely related to b_{sca}. Similarly the "haziness" associated with the atmospheric aerosol at high humidities is a result of scattering and is also closely related to b_{sca}. In actual practice (see discussion Section 12.4.2) transmissometers which measure b_{ext} and nephelometers which measure b_{sca} are both used on occasions to monitor the development of the aerosol of the polluted atmosphere.

The various contributions to the total extinction are additive. Thus we may write

$$b_{ext}^{total} = b_{sca} + b_{abs} + b_{atm}$$

where b_{sca} and b_{abs} refer to the scattering and absorption by the particulate phase of the aerosol, and b_{atm} refers to the combined scattering and absorption by the gaseous phase of the aerosol.

The visual range of a hazy atmosphere is defined by the contrast of a dark object when viewed against the distant horizon (Ref. 27). The radiative transfer equation and a somewhat arbitrary definition of contrast lead to the definition of visual range L_v to correspond to an optical depth for extinction equal to ln 50. Thus, the visual range is given by

$$L_v = \frac{3.912}{b_{ext}} \tag{8.31}$$

For air molecules the value of b_{ext} weighted by wavelength according to the standard visibility curve is 0.0123 km^{-1} (Penndorf, Ref. 32). In this case the value of the visual range is 318 km, and therefore molecular scattering does not cause a significant deterioration of visibility. The atmospheric aerosol, of either natural or anthropogenic origin, is responsible for the deterioration of visibility when a "haze" is present in the atmosphere.

The possibility of using a nephelometer or transmissometer to measure directly the total mass content of the particulate phase of the atmospheric aerosol is particularly attractive. These instruments offer the advantage of high time resolution and the potential for instantaneous readings. The relationship between mass concentration and b_{ext} depends on quantities such as the particle size distribution function, refractive index, particle density, and the particle shape if it is nonspherical. Charlson (Ref. 33) has reviewed this topic in detail and concludes that an approximate relationship between visual range and particulate concentration does exist provided relative humidity is not in excess of 70%.

8.5.3. The Chemical Composition of the Atmospheric Aerosol and Identification of Source Materials

The natural aerosol mainly consists of soil dust, the particulate products formed by natural hydrocarbons, and sea salt. The aerosol of the polluted atmosphere contains many additional substances which may be primary or secondary pollutants. For example, the primary pollutant SO_2 emitted by the combustion of fossil fuels is converted to SO_3 and finally to a nongaseous sulfate compound such as ammonium sulfate $(NH_4)_2SO_4$ through a series of atmospheric chemical reactions. The primary anthropogenic NO and reactive hydrocarbons also form particulate materials.

The chemical processes involving formation of particles from the gas phase or the atmospheric reactions of particles with gaseous species may be quite complex, and not all details of these reactions are well understood. Nevertheless the mass of each chemical element is conserved in any process that involves purely chemical transformations. Furthermore, the various natural and anthropogenic contributors to the aerosol of the polluted atmosphere each possess a distinctive composition by chemical elements. Recognition of these facts led Friedlander and colleagues (Ref. 27) to devise the material balance method to deduce quantitatively the relative distribution of source materials in the atmospheric aerosol.

The concentrations of certain elements in six dominant source materials considered by Friedlander to apply to the Los Angeles area are shown in Table 8.7. Two of the source materials (sea salt and soil dust) are natural, while the remaining four (auto exhaust, fuel oil fly ash, portland cement, and tire dust) are anthropogenic. The list of source materials in Table 8.7 is appropriate for the Los Angeles area and it will apply to other regions to some extent. Other source materials distinctive to these other regions may also be important supplements to the above list. The metallic elements listed in Table 8.7 are especially important because they are emitted by their sources as particulate species and, since they form no gaseous compounds, they remain in the particulate phase.

Examination of Table 8.7 shows that most of the source materials possess unique proportions of at least two of the metallic elements. Thus sea salt contains Na and Mg in the ratio 8.3; soil has Si and Al in the ratio 2.4; fly ash contains V and Na in the ratio 1.4; and portland cement contains Ca and Si in the ratio 4.3. Auto exhaust is essentially the role source of lead and tire dust is the sole source of elemental carbon. The uniqueness of the elemental composition of the various source materials means that the relative proportions of the latter in an arbitrary mixture of the particulate material of an aerosol can be determined by means of a chemical analysis.

For example, an aerosol sampled downwind of a cement factory in a rural location might contain only soil dust and portland cement of unknown

Table 8.7. Fractional Composition in Percent of Certain Elements in Important Source Materials of the Atmospheric Aerosol (Friedlander, Ref. 27)

Elements	Sea Salt	Soil Dust	Auto Exhaust	Fuel Oil Fly Ash	Portland Cement	Tire Dust
Elemental Carbon (C)	*n*	*n*	*n*	*u*	*n*	29.3
Carbon Compounds (C-)	*n*	*n*	40.3	*u*	*n*	58
Sodium (Na)	30.6	2.5	*u*	5	0.4	*u*
Magnesium (Mg)	3.7	1.4	*u*	0.06	0.48	*u*
Aluminum (Al)	*n*	8.2	*u*	0.8	2.4	*u*
Silicon (Si)	*n*	20	*u*	1	10.7	*u*
Sulfur (S)	2.6	*n*	*u*	*u*	*n*	*u*
Chlorine (Cl)	55.0	*n*	6.8	*n*	*n*	*u*
Calcium (Ca)	1.16	1.5	*u*	1.3	46.0	*u*
Vanadium (V)	*n*	0.006	*u*	7	*n*	*u*
Bromine (Br)	0.19	*n*	7.9	*n*	*n*	*u*
Lead (Pb)	*n*	0.02	40.0	0.07	*n*	*u*

Notes: 1. The letters *n* and *u* designate negligible and unknown fractions, respectively.
2. Only those elements are here listed which constitute more than 2.5% of the source materials; a more complete listing is given in Ref. 27.

proportions. Suppose a limited chemical analysis showed the aerosol to contain 0.58 $\mu g/m^3$ of sodium and 2.12 $\mu gm/m^3$ of aluminum. The concentration of Na and Al in the aerosol is expressed, for each individual element, as the sum of the concentrations of each source material multiplied by the fractional concentration of each element. Thus we find

$$\rho_{ei} = C_{i1}m_{s1} + C_{i2}m_{s2} \tag{8.32}$$

where ρ_{ei} is the mass concentration of the ith element, m_{sj} is the mass concentration of the jth source material, and C_{ij} are the fractional compositions of the elements in the source materials that are given in Table 8.7. In this case, with $i = 1, 2$ denoting Na and Al, respectively, the values we find from Table 8.7 are $C_{11} = 0.025$, $C_{12} = 0.004$, $C_{21} = 0.082$, and $C_{22} = 0.024$. With the above measured values of ρ_{e1} and ρ_{e2}, the soil dust concentration in the aerosol is found, by solving the two equations corresponding to Eq. 8.32, to be 20 $\mu g/m^3$, and the portland cement fraction also amounts to 20 $\mu g/m^3$. The total mass concentration of 40 $\mu g/m^3$ must also agree with a measurement of this quantity.

The method is readily extended to the case when the number of source materials is larger. In this case the mass concentration of the ith element is given by the summation over the N source materials as

$$\rho_{ei} = \sum_{j=1}^{N} C_{ij} m_{sj} \tag{8.33}$$

In the ideal case where the ρ_{ei} and C_{ij} for all source materials and elements are precisely known, a closed set of linear algebraic equations is obtained if ρ_{ei} is determined for N elements. Equations 8.33 can then be solved for the concentrations of the source materials in the aerosol.

The composition of the aerosol will differ from the composition of the aggregate contributions of the sources if one particular element is subject to disproportionate depletion from the particulate phase. This effect, called fractionalization, would occur if, for example, a particular metallic element was more likely to be found in the largest particles which might be lost from the aerosol by sedimentation near the source. Chlorine from sea salt can be liberated as a gaseous component in the form of HCl and is thus subject to fractionalization through gas phase exchange.

Gartrell and Friedlander (Ref. 27) used Na, Al, Ca, Pb, K, V, and Mg as tracer elements to determine the contributions represented by five source materials—those listed in Table 8.7, except tire dust. This study was performed at three locations in the greater Los Angeles area. With seven tracer elements and five source materials the system of equations in principle contains redundant information. However, an experimental uncertainty is present and the redundant information can be used to improve the ultimate accuracy. The method of calculation was executed to minimize the mean-square difference between the measured and the calculated values using the experimental uncertainty as a weighting factor. The results of this study show good agreement between the calculated and measured ρ_{ei} for the inferred values of m_{sj}. Additional components of the aerosol such as sulfates ($SO_4^{=}$) and nitrates (NO_3^-) were determined by direct measurement.

Studies of this type are valuable because the knowledge of the source materials facilitates the evaluation of alternative control strategies that might be applied to reduce the concentration of air pollutants.

REFERENCES

1. Argonne National Laboratory Study, reported in *Environ. Sci. & Technol.*, **6**, 679 (1972).
2. B. Dimitriades and M. Whisman, "Carbon Monoxide in Lower Atmosphere Reactions," *Environ. Sci. & Technol.*, **5**, 219 (1971).
3. E. Robinson and R. C. Robbins, "Gaseous Atmospheric Pollutants from Urban and Natural Sources," *J. Air Pollut. Control Assoc.*, **20**, 303–306, (1970).

4. A. Hoffman, "National Emission Reference File," National Air Pollution Control Administration, Durham, N. C. (1970 Edition).
5. W. G. Agnew, "Automotive Air Pollution Research," *Proc. Roy. Soc.* (*London*), A, **307**, 153–181 (1968).
6. D. O. Anderson, "The Effects of Air Pollution on Health, Parts I, II and III," *J. Can. Med. Assoc.*, **97**, 528–536, 585–593, 802–806 (1967).
7. I. T. T. Higgins and J. R. McCarroll, "Types, Ranges, and Methods for Classifying Human Pathophysiological Changes and Responses to Air Pollution," in *Development of Air Quality Standards*, A. Atkisson and R. S. Gaines, Editors (Merrill Publishing Co., Columbus, Ohio, 1970).
8. U.S.E.P.A., "Air Quality Criteria for Carbon Monoxide," *National Air Pollution Control Administration*, Publication No. AP-62 (March 1970).
9. U.S.E.P.A., "Air Quality Criteria for Sulfur Oxides," *National Air Pollution Control Administration*, Publication No. AP-50 (January 1969).
10. U.S.E.P.A., "Air Quality Criteria for Hydrocarbons," *National Air Pollution Control Administration*, Publication No. AP-64 (March 1970).
11. U.S.E.P.A., "Air Quality Criteria for Nitrogen Oxides," *National Air Pollution Control Administration*, Publication No. AP-84 (January 1971).
12. U.S.E.P.A., "Air Quality Criteria for Ozone and Other Photochemical Oxidants," EPA 600/8-78-004 (April 1978).
13. R. P. C. Wolf, "Carbon Monoxide Measuring and Monitoring in Urban Air," *Environ. Sci. & Technol.*, **5**, 212–218 (1971).
14. U.S.E.P.A., "National Primary and Secondary Ambient Air Quality Standards," *Federal Register*, Vol. 36, No. 84, Part II (30 April 1971).
15. Committee on Public Works, U.S. Senate, "Air Quality and Automobile Emission Control," Coordinating Committee on Air Quality Studies, National Academy of Sciences, National Academy of Engineering (U.S. Printing Office, September 1974).
16. A. J. Haagen-Smit, "Chemistry and Physiology of Los Angeles Smog," *Ind. and Eng. Chem.*, **44**, 1342–1346 (1952).
17. K. L. Demerjian, J. A. Kerr, and J. G. Calvert, "The Mechanism of Photochemical Smog Formation," in *Advances in Environmental Sciences and Technology*, Vol. 4 (John Wiley & Sons, New York, 1974).
18. P. A. Leighton, *Photochemistry of Air Pollution* (Academic Press, New York, 1961).
19. H. Niki, E. E. Daby, and B. Weinstock, "Mechanisms of Smog Reactions," in *Photochemical Smog and Ozone Reactions*, R. F. Gould, Editor, Advances in Chemistry Series 113 (American Chemical Society, Washington, D.C., 1972).
20. J. N. Pitts et al., "Hydrocarbon Reactivity and the Role of Hydrocarbons, Oxides of Nitrogen, and Aged Smog in the Production of Photochemical Oxidants," Proceedings of the International Conference on Photochemical Oxidant Pollution and its Control, E.P.A.-600/3-77-001, Research Triangle Park, N. C. (1977).

21. J. G. Calvert, "Test of the Theory of Ozone Generation in Los Angeles Atmosphere," *Environ. Sci. & Technol.*, **10**, 248–256 (1976).
22. G. H. Taylor and A. Q. Eschenroeder, "Tests of Difkin Photochemical Diffusion Model using Los Angeles Reactive Pollutant Program Data," Proceedings of the International Conference on Photochemical Oxidant Pollution and Its Control, E.P.A.-600/3-77-001, Research Triangle Park, N. C. (1977).
23. R. G. Lamb, "A Case Study of Stratospheric Ozone Affecting Ground-Level Oxidant Concentrations," *J. Appl. Meteorol.*, **16**, 780–794 (1977).
24. B. Dimitriades, Editor, Proceedings of the International Conference on Photochemical Oxidant Pollution and Its Control, E.P.A.-600/3-77-001, Research Triangle Park, N. C. (1977).
25. B. Dimitriades and A. P. Altshuller, "International Conference on Oxidant Problems: Analysis of Evidence/Viewpoints Presented. Part II. Evidence/Viewpoints on Key Issues," *J. Air Pollut. Control Assoc.*, **28**, 207–212 (1978).
26. C. N. Davies, Editor, *Aerosol Science* (Academic Press, New York, 1966).
27. S. K. Friedlander, *Smoke, Haze, and Dust* (Wiley Interscience, New York, 1977).
28. K. T. Whitby, R. B. Husar, and B. Y. Liu, "The Aerosol Size Distribution of Los Angeles Smog," in *Aerosols and Atmospheric Chemistry*, G. M. Hidy, Editor (Academic Press, New York, 1972).
29. H. C. Van de Hulst, *Light Scattering by Small Particles* (Wiley, New York, 1969).
30. M. Kerker, *The Scattering of Light and Other Electromagnetic Radiation* (Academic Press, New York, 1969).
31. B. A. Maguire, "Mie Scattering of Electromagnetic Radiation by a sphere: A Computer program," *J. Aerosol. Sci.*, **2**, 417–423 (1971).
32. R. Penndorf, "Tables of the Refractive Index for Standard Air and the Rayleigh Scattering Coefficient for the Spectral Region Between 0.2 and 20.0 Microns and Their Application to Atmospheric Optics," *J. Opt. Soc. Amer.*, **47**, 176–182 (1957).
33. R. J. Charlson, "Atmospheric Visibility Related to Aerosol Mass Concentration," *Environ. Sci. and Technol.*, **3**, 913–918 (1969).

PROBLEMS

8.1. We wish to estimate how long ozone, O_3, would persist after dark when a fresh injection of nitric oxide, NO, occurs from automobiles in the surface layer. For this purpose we assume that NO and O_3 in the air near the ground are well mixed at specified concentrations $[NO]_0$ and $[O_3]_0$ at sunset and we inquire on the fate of these compounds if they are allowed to achieve chemical equilibrium.

1. What chemical reaction will occur?
2. What is the equation for the disappearance of O_3 in terms of the concentrations of NO and O_3?

3. Suppose $[O_3]_0 = 0.1$ ppm and $[NO]_0 = 1.0$ ppm so that the latter is present in large excess and we may treat its concentration as constant. Give an approximate expression for the rate of disappearance of ozone under these conditions.
4. How much time will elapse before the concentration of O_3 decays to $e^{-1} = 0.368$ of its initial value? Assume the rate constant for the reaction alluded to in part 2 is 23 ppm^{-1} min^{-1}.

8.2. The National Ambient Air Quality Standard for ozone was exceeded 11.6% of all hourly observations during the summer of 1976 at a rural location in northeastern United States. At a nearby city the Standard was exceeded only 2.90% of the time during the same period. The prevailing wind blows roughly in the direction from the rural site toward the urban location. Discuss the factors which would cause a generally higher ozone concentration in the rural location in this case.

8.3. Derive the expression for the concentration of atomic oxygen O if photostationary equilibrium persists. Calculate the concentration of O if $[NO_2] = 0.020$ ppm, $[NO] = 0.080$ ppm, $K_1 = 0.48$ min^{-1}, and $K_2 = 2.0 \times 10^{-5}$ ppm^{-2} min^{-1}. Note that the third body M in reaction R-2 may be either oxygen or nitrogen and therefore its concentration is 10^{+6} ppm.

8.4. A small particle in an aerosol falls in air under the action of gravity at a terminal velocity corresponding to a low Reynolds number. The drag force on the particle is then given by the Stokes drag law, $6\pi\mu R w_0$, where μ is the absolute viscosity of air, R is the particle radius, and w_0 is the downward particle velocity. Show that the latter is given by

$$w_0 = \frac{2}{9}\frac{(\rho_p - \rho_a)R^2 g}{\mu}$$

where ρ_a and ρ_p are the air and particle densities.

8.5. Calculate the settling velocity in air at STP for water droplets whose diameters are 0.50 and 5.00 μm. What are the Reynolds numbers in both of these cases?

8.6. Show that the extinction coefficient b_{ext} given by Eq. 8.29 for a polydispersion of spherical particles depends upon the sixth moment of the particle size distribution function if all sizes are small compared to the wavelength of the incident light, $D \ll \lambda$. Show that if $D \gg \lambda$ for all sizes of a polydispersion, the scattering power is proportional to the second moment of the size distribution function.

9

DISPERSION OF AIR POLLUTANTS— INTRODUCTION

A central practical problem is the prediction of the average concentration of a pollutant, at a specified location, which results from the contributing sources whose emission rates and locations are prescribed. This problem is solved only after a variety of simplifying assumptions are made. The understanding of the limitations that are imposed by these simplifications is necessary to evaluate the circumstances under which the various methods apply. In this chapter we consider the formulation of the problem and provide the background of fundamental concepts on which the specific methods are based.

Each of the two fundamental methods of describing the motion of fluid media are valuable in addressing the problem of predicting the concentrations of air pollutants. The Lagrangian formulation gives the history of the properties of specific fluid particles and, in principle, is therefore well adapted to describing the motion of a parcel of fluid as it undergoes a complex trajectory in a turbulent flow field. The Eulerian formulation, which deals with the fluid properties at a specified location in space without recognition of the histories of the properties associated with the myriad of fluid particles which traverse the point in question, is normally advantageous because solutions are more readily obtained. In fact both of these methods make important contributions to the problem of the prediction of air pollution concentrations. In Sections 9.1, 9.2, and 9.3 we provide a background leading to the Eulerian description of dispersion of pollutants in a turbulent medium. Specific methods of calculating pollution concentrations that employ the Eulerian approach are discussed in Chapter 11. In Sections 9.4 and 9.5 we apply the Lagrangian formulation which provides valuable insight into the nature of turbulent dispersion and which leads to the widely used method of calculating air pollution concentrations that is discussed in Chapter 10.

9.1. THE CONSERVATION OF MASS IN THE PRESENCE OF MOLECULAR DIFFUSION AND CHEMICAL REACTIONS

The presence of a concentration gradient in a fluid mixture of two chemical components, a binary mixture, results in a migration or diffusion of the species that tends to eliminate the gradient. This migration, which results from the molecular motion and occurs even in the absence of pressure or temperature gradients, gravitational or other body forces, or mean fluid motion, is called molecular diffusion.

To account for the motion of the individual species it is necessary to distinguish between the velocities of each of the two components, $\mathbf{q}_A$ and $\mathbf{q}_B$. These velocities multiplied by the respective component densities will give the bulk mass flux

$$\rho\mathbf{q} = c_A\mathbf{q}_A + c_B\mathbf{q}_B$$

All three velocities in this relationship are measured with respect to a specified coordinate system and we recognize that the velocity $\mathbf{q}$ is a mass average velocity. We now define a mass flux $\mathbf{j}_A$ as the flow of component A per unit area per unit time measured with respect to the mass average velocity,

$$\mathbf{j}_A = c_A(\mathbf{q}_A - \mathbf{q}) \tag{9.1}$$

Experiments show that the difference between the velocity of a component and the velocity of the mixture, which is called the diffusion velocity of A with respect to $\mathbf{q}$, is proportional to the magnitude of the gradient of the mass fraction of the component. This statement is given mathematical form in Fick's law which is

$$\mathbf{j}_A = -\rho\mathscr{D}_{AB}\left[\frac{\partial}{\partial x}\left(\frac{c_A}{\rho}\right)\mathbf{i} + \frac{\partial}{\partial y}\left(\frac{c_A}{\rho}\right)\mathbf{j} + \frac{\partial}{\partial z}\left(\frac{c_A}{\rho}\right)\mathbf{k}\right]$$

or

$$\mathbf{j}_A = -\rho\mathscr{D}_{AB}\nabla\left(\frac{c_A}{\rho}\right) \tag{9.2}$$

where the constant of proportionality $\mathscr{D}_{AB}$ is defined to be the binary diffusion coefficient or mass diffusivity of species A in the presence of B. Since the dimensions of $\mathbf{j}_A$ are mass per unit area per unit time, the dimensions of diffusivity are length squared per unit time.

The transport of mass by molecular diffusion is related to the conduction of heat in a nonisothermal medium. In the case when ρ is constant Eq. 9.2 is akin to the Fourier heat conduction law which relates the heat flux to the temperature gradient through a constant of proportionality which is defined to be the thermal conductivity. It is possible to determine both the binary diffusion coefficient and the thermal conductivity of a gas by the application of kinetic theory. The conservation of species A is formulated by considering the processes of transport, and production or consumption by chemical reactions as they apply to a control volume which is fixed in space and through which the fluid moves. This conservation principle is expressed by the statement

$$\underset{(1)}{\iiint_{\mathscr{V}} \frac{\partial c_A}{\partial t} d\mathscr{V}} = \underset{(2)}{-\iint_{S} (c_A \mathbf{q}_A) \cdot d\mathbf{S}} + \underset{(3)}{\iiint_{\mathscr{V}} r_A d\mathscr{V}} \tag{9.3}$$

The three terms represent (1) the net rate of accumulation of species A within the control volume, (2) the rate of flow of species A through the boundaries of the control surface, and (3) the total production of A within the control volume as a result of a volumetric chemical production rate r_A possessing the units of mass per unit volume per unit time.

The surface integral in Eq. 9.3 is converted into a volume integral by Gauss' theorem and all terms are collected under a common volume integral. The result is

$$\iiint_{\mathscr{V}} \left[\frac{\partial c_A}{\partial t} + \nabla \cdot (c_A \mathbf{q}_A) - r_A\right] d\mathscr{V} = 0$$

Because the above integral vanishes for any arbitrarily chosen volume and the integrand is continuous, the latter vanishes at every point within the control volume and we find

$$\frac{\partial c_A}{\partial t} + \nabla \cdot (c_A \mathbf{q}_A) = r_A \tag{9.4}$$

This equation is a continuity equation for species A which expresses the change of concentration of this component at each point in space when motion and chemical reactions are present. A corresponding continuity equation for species B also can be written and, if the mixture consists solely of these two components, then $r_A = -r_B$ in any chemical reaction by the conservation of mass. The addition of the two species continuity equations then yields the customary global continuity equation in view of the above definition of the bulk mass flux and the fact that $\rho = c_A + c_B$.

Equation 9.4 is rendered more useful if the mass average velocity is used in place of $\mathbf{q}_A$. The quantity $c_A \mathbf{q}_A$ is eliminated through Eq. 9.1, and Fick's law is used to eliminate $\mathbf{j}_A$. We find

$$\frac{\partial c_A}{\partial t} + \nabla \cdot (c_A \mathbf{q}) = +\nabla \cdot \left[\rho \mathscr{D}_{AB} \nabla\left(\frac{c_A}{\rho}\right)\right] + r_A \tag{9.5}$$

We now assume the density of the mixture is constant and therefore $\nabla \cdot \mathbf{q} = 0$. The second term then reduces to $\mathbf{q} \cdot \nabla c_A$ and the diffusion term also simplifies. We find

$$\frac{\partial c_A}{\partial t} + \mathbf{q} \cdot \nabla c_A = \nabla \cdot (\mathscr{D}_{AB} \nabla c_A) + \mathrm{r}_A \tag{9.6a}$$

In expanded form this equation becomes

$$\frac{\partial c_A}{\partial t} + u\frac{\partial c_A}{\partial x} + v\frac{\partial c_A}{\partial y} + w\frac{\partial c_A}{\partial z} = \frac{\partial}{\partial x}\left(\mathscr{D}_{AB}\frac{\partial c_A}{\partial x}\right) + \frac{\partial}{\partial y}\left(\mathscr{D}_{AB}\frac{\partial c_A}{\partial y}\right) + \frac{\partial}{\partial z}\left(\mathscr{D}_{AB}\frac{\partial c_A}{\partial z}\right) + r_A \tag{9.6b}$$

We recognize the first four terms to constitute the particle derivative and therefore we may also write, in vector form,

$$\frac{Dc_A}{Dt} = \nabla \cdot (\mathscr{D}_{AB} \nabla c_A) + r_A \tag{9.6c}$$

In general the molecular diffusion coefficient is a function of concentration which is itself influenced by the presence of diffusion. For our purposes it is adequate to consider species A to be a minor component in a mixture of gases. In this case the relevant diffusion coefficient will be that for species A in the presence of the mixture which we designate as $\mathscr{D}_A$. If temperature variations are not large then $\mathscr{D}_A$ is constant and the Eqs. 9.6 reduce to

$$\frac{\partial c_A}{\partial t} + u\frac{\partial c_A}{\partial x} + v\frac{\partial c_A}{\partial y} + w\frac{\partial c_A}{\partial z} = \mathscr{D}_A\left(\frac{\partial^2 c_A}{\partial x^2} + \frac{\partial^2 c_A}{\partial y^2} + \frac{\partial^2 c_A}{\partial z^2}\right) + r_A \tag{9.7}$$

Finally we may divide the above by molecular weight to obtain the form of the diffusion equation expressed in molar concentration which is advantageous for accounting for chemical production.

Equations 9.5, 9.6, and 9.7 assert that the rate of increase of concentration within a small volume that moves with the fluid is equal to the net diffusional influx plus the rate of generation by chemical reactions. A more detailed discussion of molecular diffusion is provided in Ref. 1.

9.2. SOME EXACT SOLUTIONS FOR THE DIFFUSION EQUATION

Molecular diffusion is not an important factor in the spreading of chemical species in the atmosphere. Nevertheless some of the solutions that are used in the determination of concentrations of atmospheric pollutants borrow mathematical concepts which originate in the exact solutions for molecular diffusion. A digression is therefore profitable in order to examine the properties of several of these solutions.

9.2.1. Unsteady Diffusion from an Instantaneous Point Source in a Quiescent Fluid

We restrict the diffusion equation to apply to a fluid at rest, $\mathbf{q} = 0$, without chemical reactions. Equation 9.7 reduces to

$$\frac{\partial c}{\partial t} = \mathscr{D}\nabla^2 c \tag{9.8}$$

where c is the molar concentration of the species of interest and $\mathscr{D}$ is its diffusion coefficient which is assumed constant throughout all space. A finite amount of material is released instantaneously at the time $t = 0$ at the origin. Since the diffusion from an isolated point source is uniform in all directions, the concentration is a function of the radial distance from the origin and time, $c = c(r, t)$. The boundary conditions for the problem are:

1. The concentration is zero throughout all space at the initial time, $c(r, 0) = 0$.
2. The concentration tends to zero at large values of distance from the origin for all values of time, $c(r \to \infty, t) \to 0$.
3. The total amount of material distributed throughout all space at any time is equal to the amount released Q at the initial moment,

$$4\pi \int_0^\infty c(r, t) r^2 dr = Q \tag{9.9}$$

The spherical symmetry of the problem requires the Laplacian operator in the diffusion equation to be formulated in spherical coordinates (see Appendix I),

$$\frac{\partial c}{\partial t} = \mathscr{D} \frac{1}{r^2} \frac{\partial}{\partial r}\left(r^2 \frac{\partial c}{\partial r}\right) \tag{9.10}$$

The solution to this problem is facilitated by the recognition that r must be made dimensionless with $\mathscr{D}$, there being no other length scale, and this requires the inclusion of t as well. A new variable $\eta = r/2\sqrt{\mathscr{D}t}$ is of the traditional form although the value of the numerical constant is arbitrary. If this variable is used to make the boundary condition given by Eq. 9.9 dimensionless, we deduce that the dimensionless concentration $\bar{c}$ must be of the form $c(\mathscr{D}t)^{3/2}/Q$. The inclusion of the numerical factor of $(4\pi)^{3/2}$ will reduce the boundary condition to the form

$$\int_0^\infty \bar{c}(\eta)\eta^2 d\eta = \frac{\sqrt{\pi}}{2} \tag{9.11}$$

and therefore $\bar{c} = \exp(-\eta^2)$ or

$$c = \frac{Q}{(4\pi\mathscr{D}t)^{3/2}} \exp\left(-\frac{r^2}{4\mathscr{D}t}\right) \tag{9.12}$$

where $r^2 = x^2 + y^2 + z^2$.

Equation 9.12, which satisfies all of the boundary conditions and the differential equation, is the familiar normal distribution. The concentration is always greatest at the origin, $r = 0$, where the point release initially occurred. The width of the curve increases with time, and the position of the location $r_{1/2}$ where the concentration is always 50% of the value at the origin is given by

$$r_{1/2} = 2(ln\ 2)^{1/2}(\mathscr{D}t)^{1/2} \tag{9.13}$$

We also evaluate the spread of material by molecular diffusion as the mean-squared radial displacement σ_r^2 by summing over all volume elements the product of the concentration of each volume element times the square of its distance from the origin. Dividing by the total mass released, we find

$$\sigma_r^2 = \frac{1}{Q}\sum_i c(r_i)r_i^2 d\mathscr{V}_i = \frac{4\pi}{Q}\int_0^\infty cr^4 dr \tag{9.14}$$

Converting c and r to their dimensionless counterparts c and η, we find the mean-square displacement given by

$$\sigma_r^2 = \frac{16}{\pi^{1/2}}\int_0^\infty \bar{c}\eta^4 d\eta \tag{9.15}$$

Finally, if the above integral is evaluated using the known $\bar{c}(\eta)$, we find

$$\sigma_r^2 = 6\mathscr{D}t$$

The mean-square displacement is often expressed as the sum of three components in the x, y, and z directions

$$\sigma_r^2 = \sigma_x^2 + \sigma_y^2 + \sigma_z^2 = 3\sigma^2$$

and therefore

$$\sigma^2 = 2\mathscr{D}t \tag{9.16}$$

and Eq. 9.12 becomes

$$c = \frac{Q}{(2\pi)^{3/2}\sigma^3} \exp\left(-\frac{r^2}{2\sigma^2}\right) \tag{9.17}$$

In this form the spread of the material by molecular diffusion is measured by the mean-square displacement which varies as the square root of time.

9.2.2. Diffusion from an Instantaneous Point Source in an Anisotropic Medium

In this case the form of the diffusion equation is

$$\frac{\partial c}{\partial t} = \frac{\partial}{\partial x}\left(\mathscr{D}_x \frac{\partial c}{\partial x}\right) + \frac{\partial}{\partial y}\left(\mathscr{D}_y \frac{\partial c}{\partial y}\right) + \frac{\partial}{\partial z}\left(\mathscr{D}_z \frac{\partial c}{\partial z}\right) \tag{9.18}$$

where $\mathscr{D}_x$, $\mathscr{D}_y$, and $\mathscr{D}_z$ represent the values of the binary diffusion coefficient in the three directions which may not necessarily be equal to one another. For an instantaneous release at the origin the solution is given by

$$c = \frac{Q}{(4\pi t)^{3/2}(\mathscr{D}_x\mathscr{D}_y\mathscr{D}_z)^{1/2}} \exp\left[-\frac{x^2}{4\sigma_x t} - \frac{y^2}{4\sigma_y t} - \frac{z^2}{4\sigma_z t}\right] \tag{9.19}$$

as can be confirmed by substitution. This solution is clearly the product of three solutions for unidirectional diffusion from the three planes which form the coordinate axes. The solution reduces to Eq. 9.12 for isotropic diffusion from a point source when $\mathscr{D}_x = \mathscr{D}_y = \mathscr{D}_z$. We may define the quantities

$$\sigma_x^2 = 2\mathscr{D}_x t \qquad \sigma_y^2 = 2\mathscr{D}_y t \qquad \sigma_z^2 = 2\mathscr{D}_z t \tag{9.20}$$

with which the concentration can be expressed as

$$c = \frac{Q}{(2\pi)^{3/2}\sigma_x\sigma_y\sigma_z} \exp\left[-\frac{x^2}{2\sigma_x^2} - \frac{y^2}{2\sigma_y^2} - \frac{z^2}{2\sigma_z^2}\right] \tag{9.21}$$

The directional mean-square displacements σ_x, σ_y, and σ_z are more frequently referred to as the dispersion coefficients. Equation 9.21 reduces to Eq. 9.17 when the diffusion is isotropic and $\sigma_x = \sigma_y = \sigma_z$.

9.2.3. The Instantaneous Point Source above an Impenetrable Ground Plane

We convert the solution for an instantaneous release at the origin, Eq. 9.21, to apply to a release of material at $(0, 0, H)$ by a translation of coordinates wherein z is replaced by $z - H$. The result, assuming the diffusion is isotropic, is

$$c = \frac{Q}{(2\pi)^{3/2}\sigma^3} \exp\left[-\frac{x^2 + y^2 + (z - H)^2}{2\sigma^2}\right] \tag{9.22}$$

Since the diffusion equation is a linear differential equation, the property of superposition is applicable. To obtain the solution for an instantaneous release above an impenetrable ground plane, we add an image source at $z = -H$ to Eq. 9.22. Because the source at $z = +H$ and its image at $z = -H$ will combine to produce a zero net flux at $z = 0$ for all x and y, this plane may be considered to be impenetrable. The solution in this case, for isotropic diffusion, is

$$c = \frac{Q}{(2\pi)^{3/2}\sigma^3} \exp\left[-\frac{x^2+y^2}{2\sigma^2}\right]\left\{\exp\left[-\frac{(z-H)^2}{2\sigma^2}\right] + \exp\left[-\frac{(z+H)^2}{2\sigma^2}\right]\right\} \tag{9.23}$$

where, as previously, $\sigma^2 = 2\mathcal{D}t$. This equation clearly satisfies the boundary conditions that $c = 0$ for any $x, y, z > 0$ at $t = 0$, that $c \to 0$ as $x, y, z \to \infty$ for any finite t, and that the diffusional flux $(-\mathcal{D}(\partial c/\partial z))$ at the ground plane is zero, $z = 0$ for all x and y.

9.2.4. Diffusion from an Instantaneous Release in a Uniform Laminar Flow

The solution for an instantaneous release in a uniform laminar flow is obtained by again adapting the solution for a point source in a quiescent fluid. In this case we translate the location of the point of release at a uniform velocity. Thus, if Eq. 9.21 with x, y, and z replaced by x', y', and z' represents the solution in the primed coordinates that move at uniform velocity u_0 in the $+x$ direction, then by setting $x = x' + u_0 t$, $y = y'$, and $z = z'$, we find

$$c = \frac{Q}{(2\pi)^{3/2}\sigma_x\sigma_y\sigma_z} \exp\left[-\frac{(x-u_0 t)^2}{2\sigma_x^2} - \frac{y^2}{2\sigma_y^2} - \frac{z^2}{2\sigma_z^2}\right] \tag{9.24}$$

In this case the maximum concentration is at the point of release $(u_0 t, 0, 0)$ which moves at constant velocity.

9.2.5. Diffusion from an Instantaneous Elevated Release in a Laminar Flow above an Impenetrable Ground Plane

The solution for the instantaneous release from a source at the height H above an impenetrable ground plane is obtained by translating the point of release to the point $(u_0 t, 0, +H)$ and adding an image source of equivalent strength at the point $(u_0 t, 0, -H)$. The result is

$$c = \frac{Q}{(2\pi)^{3/2}\sigma_x\sigma_y\sigma_z} \exp\left[-\frac{(x-u_0 t)^2}{2\sigma_x^2} - \frac{y^2}{2\sigma_y^2}\right]\left\{\exp\left[-\frac{(z-H)^2}{2\sigma_z^2}\right] + \exp\left[-\frac{(z+H)^2}{2\sigma_z^2}\right]\right\} \tag{9.25}$$

9.3. THE EULERIAN DESCRIPTION OF TURBULENT DISPERSION

The conservation of each species in a moving fluid mixture is expressed by Eq. 9.7 which is the continuity equation including the Fickian diffusion and chemical reactions. When the motion of the fluid is turbulent there is usually a significant augmentation of mass transfer from regions of high to low mass fraction of a particular component. In actuality the influence of this mechanism, which is frequently referred to as turbulent dispersion, is entirely dominant over molecular diffusion in the lower troposphere. In deriving the above Eq. 9.7 we employed an Eulerian formulation wherein the fluid properties—density or concentration, velocity, and so on—refer to the instantaneous values at a particular point in space. The coordinates of this point are said to be the Eulerian coordinates and the specified properties apply to the particular fluid particle that is momentarily present at the specified time. In this section we will explore the Eulerian description of dispersion in a turbulent medium using the species continuity equation as a point of departure.

To adapt the continuity-diffusion equation to describe turbulent dispersion, we must obtain its time-averaged form which is analogous to the similar forms of the momentum equations obtained in Chapter 6. Thus we define the time-averaged concentration of a particular species at a specified point in space as

$$\bar{c} = \frac{1}{\Delta\tau} \int_{t-(\Delta\tau/2)}^{t+(\Delta\tau/2)} c(t)dt \tag{9.26}$$

We now proceed to time-average Eq. 9.7 by using the rules that apply to this operator that were previously exploited. The time average of the derivative terms is equal to the derivative of the time-averaged concentration and thus

$$\overline{\left(\frac{\partial c}{\partial t}\right)} = \frac{\partial \bar{c}}{\partial t}, \qquad \overline{\left(\frac{\partial^2 c}{\partial x^2}\right)} = \frac{\partial^2 \bar{c}}{\partial x^2}, \qquad \text{and so on}$$

The more interesting result arises when the convective terms are time-averaged. Thus, expressing the instantaneous values as the sum of their time-averaged value plus their fluctuation, $c = \bar{c} + c'$ and $u = \bar{u} + u'$, we find, for example,

$$\overline{\left(u\frac{\partial c}{\partial x}\right)} = \bar{u}\frac{\partial \bar{c}}{\partial x} + \overline{\left(u'\frac{\partial \bar{c}}{\partial x}\right)} + \overline{\left(\bar{u}\frac{\partial c'}{\partial x}\right)} + \overline{\left(u'\frac{\partial c'}{\partial x}\right)}$$

The second and third terms on the right side vanish because $\overline{u'} = 0$ and $\overline{c'} = 0$. Accordingly three new terms

$$\overline{\left(c'\frac{\partial u'}{\partial x}\right)}, \qquad \overline{\left(c'\frac{\partial v'}{\partial y}\right)}, \qquad \text{and } \overline{\left(c'\frac{\partial w'}{\partial z}\right)}$$

arise as a result of the time-averaging operation. These terms can be written in a different form by noting

$$\frac{\partial(c'u')}{\partial x} + \frac{\partial(c'v')}{\partial y} + \frac{\partial(c'w')}{\partial z} = u'\frac{\partial c'}{\partial x} + v'\frac{\partial c'}{\partial y} + w'\frac{\partial c'}{\partial z} + c'\left(\frac{\partial u'}{\partial x} + \frac{\partial v'}{\partial y} + \frac{\partial w'}{\partial z}\right)$$

The sum of the three terms in the parenthesis vanishes and the remaining terms are time-averaged to yield

$$\overline{\left(u'\frac{\partial c'}{\partial x}\right)} + \overline{\left(v'\frac{\partial c'}{\partial y}\right)} + \overline{\left(w'\frac{\partial c'}{\partial z}\right)} = \frac{\partial}{\partial x}\overline{(c'u')} + \frac{\partial}{\partial y}\overline{(c'v')} + \frac{\partial}{\partial z}\overline{(c'w')}$$

The conservation equation then becomes

$$\frac{\partial \bar{c}}{\partial t} + \bar{u}\frac{\partial \bar{c}}{\partial x} + \bar{v}\frac{\partial \bar{c}}{\partial y} + \bar{w}\frac{\partial \bar{c}}{\partial z} = \frac{\partial}{\partial x}\left(\mathscr{D}\frac{\partial \bar{c}}{\partial x}\right) + \frac{\partial}{\partial y}\left(\mathscr{D}\frac{\partial \bar{c}}{\partial y}\right) + \frac{\partial}{\partial z}\left(\mathscr{D}\frac{\partial \bar{c}}{\partial z}\right)$$
$$+ \left[\frac{\partial(-\overline{u'c'})}{\partial x} + \frac{\partial(-\overline{v'c'})}{\partial y} + \frac{\partial(-\overline{w'c'})}{\partial z}\right] + \bar{R} \tag{9.27}$$

where the reaction rate $\bar{R}$ is a suitably time-averaged value. This term will depend on the concentration of the various species which participate in the formation of the particular component, but it is not at all apparent that $\bar{R}$ depends solely on the time-averaged values of these species; on some occasions chemical production terms may be strongly influenced by the concentration fluctuation as well as the mean concentration of the reactants.

Equation 9.27 has three new terms that are of similar origin to the Reynolds stress terms in the momentum equation. By writing the equation in the above form we recognize that a term such as $-\overline{u'c'}$ is a counterpart of the diffusional flux term $\mathscr{D}(\partial\bar{c}/\partial x)$ and possesses identical units. We conclude that the correlations $-\overline{u'c'}$, $-\overline{v'c'}$, and $-\overline{w'c'}$ represent a type of diffusion process resulting from turbulence which is called turbulent dispersion. We designate these terms the Reynolds dispersion terms.

The laminar flow diffusion Eq. 9.7 is a linear equation in c_A if the diffusion coefficient is constant. If, in a laminar flow, the velocity components are known and appropriate boundary conditions are prescribed, then Eq. 9.7 represents a single equation in a single unknown whose spatial and temporal dependence can then be determined. The conservation-dispersion Eq. 9.27 for a turbulent flow is of an entirely different nature because it contains not only the mean concentration, but also three additional unknown Reynolds dispersion terms about which no information is supplied. Thus Eq. 9.27, with velocity components specified, represents one equation with four unknown quantities and cannot be solved without additional information. We are again confronted with the closure problem of turbulence.

In studying the transfer of momentum in the turbulent atmospheric boundary layer, we found the Reynolds stress terms to be as much as 10^{+5} times the molecular stress terms. A corresponding relationship holds for mass transfer in the turbulent atmospheric boundary layer. Molecular diffusion is entirely negligible compared with turbulent transport, $\mathscr{D}(\partial\bar{c}/\partial x) \ll -\overline{u'c'}$, and the above Eq. 9.27 can be simplified to

$$\frac{\partial\bar{c}}{\partial t} + \bar{u}\frac{\partial\bar{c}}{\partial x} + \bar{v}\frac{\partial\bar{c}}{\partial y} + \bar{w}\frac{\partial\bar{c}}{\partial z} = \frac{\partial(-\overline{u'c'})}{\partial x} + \frac{\partial(-\overline{v'c'})}{\partial y} + \frac{\partial(-\overline{w'c'})}{\partial z} + \bar{R} \quad (9.28)$$

The closure problem of turbulence still confronts this equation which represents an exact description of turbulent dispersion using an Eulerian formulation. The methods of dealing with the closure problem will be discussed in Chapter 11.

9.4. THE LAGRANGIAN DESCRIPTION OF TURBULENT DISPERSION

The Lagrangian description of motion in a fluid possesses advantages in the formulation of turbulent dispersion because the method focuses attention on the behavior of individual fluid particle. The latter is designated by its position $\mathbf{r}_0$ at a specified initial time t_0. The subsequent motion of the particle is represented by the time dependent position vector $\mathbf{r}$. Thus the expression $\mathbf{r} = \mathbf{r}(\mathbf{r}_0, t_0; t)$ gives the position of the particle at time t whose initial position is $\mathbf{r}_0$ at the time t_0. This method is closely related to the formulation of problems in particle mechanics when the position vector of the particle is the principle independent variable. In the Lagrangian formulation the coordinates x, y, and z are the components of the position vector of the particle and are referred to as the Lagrangian coordinates. The velocity of the particle, as in the case of particle mechanics, is found by differentiating the position vector. The application of the Lagrangian formulation to the statistical description of turbulent dispersion is illustrated by the following discussion. A more detailed development of the Lagrangian approach is found in Refs. 2 to 5.

We imagine conducting an experiment in which a series of marked particles are released at a fixed point in space in a turbulent flow field which is homogeneous and stationary (Fig. 9.1). The marked particles are small in size and neutrally buoyant so that they follow the excursions of the turbulent flow field in precise detail. Each particle will pursue a unique tortuous trajectory which does not duplicate the motion of the other particles that are

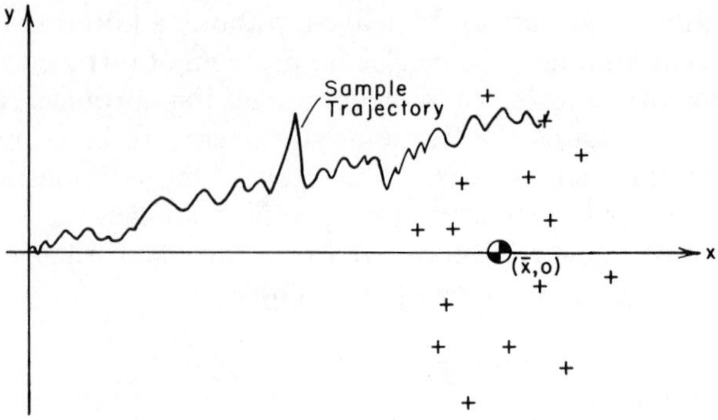

Figure 9.1. The dispersion of particles in a turbulent flow. Many neutrally buoyant particles are released at the origin at the times $t_{10}, t_{20}, \ldots, t_{NO}$ and their positions are noted after a fixed time interval τ_0 has elapsed.

released at earlier or later times. The Eulerian flow field is found to be made up of a constant x component $\bar{u}_0$ plus fluctuating components in the x, y, and z directions. Even though the mean flow is in the x direction, the various particles do not arrive at the same point after a specified travel time because of their distinctive turbulent trajectories. The final particle positions form a scattered array that is depicted by the crosses in Fig. 9.1. We wish to quantify the scatter in this array resulting from turbulent dispersion. For this purpose we will assume that the (Lagrangian) velocity of each particle is a known function of time. The y velocity component of the ith particle is $v_{yi}(t)$ and the y position of the particle after a specified travel time τ_0 after its release at time t_{io} is given by

$$y_i(t_i) = \int_0^{\tau_0} v_{yi}(t_{io} + \tau)d\tau \tag{9.29}$$

We wish to calculate the average y coordinate for N particles released at various times $t_{10}, t_{20}, \ldots, t_{NO}$ which would be given by

$$\bar{y} = \frac{1}{N} \sum_{i=1}^{N} \int_0^{\tau_0} v_{yi}(t_{io} + \tau)d\tau \tag{9.30}$$

The integration and summation processes indicated in Eq. 9.31 can be interchanged and therefore

$$\bar{y} = \int_0^{\tau_0} \left[\frac{1}{N} \sum_{i=1}^{N} v_{yi}(t_{io} + \tau)\right] d\tau \tag{9.31}$$

The averaging process indicated within the square brackets, wherein the average is calculated for a group of identical experiments, is called the ensemble average and is represented by $\langle v_y(t_{io} + \tau)\rangle$. The quantity $\langle v_y(t_{io} + \tau)\rangle$ is the average Lagrangian y component of velocity formed by observing a large number of particles at time τ after they leave the origin. The momentary velocity at each point in space is the instantaneous Eulerian velocity and $\langle v_y(t_{io} + \tau)\rangle$ can be expressed as an average of Eulerian velocities which are sampled, in space and time, in the highly selective way that is necessary to determine the particular particle velocities. In stationary, homogeneous turbulence it seems plausible that the average value of a large number of selectively sampled Eulerian velocities will equal the Eulerian velocities, sampled at one time, averaged over a suitable volume. In addition, we believe the volume-averaged Eulerian velocity will equal the more conventional time-averaged Eulerian velocity at any specified point in space. The presumed equivalence between time-averaged and volume-averaged quantities in stationary, homogeneous turbulence is called the ergodic hypothesis. Accordingly, we conclude

$$\langle v_y(t_{io} + \tau)\rangle = \overline{[v(t)]} = \overline{v'(t)} = 0 \tag{9.32}$$

where $\bar{v}$, the time mean y directed Eulerian velocity, is zero and $v'(t)$ is the corresponding velocity fluctuation. From Eqs. 9.32 and 9.31 it follows that $\bar{y} = 0$.

Even though the average y coordinate is equal to zero for a large number of particles after a travel time τ_0, no single particle arrives precisely at the mean position of the ensemble $(\bar{x}, 0)$. The particle positions scatter about the mean position and we wish to quantify the lateral spread. The mean-square displacement which was introduced earlier provides a convenient tool to measure the lateral spread. In this case the mean-square displacement in the y direction is given by

$$\sigma_y^2 = \frac{1}{N}\sum_{i=1}^{N} y_i^2(t_{io} + \tau_0) \tag{9.33}$$

Even though the turbulence is stationary σ_y^2 will increase with time and is thus said to be an evolutionary quantity. By differentiating Eq. 9.33 with respect to τ_0 we find, since $dy_i/d\tau_0 = v_{yi}$,

$$\frac{d(\sigma_y^2)}{d\tau_0} = \frac{2}{N}\sum_{i=1}^{N} y_i(t_{io} + \tau_0)v_{yi}(t_{io} + \tau_0) \tag{9.34}$$

The latter equation may be rewritten using Eq. 9.29 as

$$\frac{d(\sigma_y^2)}{d\tau_0} = \frac{2}{N}\sum_{i=1}^{N} v_{yi}(t_{io} + \tau_0)\int_0^{t_0} v_{yi}(t_{io} + \tau)d\tau \tag{9.35}$$

At this point it is advantageous to measure time backward from τ_0 by the introduction of the time scale $\xi = \tau_0 - \tau$. The quantity within the summation sign can be rewritten as

$$v_{yi}(t_{i0} + \tau_0) \int_0^{\tau_0} v_{yi}(t_{i0} + \tau) d\tau = -v_{yi}(t_i) \int_{\tau_0}^{0} v_{yi}(t_i - \xi) d\xi$$

$$= \int_0^{\tau_0} v_{yi}(t_i) v_{yi}(t_i - \xi) d\xi$$

Equation 9.35 becomes

$$\frac{d}{d\tau}(\sigma_y^2) = \frac{2}{N} \sum_{i=1}^{N} \int_0^{\tau_0} v_{yi}(t_i) v_{yi}(t_i - \xi) d\xi \tag{9.36}$$

We interchange the order of integration and summation and recover an ensemble average, designated by the symbols introduced above, in the form

$$\frac{d}{d\tau_0}(\sigma_y^2) = 2 \int_0^{\tau_0} \langle v_{yi}(t_i) v_{yi}(t_i - \xi) \rangle d\xi \tag{9.37}$$

The concept of the Lagrangian correlation function provides a useful way of expressing the statistical behavior of particles in a turbulent flow. We define the Lagrangian autocorrelation as

$$R_y(\xi) = \frac{\langle v_y(t) v_y(t + \xi) \rangle}{\langle v_y^2(t) \rangle} \tag{9.38}$$

For values of the delay time $\xi \to 0$, $R_y(\xi) \to 1$ and for large ξ the y component of velocity of the particle bears no relationship to its earlier value and $R_y(\xi) \to 0$. In stationary turbulence $R_y(\xi) =$ constant for all time t and thus

$$\frac{\langle v_y(t) v_y(t + \xi) \rangle}{\langle v_y^2(t) \rangle} = \frac{\langle v_y(0) v(\xi) \rangle}{\langle v_y^2(0) \rangle} = \frac{\langle v_y(-\xi) v_y(0) \rangle}{\langle v_y^2(-\xi) \rangle}$$

Therefore we conclude that $R_y(\xi) = R_y(-\xi)$ in a stationary turbulent flow field. A function possessing this property is said to be an even function and is symmetric about the line $\xi = 0$. It can also be demonstrated that the slope of $R_y(\xi)$ is zero at $\xi = 0$ and the general character of the function is shown in Fig. 9.2. The Lagrangian autocorrelation is difficult to determine because

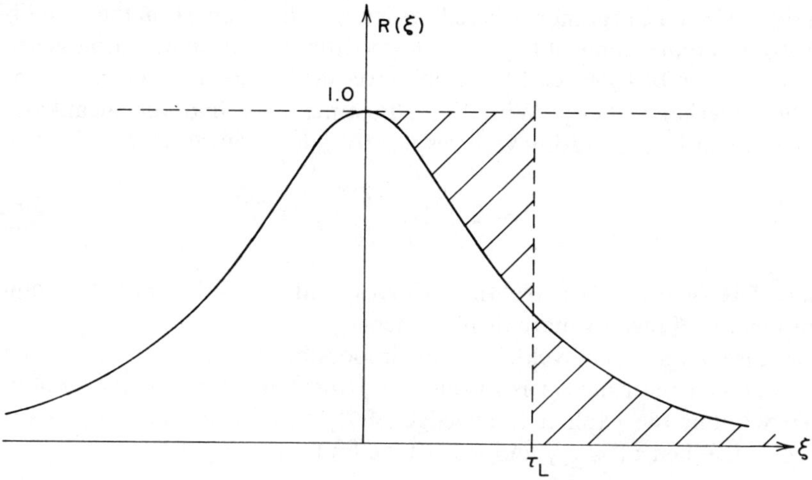

Figure 9.2. The Lagrangian autocorrelation (schematic). The Lagrangian time scale τ_L is defined by equating the shaded areas.

Lagrangian velocities are not easily observed. Clearly one of the most important characteristics of the $R_y(\xi)$ curve is its width which is defined by the value of $\xi = \tau_{LY}$ that equates the shaded areas above and below the $R(\xi)$ curve as depicted in Fig. 9.2. This time interval, designated the Lagrangian time scale τ_{LY}, is thus defined by

$$\int_0^{\tau_{LY}} [1 - R_y(\xi)]d\xi = \int_{\tau_{LY}}^{\infty} R_y(\xi)d\xi$$

or

$$\tau_{LY} \equiv \int_0^{\infty} R_y(\xi)d\xi \tag{9.39}$$

The statistical behavior of the particles can now be restated in terms of the Lagrangian time scale; for $\xi \ll \tau_{LY}$ the particle motion is perfectly self-correlated, and for $\xi \gg \tau_{LY}$ the autocorrelation vanishes.

The mean-square displacement can be formulated in terms of the Lagrangian autocorrelation. By inserting Eq. 9.38 into 9.37 we find

$$\frac{d}{d\tau_0}(\sigma_y^2) = 2\langle v_{yi}^2(t_i)\rangle \int_0^{\tau_0} R_y(\xi)d\xi \tag{9.40}$$

On the basis of the argument offered earlier, we will assume that the ensemble-averaged y components of Lagrangian velocities in stationary, homogeneous turbulence can be replaced by the time-averaged mean-square y component of the Eulerian velocity. With this adjustment, we find the mean-square displacement by integrating with respect to τ_0 for a duration T

$$\sigma_y^2 = 2(\overline{v'^2}) \int_0^T \int_0^{\tau_0} R_y(\xi) d\xi d\tau_0 \tag{9.41}$$

where T is an arbitrary travel time. Corresponding results apply for x and z components of mean-square displacements.

Because Lagrangian velocities are impossible to measure at this time, little detailed information is available on the Lagrangian autocorrelation. Even without the explicit knowledge of $R_y(\xi)$, two limiting cases can be discussed. When $\tau_0 \ll \tau_{LY}$ and $R_y \to 1$ we find, from Eq. 9.41,

$$\sigma_y^2 = \overline{v'^2}\, T^2 \tag{9.42}$$

and the mean-square displacement varies directly with travel time. On the other hand, when $\tau_0 \gg \tau_{LY}$ we find

$$\sigma_y^2 = \overline{2v'^2}\, \tau_{LY} T \tag{9.43}$$

and the mean-square displacement varies as the square root of travel time.

The significance of the results given by Eqs. 9.42 and 9.43, which were originally enunciated by Taylor (Ref. 6), can be best appreciated by recalling that the mean-square displacement for molecular diffusion given by Eq. 9.16 or 9.20 varies as the square root of the travel time from the source for all values of time. Turbulent dispersion, in contrast, has an early time behavior which is dramatically different from the later behavior that is of a similar algebraic form to molecular diffusion. To dramatize the difference between σ_y in molecular diffusion and turbulent dispersion, we may visualize a second source of marked particles slightly upstream of the point $(\bar{x}, 0)$ in Fig. 9.1. If the travel time from the second source is short compared with the Lagrangian time scale τ_{LY} while the travel time from the first source $\gg \tau_{LY}$, then the mean squared displacements for the material released at the two sources will have radically different dependence on the time required to arrive at the point $(\bar{x}, 0)$. Thus no property of the turbulent flow can be defined which describes the rate of spreading of the effluent from both sources; turbulent dispersion in this case depends on $\overline{v'^2}$, $R_y(\xi)$ and the two unequal travel times. These observations have important implications on the prediction of the spreading of material in a turbulent flow field.

9.5. SOME IMPORTANT EXPERIMENTAL RESULTS ON TURBULENT DISPERSION

The role of experimental observation of turbulent dispersion is especially important because of the limitations of the theoretical methods that are presently available. The dispersion of a marked species or contaminant released in a turbulent flow has been the object of a series of experimental investigations. Often the "contaminant" selected for study has been the heat generated by ohmic heating of a fine wire which can simulate a line source of negligible diameter with great fidelity. The unwanted buoyancy effects associated with the heat from the wire are minimized by limiting the temperature rise and using sensitive detection equipment. The heat released by the wire then serves as a passive contaminant to tag the fluid particles which pass its immediate vicinity. The distribution of temperature in the region downstream of the wire serves as a quantitative indicator of the role of turbulence in causing the spread of material provided that turbulent dispersion is dominant over molecular conduction.

A recent study of this type was performed by Shlien and Corrsin (Ref. 7) in a wind tunnel whose test section was 1.0 × 1.3 m in cross section and 10 m in length. The wind tunnel possessed a design feature, a grid of special construction followed by a contraction, to produce turbulence of the desired characteristics. Square rods, 9.7 mm on each side, were arranged in a parallel row at centerline distances of 50.8 mm. A second similar arrangement of rods was overlaid on the first set so as to form a rectangular biplanar grid possessing a 34% solid area. The contraction located 0.91 m downstream of the grid was used to reduce the area by the ratio of 1.27 to 1 in order to amplify relatively the transverse velocity fluctuations. The resulting turbulent flow is then nearly isotropic, is homogeneous in planes parallel to the grid, and is slowly decaying in the streamwise direction. The tagging wire, either of 0.0127 or 0.127 mm diameter, was located 2.13 m downstream of the grid and it was heated to a uniform temperature by an electric current. The mean velocity in the turbulent flow downstream of the grid was 12.7 m/s. The air temperature downstream of the tagging wire was measured by resistance thermometers of 0.63 μm diameter which could detect the very rapid oscillations of temperature that are associated with small-scale turbulent fluctuations of temperature. The output of the resistance thermometer was integrated by electrochemical means in order to obtain the mean temperature.

The main object of these experiments was to derive information on certain statistical properties of the flow field. An additional result is shown in Fig. 9.3 where we reproduce the mean temperature measurements that were observed in the y direction mutually perpendicular to both the mean flow

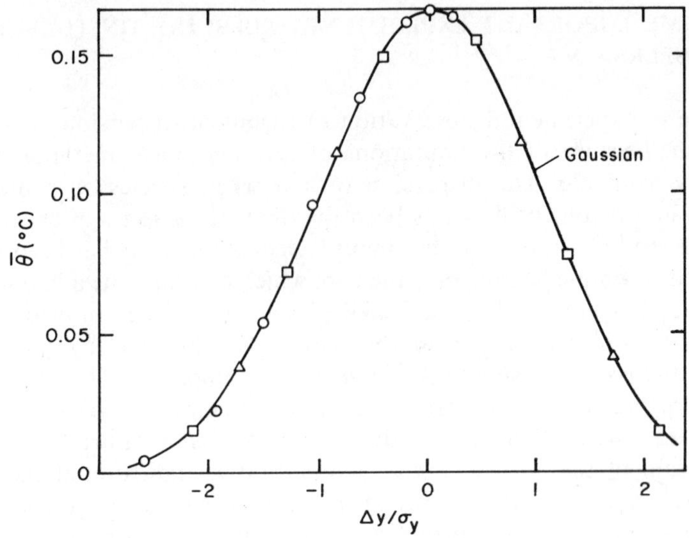

Figure 9.3. A typical mean temperature profile downstream of a heated wire in a nearly isotropic turbulent flow. The Gaussian curve has been fitted to the experimental data. [From D. J. Shlien and S. Corrsin, J. *Fluid Mech.*, **62**, 261 (1974). Reprinted with permission of the Cambridge University Press and D. J. Shlien.]

and the tagging wire at a distance 2.44 m downstream of the latter. A Gaussian curve was fitted to the data and is shown as the continuous line in Fig. 9.3. The remarkably good fit between the measured mean temperature profile and the Gaussian curve, a result that is not predicted by any theory, is a useful discovery that is the basis of the method of predicting the turbulent dispersion of air pollutants that is discussed in the next chapter.

REFERENCES

1. R. B. Bird, W. E. Stewart, and E. N. Lightfoot, *Transport Phenomena* (Wiley, New York, 1960).
2. G.T. Csanady, *Turbulent Diffusion in the Environment* (D. Reidel Publishing Co., Dordrecht, Holland, 1973).
3. F. Pasquill, *Atmospheric Diffusion* (Halsted Press of John Wiley & Sons, New York, 1974), Second Edition.
4. J. H. Seinfeld, *Air Pollution—Physical and Chemical Fundamentals* (McGraw-Hill, New York, 1975).
5. R. G. Lamb and J. H. Seinfeld, "Mathematical Modeling of Urban Air Pollution," *Env. Sci. and Technol.*, 7, 253–261 (1973).

6. G. I. Taylor, "Diffusion By Continuous Movements," *Proc. London Math. Soc.*, **20**, 196–212 (1921).
7. D. J. Shlien and S. Corrsin, "A Measurement of Lagrangian Velocity Autocorrelation in Approximately Isotropic Turbulence," *J. Fluid Mech.*, **62**, 255–271 (1974).

PROBLEMS

9.1. Equation 9.28 reduces to

$$\frac{\partial \bar{c}}{\partial t} + \bar{U}\frac{\partial \bar{c}}{\partial x} = \frac{\partial}{\partial x}(-\overline{u'c'}) + \frac{\partial}{\partial y}(-\overline{v'c'}) + \frac{\partial}{\partial z}(-\overline{w'c'})$$

when the x coordinate is chosen to coincide with the direction of the mean wind $\bar{U}$ and the species is nonreactive. If we let

$$-\overline{u'c'} = K_x \frac{\partial \bar{c}}{\partial x}, \quad -\overline{v'c'} = K_y \frac{\partial \bar{c}}{\partial y} \quad \text{and } -\overline{w'c'} = K_z \frac{\partial \bar{c}}{\partial z}$$

then K_x, K_y, and K_z represent the so-called "apparent eddy diffusivities." The above equation, if K_x, K_y, and K_z are constants, then reduces to a diffusion type equation,

$$\frac{\partial \bar{c}}{\partial t} + \bar{U}\frac{\partial \bar{c}}{\partial x} = K_x \frac{\partial^2 \bar{c}}{\partial x^2} + K_y \frac{\partial^2 \bar{c}}{\partial y^2} + K_z \frac{\partial^2 \bar{c}}{\partial z^2}$$

In this case K_x, K_y, and K_z can be related to the mean-square displacements σ_x, σ_y, and σ_z, which in general are given by Eq. 9.41.

1. Show that Eq. 9.24, which was derived for molecular diffusion but which may be applied to turbulent dispersion that also produces a Gaussian concentration distribution, yields

$$\frac{\partial \bar{c}}{\partial t} + \bar{U}\frac{\partial \bar{c}}{\partial x} = \bar{c}\frac{d\sigma_x}{dt}\left[\frac{(x - \bar{U}t)^2}{\sigma_x^3} - \frac{1}{\sigma_x}\right] + \bar{c}\frac{d\sigma_y}{dt}\left(\frac{y^2}{\sigma_y^3} - \frac{1}{\sigma_y}\right) + \bar{c}\frac{d\sigma_z}{dt}\left(\frac{z^2}{\sigma_z^3} - \frac{1}{\sigma_z}\right)$$

where σ_x, σ_y, and σ_z are functions of time.

2. Show that Eq. 9.24 also gives

$$\sigma_x \frac{d\sigma_x}{dt}\frac{\partial^2 \bar{c}}{\partial x^2} + \sigma_y \frac{d\sigma_y}{dt}\frac{\partial^2 \bar{c}}{\partial y^2} + \sigma_z \frac{d\sigma_z}{dt}\frac{\partial^2 \bar{c}}{\partial z^2}$$
$$= \bar{c}\frac{d\sigma_x}{dt}\left[\frac{(x - \bar{U}t)^2}{\sigma_x^3} - \frac{1}{\sigma_x}\right] + \bar{c}\frac{d\sigma_y}{dt}\left(\frac{y^2}{\sigma_y^3} - \frac{1}{\sigma_y}\right) + \bar{c}\frac{d\sigma_z}{dt}\left(\frac{z^2}{\sigma_z^3} - \frac{1}{\sigma_z}\right)$$

3. With the above results show that

$$K_x = \frac{1}{2}\frac{d\sigma_x^2}{dt}, \qquad K_y = \frac{1}{2}\frac{d\sigma_y^2}{dt} \qquad \text{and } K_z = \frac{1}{2}\frac{d\sigma_z^2}{dt}$$

9.2 Use the relation from Problem 9.1.3 for K_y and Eqs. 9.42 and 9.43 to show:

1. If $T \ll \tau_{LY}$, then $K_y = \overline{v'^2}\,T$
2. If $T \gg \tau_{LY}$, then $K_y = \overline{v'^2}\tau_{LY}$
3. What important assumptions about the character of the turbulent flow field must be fulfilled in order that Eqs. 9.42 and 9.43 are applicable?

9.3. In an experiment reported in Ref. 7, the mean-square displacement as a result of turbulent dispersion is $\sigma_y = 5.60$ mm at a position 7.0 m downstream of a source when the mean velocity in the x direction $\overline{U}$ is 12.7 m/s and the Lagrangian time scale $\tau_{LY} = 110$ ms.

1. Assuming that we may consider the travel time T to be much larger than the Lagrangian time scale τ_{LY}, then

$$\sigma_y^2 = 2\overline{v'^2}\tau_{LY}\mathrm{T}$$

Find the $\overline{v'^2}$ and the intensity of turbulence in the y direction defined by $i_v = (\overline{v'^2})^{1/2}/\overline{U}$.

2. When $T \gg \tau_{LY}$ we find $K_y = \overline{v'^2}\tau_{LY}$. Determine the "apparent eddy diffusivity" K_y in the case of the experiment cited above and find the ratio of $K_y/\mathscr{D}_{ab}$ where $\mathscr{D}_{ab}$, the binary diffusion coefficient for molecular diffusion, is of the order of 15×10^{-6} m^2/s.

10

DISPERSION OF AIR POLLUTANTS—THE GAUSSIAN PLUME MODEL

A calculational procedure for the determination of the concentration of reactive pollutants that attempts to incorporate the details of the physics of the turbulent flow and mixing processes and the chemical interaction of reacting gaseous species must necessarily be a numerical analysis. Solutions of this type for just a single set of meteorological conditions are achieved only by the expenditure of a large computational effort. The prediction of the long-term averages compounds even further the effort and cost. The Gaussian plume model, which is the subject of this chapter, is a simple method to apply that can give useful results provided its limitations are properly observed.

A source of complexity in atmospheric dispersion calculations is the departure of the winds from quasisteadiness. The occurrence of a diurnal cycle of the atmospheric boundary layer during clear weather implies that a time dependence of the winds near the surface is the rule rather than the exception. Nevertheless, for limited time periods and for limited geographical distances the assumption of quasisteadiness of the boundary layer winds, which is the first prerequisite for the Gaussian plume model, may be a reasonably satisfactory approximation.

The occurrence of chemical reactions that involve the formation of one or more secondary pollutants also increases the mathematical complexity of the atmospheric dispersion problem. The chemical inertness of the pollutants in question is a second requirement of the Gaussian plume model.* A variety of pollutants are either chemically inert or at least slowly reacting on the time scale of the validity of the quasisteadiness of the wind field. All gaseous compounds in the OH reactivity classes I, II, and some of the less reactive members of class III (Table 8.5) qualify as being inert for this purpose. Methane (CH_4), carbon monoxide (CO), ethane (C_2H_6), acetylene (C_2H_2), plus probably propane (C_3H_8), and benzene (C_6H_6) possess low reactivity and can usually be considered chemically inert for this purpose. These same species may, nevertheless, participate in the atmospheric reactions on a longer time scale. Particulate material in forms such as fly ash and soot can also be considered inert and its dispersion can be calculated by the methods applicable to gases, if the particle size is small enough to allow the material to be carried with the wind. Carbon monoxide, sulfur dioxide, and particulate material are the community pollutants whose concentrations are commonly analyzed by the Gaussian plume model.

Finally, a third assumption of the model is that the boundary layer wind profile can be approximated by a single vector whose magnitude $\bar{u}$ and direction are constant at all levels. The inability of the model to include the vertical variation of wind speed is a shortcoming which is sometimes limiting in its application to ground-level sources.

10.1. THE GAUSSIAN PLUME MODEL FOR THE TURBULENT DISPERSION FROM AN ISOLATED POINT SOURCE

Experiments described in Section 9.5 show that the concentration downwind of a steady line source in an isotropic, homogeneous, turbulent flow possesses a Gaussian distribution. The Gaussian plume model is based on the approximation that the concentration downwind of a point source in the atmospheric boundary layer is also Gaussian but with unequal dispersion coefficients in the horizontal and vertical directions. Throughout this chapter a wind-oriented coordinate system is employed in which the x axis is in the direction of the wind (see Fig. 10.1 below). The geometric factor that represents the ratio of the concentration at an arbitrary off-axis location (x, y, z) to the

* The Gaussian plume model is capable of describing the time-dependent depletion that is simply related to the travel time from the source. This capability is useful in accounting for the decay of radioactive species, and also is often used to account for the chemical depletion of SO_2 (see p. 243). The restriction that the rate process be related to time after release is, however, a severe one. Thus the influences of chemically interacting sources or source components, of formation reactions involving several primary of secondary pollutants, or of photochemical effects are beyond the scope of this method as it is normally implemented.

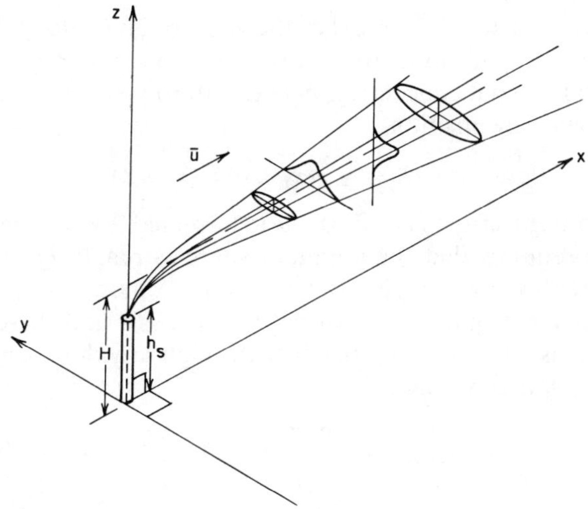

Figure 10.1. An elevated point source located at (0, 0, H) in a wind-oriented coordinate system.

concentration at the centerline of the plume is therefore given by

$$\mathscr{F}(x, y, z) = \exp\left(-\frac{y^2}{2\sigma_y^2} - \frac{z^2}{2\sigma_z^2}\right) \tag{10.1}$$

where σ_y and σ_z are the x dependent lateral (or crosswind) and vertical dispersion coefficients. The Gaussian distribution is used in the atmospheric boundary layer even though the turbulence there is not isotropic (e.g., $\overline{u'^2} \neq \overline{w'^2}$) or homogeneous (e.g., $\overline{w'^2}$ is a function of z). Unequal lateral and vertical dispersion coefficients are used to compensate, at least in part, for these departures.

The quantities σ_y and σ_z, here referred to as dispersion coefficients, are synonymous with the quantities referred to in Section 9.4 as the mean-square displacements in the y and z directions. This traditional switch in terminology tends to obscure the equivalence of these quantities. In isotropic, homogeneous turbulence the value of the dispersion coefficient or mean-square displacement was found, Section 9.4, to be determined by the travel time or distance of the particle from the source, by the Lagrangian time scale of the turbulent flow, and by the intensity of turbulence. In atmospheric calculations, where the turbulence is anisotropic and nonhomogeneous, we are obliged to rely on values of the dispersion coefficients that have been determined by field observations as discussed below.

The geometric factor $\mathscr{F}(x, y, z)$ has the value of unity along the plume axis where $y = z = 0$ and must therefore represent the quantity which is multiplied by the axial concentration $\bar{c}(x, 0, 0)$ to yield the mean concentration at any arbitrary position,

$$\bar{c}(x, y, z) = \bar{c}(x, 0, 0)\mathscr{F}(x, y, z) \tag{10.2}$$

The axial concentration $\bar{c}(x, 0, 0)$ is determined by noting that species conservation requires that the pollutant source strength Q_p (gm/s) is equal to its flow rate through any plane normal to the x axis. The species flow rate through an element of area dA normal to $\bar{u}$ is $\bar{c}\bar{u}dA$ and therefore the conservation of mass requires that the flow through any downwind $y-z$ plane is equal to the source strength,

$$Q_p = \bar{u}\bar{c}(x, 0, 0) \int_{-\infty}^{+\infty} \int_{-\infty}^{+\infty} \mathscr{F}(x, y, z)\, dydz \tag{10.3}$$

The latter integral is readily evaluated,

$$\int_{-\infty}^{+\infty} \int_{-\infty}^{+\infty} \mathscr{F}(x, y, z) dydz = 2\pi\sigma_y(x)\sigma_z(x)$$

since it can be reduced to two successive single integrations of the form

$$\int_{-\infty}^{+\infty} \exp\left(-\frac{t^2}{2}\right) dt = \sqrt{2\pi} \tag{10.4}$$

Therefore, we find the centerline concentration given by

$$\bar{c}(x, 0, 0) = \frac{Q_p}{2\pi\bar{u}\sigma_y\sigma_z} \tag{10.5}$$

We then determine the spatial variation of concentration downwind of the source by inserting Eqs. 10.1 and 10.5 into Eq. 10.2 to obtain

$$\bar{c}(x, y, z) = \frac{Q_p}{2\pi\bar{u}\sigma_y\sigma_z} \exp\left(-\frac{y^2}{2\sigma_y^2} - \frac{z^2}{2\sigma_z^2}\right) \tag{10.6}$$

which is the basic Gaussian plume formula. Equation 10.6 gives the downwind concentration from an isolated point source located at the origin (0, 0, 0) in a medium of infinite extent if the values of the dispersion coefficients are properly chosen.

The concentration from an elevated point source above a ground plane can be found from Eq. 10.6 by the application of the concepts discussed in Chapter 9. The source is relocated at the point $(0, 0, H)$ by a translation of the coordinate system where z is replaced by $z - H$. An image source is also added at the location (0, 0, $-H$) in order that the ground plane should be, by symmetry, a no flux or impenetrable surface, $\partial\bar{c}/\partial z = 0$ at $z = 0$. The result is

$$\bar{c}(x, y, z) = \frac{Q_p}{2\pi\bar{u}\sigma_y\sigma_z}\exp\left(-\frac{y^2}{2\sigma_y^2}\right)\left\{\exp\left[-\frac{(z-H)^2}{2\sigma_z^2}\right] + \exp\left[-\frac{(z+H)^2}{2\sigma_z^2}\right]\right\} \tag{10.7}$$

Figure 10.1 shows a sketch of an elevated point source in the wind-oriented coordinate system. We find the surface level concentrations that are of greater interest by setting $z = 0$ in Eq. 10.7 and obtain

$$\bar{c}(x, y, 0) = \frac{Q_p}{\pi\bar{u}\sigma_y\sigma_z}\exp\left(-\frac{y^2}{2\sigma_y^2} - \frac{H^2}{2\sigma_z^2}\right) \tag{10.8}$$

The dispersion coefficients are determined by the mixing or dispersive power of the turbulent flow within the atmospheric boundary layer. For this purpose the state of the boundary layer was characterized by Gifford (Ref. 1), based on research by F. Pasquill, as being in one of nine classes which are summarized in Table 10.1. The determinants of the dispersion classes

Table 10.1. Pasquill Dispersion Classes (Adapted from Ref. 2)

		Surface Wind Speed (m/s)				
Insolation/Cloud Cover		<2.0	2 to <3	3 to <5	5 to <6	≥6
Day	Strong Insolation	A	A–B	B	C	C
	Moderate Insolation	A–B	B	B–C	C–D	D
	Slight Insolation	B	C	C	D	D
Day or Night	Overcast	D	D	D	D	D
Night	Thin overcast or ≥0.5 cloud cover	—	E	D	D	D
	≤0.4 cloud cover	—	F	E	D	D

Notes: 1. Strong insolation corresponds to a solar elevation angle of 60° or more above the horizon. Slight insolation corresponds to a solar elevation angle of 15° to 35°.

2. Pollutants emitted under clear nighttime skies with winds less than 2.0 m/s, more recently defined to be class G, may be subject to unsteady meandering which renders the prediction of concentrations at downwind locations unreliable.

are wind speed and the state of incoming or outgoing radiation, which affect the lapse rate, the presence or absence of convective activity, and the dynamics of the mixed layer. Thus a neutral stratification when the net radiation flux is zero under conditions of heavy cloudiness by day or night, regardless of wind speed, is defined to be dispersion class *D*. A radiation inversion may form during nighttime and will give rise to poor dispersion under either classes *E* or *F* according to the state of partial cloudiness and wind speed. Dispersion classes *A*, *B*, and *C* represent the daytime conditions with varying amounts of incoming solar and sky radiation. With low wind speeds and strong incoming radiation, class *A*, there is vigorous convective activity and strong mixing. The latter produces a well-developed mixed layer of finite height often capped by an elevated inversion. Under these conditions the atmospheric boundary layer is described by the model of Tennekes discussed in Section 6.4.1. Lower amounts of incoming radiation flux or higher wind speeds correspond to moderate convective activity, class *B*, or to slight convective activity, class *C*, or to one of the intermediate classes *A–B*, *B–C*, or *C–D*.* The values of σ_y and σ_z are determined from graphs, Figs. 10.2*a* and 10.2*b*, originally provided by Gifford (Ref. 1). Separate curves for σ_y and σ_z are given for each dispersion class *A* through *F*. The values of the dispersion coefficients for the double letter classes, *A–B*, *B–C*, or *C–D* are found by averaging the values of σ_y and σ_z for both classes.

Throughout our discussions we will use the Pasquill–Gifford dispersion classes and their values of the dispersion coefficients given by Figs. 10.2. According to a later survey article by Gifford (Ref. 3) these values of σ_y and σ_z are appropriate "... for diffusion from low-level, nonbuoyant sources over open country." There exist several other sets of recommendations that define the dispersion classes and their corresponding dispersion coefficients. An early set of recommendations was derived from studies conducted at the Brookhaven National Laboratory in New York on nonbuoyant plumes released at a height of 108 m above ground. Another set of recommendations resulted from the experience with buoyant plumes gathered over a 20 year time period by the Tennessee Valley Authority. Yet a fourth set of recommendations has been the outcome of studies in an urban area by McElroy and Pooler. All of these recommendations are described in detail in the survey article by Gifford (Ref. 3).

* The term "unstable conditions" that is widely used to describe the dispersion classes *A*, *B*, and *C* is misleading to the extent that it suggests the existence of a large scale superadiabatic lapse rate. The conditions of light winds and strong insolation of dispersion class *A* clearly correspond to periods of strong convective activity when $d\theta/dz \approx 0$ throughout most of the mixed layer. The existence of a superadiabatic lapse rate in the surface layer under these conditions would be strongly dependent upon the topographical features.

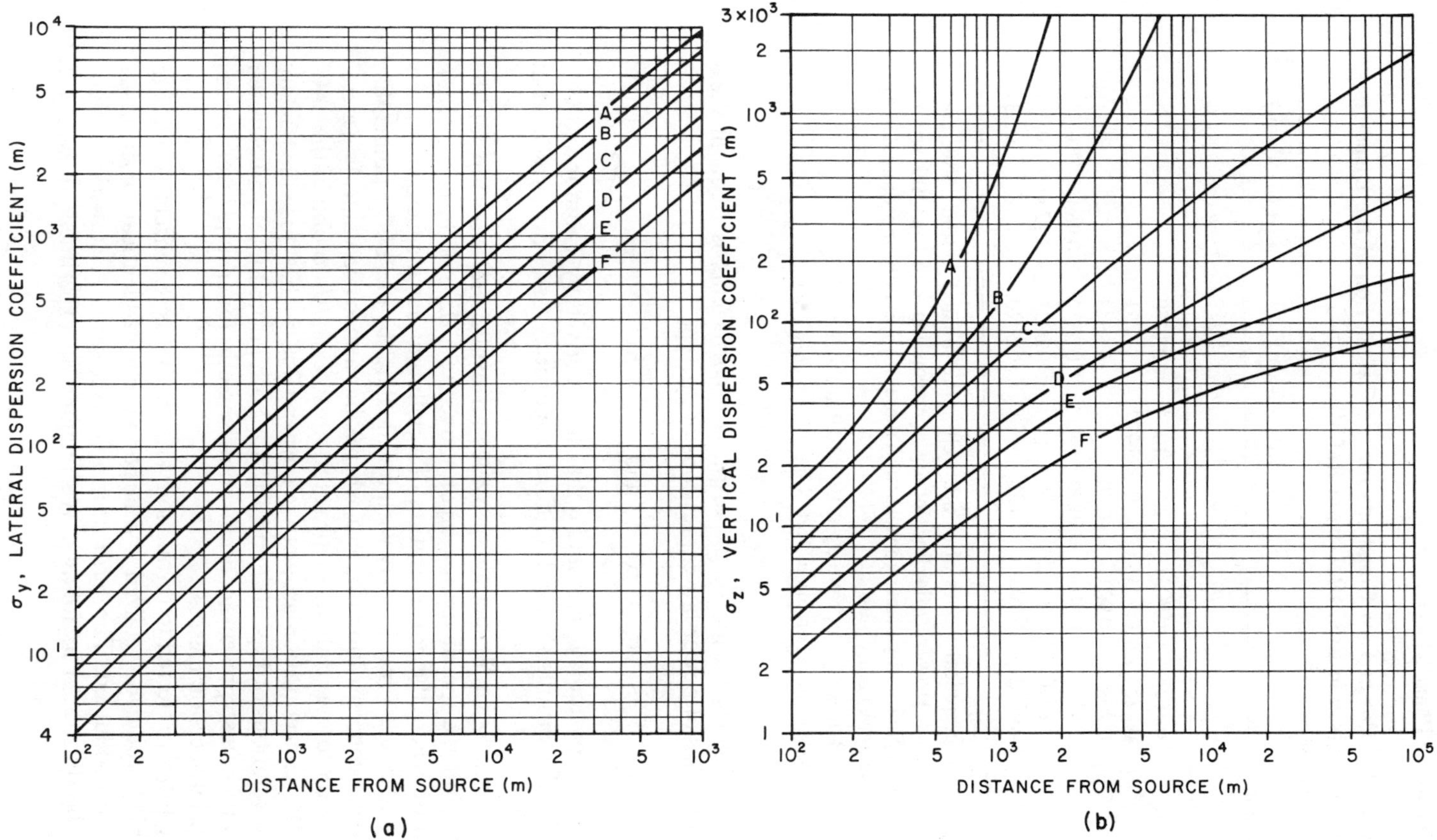

Figure 10.2. The Pasquill–Gifford dispersion coefficients versus downwind distance for various dispersion classes. (*a*) The lateral dispersion coefficient and (*b*) the vertical dispersion coefficient are plotted against x.

Table 10.2. Power Law Constants for the Calculation of the Dispersion Coefficients given in Eqs. 10.9 (from Ref. 4)

Dispersion Class	σ_y		σ_z(0.5–5 km)		σ_z(5–50 km)	
	a	p	b	q	b	q
A	0.3658	0.9031	2.5×10^{-4}	2.1250	—	—
B	0.2751	0.9031	1.9×10^{-3}	1.6021	—	—
C	0.2089	0.9031	0.20	0.8543	0.5742	0.7160
D	0.1474	0.9031	0.30	0.6532	0.9605	0.5409
E	0.1046	0.9031	0.40	0.6021	2.1250	0.3979
F	0.0722	0.9031	0.20	0.6020	2.1820	0.3310

Notes: 1. The graphical values of σ_y are well approximated by the above coefficients. However, the lines of log σ_z versus log x in Fig. 10.2 are not straight for any dispersion class and therefore σ_z is only approximated by the above constants.
2. The Units of x, σ_y, and σ_z in Eqs. 10.9 are meters.
3. Typographical errors in Ref. 4 have been corrected.

Algebraic representation of the dispersion coefficients is frequently expressed in terms of power law expressions of the type

$$\sigma_y = ax^p \quad \text{and} \quad \sigma_z = bx^q \tag{10.9}$$

Values of the constants a, p, b, and q that were derived by Tadmor and Gur (Ref. 4) for the various Pasquill–Gifford dispersion classes are presented in Table 10.2.

The Gaussian plume method provides an explicit relationship for the concentration at a specified point on or above the surface of the earth. Often the location where the concentration of a pollutant has a specified value is of interest. For example, we may wish to calculate the loci of the points that form a line, a concentration isopleth, where $\bar{c}$ has a specified value. In general it is necessary to use trial and error solutions to Eq. 10.8 to obtain the position of the concentration isopleths. Turner (Ref. 5) has provided many graphical solutions of Eqs. 10.7 and 10.8 for both ground level and elevated sources, and he also presents concentration isopleths for ground level sources for various dispersion classes.

The quantity $\bar{c}$ referred to throughout this chapter is the ensemble average value which would be realized if a large number of identical experiments were to be conducted. The term ensemble average is of statistical origin and can be illustrated as follows. We imagine an experiment in which a cloud of marked, neutrally buoyant particles is released from a point in the turbulent flow. The number of these particles found in a specified small volume at a downwind location is determined at a specified time interval after the release has occurred. The experiment is repeated a number of times under identical conditions, and the observed concentrations are found to display a range of values. No two readings of particle concentration are necessarily the same, and no single reading need coincide with the average of all values. The set of identical tests is called the ensemble of experiments and the average concentration is said to be the ensemble average. In practice, we accept the time-average concentration that is measured downwind of a continuous source as being completely equivalent to the ensemble average if two conditions are simultaneously fulfilled. The time period over which the averaging process is performed must be long enough so as to contain a full period of the slowest frequency component and, at the same time, short enough to guarantee that the time-averaged wind speed components and all other meteorological variables have not changed. The determination that these two conditions are fulfilled may be difficult to achieve in practice.

10.2. MULTIPLE POINT SOURCES WITH AN ARBITRARY WIND DIRECTION

The concentration at a location of interest, designated as a receptor, downwind of a number of sources whose coordinates are given in a geographic coordinate system can be found from the solutions given in the previous section. A convenient set of geographic coordinates in the United States is the metric cartesian Universal Transverse Mercator (UTM) system. We designate the easterly and northerly coordinates in the UTM system by the symbols E and N, and use the subscript si to denote the ith source and the subscript r for the receptor. We then wish to calculate the concentration of a pollutant at a surface level receptor located at (E_r, N_r) owing to the source at (E_{si}, N_{si}) as depicted in Fig. 10.3. The solution will be given by Eq. 10.8 in the wind-oriented $(x-y)$ coordinate system. The intermediate coordinates, $x' - y'$, shown in Fig. 10.3 are first determined by

$$x'_i = E_r - E_{si} \qquad y'_i = N_r - N_{si}$$

We now relate the $x'-y'$ coordinates to the wind-oriented x–y coordinate system. According to meteorological convention, the wind angle ψ_w is defined as the angle measured clockwise from north to a ray pointing to the direction of origin of the wind. Thus a southwest wind, a wind vector heading

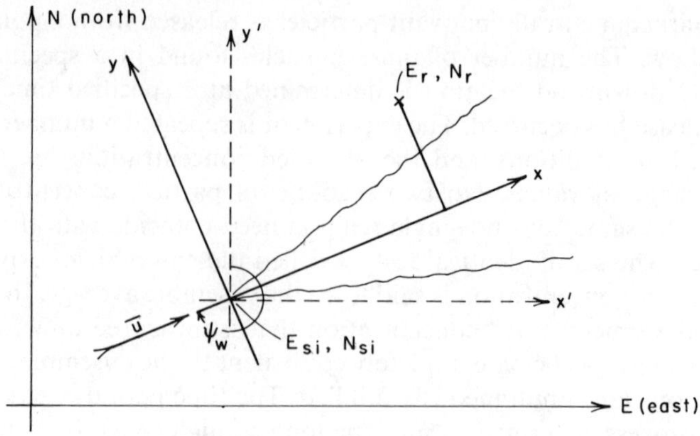

Figure 10.3. The dispersion from a point source in geographic coordinates. The locations of the ith source (E_{si}, N_{si}) and the receptor are arbitrary and the wind angle (ψ_w) is measured clockwise from north to the direction from which the wind originates.

toward the northeast, has a wind angle $\psi_w = 225°$. To obtain the (x_i, y_i) coordinates of the receptor with respect to the ith source, we use the relations

$$x_i = -x_i' \sin \psi_w - y_i' \cos \psi_w \tag{10.10a}$$

$$y_i = +x_i' \cos \psi_w - y_i' \sin \psi_w \tag{10.10b}$$

Equation 10.8 is then applicable with σ_{yi} and σ_{zi} based on the downwind distance x_i given by Eq. 10.10a.

The concentration at the receptor owing to a series of point sources is now readily formulated by adding the contributions of all sources. We thereby employ the concept of superposition, which asserts that the concentration from two or more sources is the sum of the concentrations that each source would produce in the absence of all others. We find, accordingly,

$$\bar{c}(E_r, N_r, 0) = \frac{1}{\pi \bar{u}} \sum_{i=1}^{N} \frac{Q_{pi}}{\sigma_{yi}\sigma_{zi}} \exp\left(-\frac{y_i^2}{2\sigma_{yi}^2} - \frac{H_i^2}{2\sigma_{zi}^2}\right) \tag{10.11}$$

where

$$\begin{aligned} \sigma_{yi} &= \sigma_{yi}(x_i), \qquad \sigma_{zi} = \sigma_{zi}(x_i), \\ x_i &= +(E_{si} - E_r) \sin \psi_w + (N_{si} - N_r) \cos \psi_w, \\ y_i &= -(E_{si} - E_r) \cos \psi_w + (N_{si} - N_r) \sin \psi_w, \\ H_i &= \text{the effective height of release of the } i\text{th source,} \end{aligned} \tag{10.12}$$

and

Q_{pi} = the emission rate of pollutant by the ith source located at (E_{si}, N_{si})

The summation indicated in Eq. 10.11 should be extended over those sources which lie in the upwind sector from which a significant contribution is received at the receptor. The usefulness of Eq. 10.11 is extended by noting that line and area sources can be divided into discrete elements which may be considered point sources. (The contribution from these sources is more easily evaluated by the means described below.) The lack of an adequate knowledge of source strengths frequently hampers the application of Eq. 10.11.

Calder (Ref. 6) has pointed out that the use of Eq. 10.11 implies both a horizontal homogeneity of the topography and an isotropy with respect to wind direction. Horizontal homogeneity means that the solution for a given source-receptor pair is not influenced by an arbitrary horizontal translational displacement. Isotropy implies that the concentration at a given distance downwind of a source is the same when the wind direction changes arbitrarily, with no change in wind speed or other meteorological variables. Homogeneity and isotropy may be violated if topographical features such as hills or a high density of tall buildings are present.

10.3. POLLUTANT CONFINEMENT WITHIN THE MIXED LAYER

The convective conditions that correspond to classes A, B, and C in Table 10.1 are normally associated with a mixed layer of finite height that is frequently capped by an elevated inversion. A nocturnal radiation inversion may confine pollutants within a lower stratum. Pollutants released at the surface are dispersed throughout the mixed layer at a rate that varies according to the intensity of the convective activity but are confined by the inversion which serves as an impenetrable barrier. The Gaussian plume model can be adapted to include the limitation on mixing that is imposed by the finite height of the mixed layer by one of three methods.

The method of images can again be used to produce a zero mass flux condition at the ground surface, $z = 0$ and also at the surface $z = H_m$ as indicated in Fig. 10.4. Thus to confine the pollutants at $z = +H_m$ it is necessary to add a source at $z = 2H_m - H$. The latter, in turn, must be confined by a source at $z = 2H_m + H$ which is also restrained by a series of pairs of sources extending to $z = +\infty$. The image source at $z = -H$ is similarly confined by pairs of image sources extending to $z = -\infty$. The concentration of the pollutant in the level $0 < z < H_m$ is found by adding the contribution from all sources as

$$\bar{c}(x, y, z) = \frac{Q_p}{2\pi\bar{u}\sigma_y\sigma_z} \exp\left(-\frac{y^2}{2\sigma_y^2}\right) \times S\left(\frac{z}{\sigma_z}, \frac{H}{\sigma_z}, \frac{H_m}{\sigma_z}\right)$$

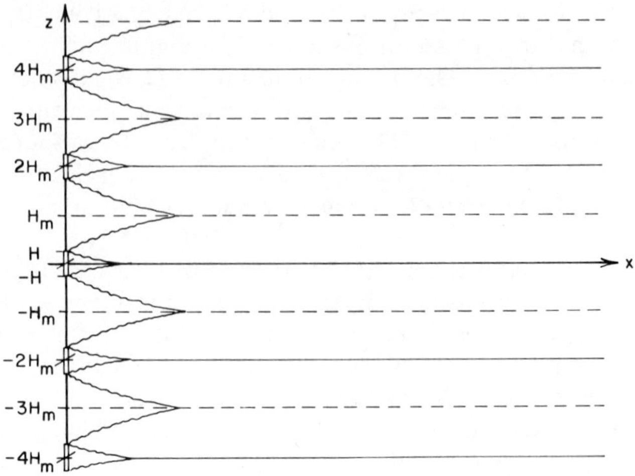

Figure 10.4. Dispersion from the source at $z = +H$ confined by an infinite array of image sources to the layer $0 < z < H_m$.

where

$$S\left(\frac{z}{\sigma_z}, \frac{H}{\sigma_z}, \frac{H_m}{\sigma_z}\right) = \sum_j \left\{\exp\left[-\frac{(z + 2jH_m - H)^2}{2\sigma_z^2}\right] + \exp\left[-\frac{(z + 2jH_m + H)^2}{2\sigma_z^2}\right]\right\} \quad (10.13)$$

with $j = 0, \pm 1, \pm 2 \ldots \pm \infty$.

The terms within the summation for S represent, with $j = 0$, the original source at $z = +H$ and its image at $z = -H$. All image sources for $z > +H_m$ are represented by the same terms with positive values of j; all image sources at $z < -H_m$ are represented by negative values of j. The effect of the confining influence of the finite mixing height can always be found by use of Eq. 10.13 and the series converges rapidly. Yamartino (Ref. 7) has shown that a practical computational scheme is to use Eq. 10.13 with $j = 0$, and ± 1 for $\sigma_z/H_m \leq 0.63$. For values of $\sigma_z/H_m > 0.63$ he recommends, for an accuracy of 1.3%, the approximation,

$$S\left(\frac{z}{\sigma_z}, \frac{H}{\sigma_z}, \frac{H_m}{\sigma_z}\right) = \frac{\sqrt{2\pi}\sigma_z}{H_m}(1 - \beta^2) \times \left[1 + \beta^2 + 2\beta \cos\left(\frac{\pi z}{H_m}\right) \cos\left(\frac{\pi H}{H_m}\right)\right] \quad (10.14)$$

where

$$\beta \equiv \exp\left[-\frac{1}{2}\left(\frac{\pi\sigma_z}{H_m}\right)^2\right]$$

Yamartino reports $\beta = 0$ for $\sigma_z/H_m > 1.08$ is consistent with the above accuracy.

A second method of calculating the influence of the confinement by an elevated inversion is discussed by Turner (Ref. 5). Equation 10.8 is used for downwind distances, $x < x_{hm}$, which are sufficiently close to the source to be free of influence of the inversion. The value of x_{hm} is defined as the distance at which the concentration at $z = H_m$ is estimated to be 10% of its centerline value. This estimate of x_{hm} ignores the existence of both the image source and also the inversion itself. Thus

$$\exp\left[-\frac{(H_m - H)^2}{2\sigma_z}\right] = 0.1$$

or $\sigma_z(x_{hm}) = 0.466\,(H_m - H)$ is the implicit equation for x_{hm} which is determined graphically by means of Fig. 10.2*b*.* For values of $x > 2x_{hm}$ it is assumed that the pollutant is mixed perfectly in the vertical direction for z such that $0 < z < H_m$. In this case the concentration is given for $0 \leq z < H_m$

$$\bar{c}(x, y, z) = \frac{Q_p}{\sqrt{2\pi}\,\bar{u}H_m\sigma_y}\exp\left(-\frac{y^2}{2\sigma_y^2}\right) \tag{10.15}$$

and further turbulent dispersion occurs only in the crosswind direction parallel to the y axis. The value of σ_y is assumed to be uninfluenced by the limit on the vertical extent of mixing. The suitability of Eq. 10.15 is demonstrated by proving that the mass flow of the pollutant through any plane parallel to the $y-z$ axes is equal to Q_p (see Problem 10.2). For intermediate values of x such that $x_{hm} < x < 2x_{hm}$ a graphical interpolation between the calculated values on a logarithmic graph is recommended by Turner.

Equation 10.15 shows that the concentration downwind of a source of specified strength is inversely proportional to the product $\bar{u}H_m$. This quantity is the ventilation factor discussed in Section 6.5.1 if the wind is independent of height. The appearance of this quantity in the above expression is the basis for its use as a significant parameter that characterizes atmospheric dispersion.

A third adaptation of the Gaussian plume method to the calculation of dispersion in the presence of an elevated inversion was used by Smith and Singer (Ref. 8) and Shieh et al. (Ref. 9). In their procedure the value of the vertical dispersion coefficient σ_z was chosen equal to the lesser of either its

* Turner actually takes H to be small compared with H_m in finding x_{hm}.

graphical value or the height of the mixed layer H_m (or a fraction of it such as $H_m/1.25$ or even $H_m/2$). In effect this method allows for a small fixed fraction of the pollutant to penetrate upward through the inversion, while the vertical variation of concentration within the mixed layer is relatively uniform. The vertical shape of the concentration profile $\bar{c}(x, 0, z)$ is thus held constant when H_m replaces σ_z in Eq. 10.8 and $\bar{c}$ decreases with increasing downwind distance x because σ_y continues to increase. Calder (Ref. 6) discusses a later variation of this third method which is easy to apply.

10.4. OTHER SOURCE GEOMETRIES

A tall stack emitting combustion products is a good example of a point source whose strength is specified by Q_p(gm/s) as discussed in the previous sections. It would be impractical to account for the individual emissions of each vehicle on a busy highway. A more suitable procedure here is to attribute the emissions from many vehicles as occurring along a line source of strength Q_l (gm/m s) which is an aggregate of the emissions of all vehicles. In a similar fashion, the pollutants, such as CO and SO_2, that are released from the combustion of hydrocarbon fuels in home heating plants in a residential area are best treated as resulting from an area source of uniform strength Q_a(gm/m^2 s) which represents the sum of all individual average emission rates divided by the area of the region. In this section we apply the Gaussian plume model to the description of the concentrations downwind of line and area sources.

10.4.1. Line Sources

The concentration downwind of a line source which is normal to the mean wind vector can be found by an extension of the concepts presented in Section 10.1 as discussed by Csanady (Ref. 10). We assume the line source is at the surface as is the case for a grade-level highway, but the theory is easily modified to apply to an elevated line source. The emissions from a differential length dy' of a line source will be $Q_l dy'$ where Q_l is the line source strength. The line source is located on the y axis, see Fig. 10.5, and the receptor is at the arbitrarily positioned point (x, y) downwind.

The concentration at the receptor caused by emissions from the element dy' located at $(0, y')$, which is effectively a point source, is found from Eq. 10.8 with $H = 0$ as

$$d\bar{c} = \frac{Q_l dy'}{\pi \bar{u} \sigma_y \sigma_z} \exp\left[-\frac{(y' - y)^2}{2\sigma_y^2}\right]$$

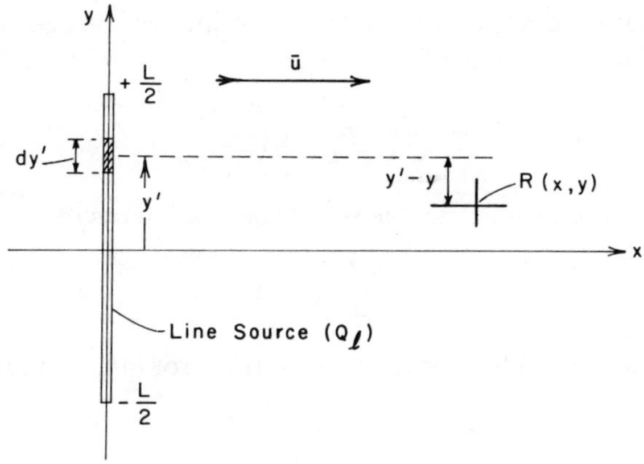

Figure 10.5. A line source of length L and strength Q_l normal to the wind vector. The receptor is located at the point (x, y).

with σ_y and σ_z evaluated at the downwind distance x to the receptor. The effect of an impenetrable ground plane is included by the use of Eq. 10.8. The concentration at the receptor caused by the entire line source is therefore found by integration over y' to be

$$\bar{c} = \frac{Q_l}{\pi \bar{u} \sigma_y \sigma_z} \int_{-L/2}^{+L/2} \exp\left[-\frac{(y'-y)^2}{2\sigma_y^2}\right] dy'$$

This integral is expressed as

$$\bar{c} = \frac{Q_l}{\sqrt{2\pi}\,\sigma_z \bar{u}} \left[\operatorname{erf}\left(\frac{L/2 - y}{\sqrt{2}\sigma_y}\right) + \operatorname{erf}\left(\frac{L/2 + y}{\sqrt{2}\sigma_y}\right)\right] \qquad (10.16)$$

where the error function is defined by

$$\operatorname{erf} t = \frac{2}{\sqrt{\pi}} \int_0^t e^{-u^2} du$$

and is a widely tabulated quantity. For small values of the argument t the error function is approximated by

$$\operatorname{erf} t = \frac{2t}{\sqrt{\pi}} \left(1 - \frac{t^2}{3} + \cdots\right)$$

while $\operatorname{erf} t \to 1$ as $t \to \infty$.

Several limiting cases of Eq. 10.16 are of interest. When the receptor is downwind of the centerline of the source, we set $y = 0$ in Eq. 10.16 and find

$$\bar{c} = \sqrt{\frac{2}{\pi}} \frac{Q_l}{\sigma_z \bar{u}} \operatorname{erf}\left(\frac{L}{2\sqrt{2}\sigma_y}\right) \tag{10.17}$$

If the length of the symmetric line source is small, namely,

$$\frac{L^2}{24} \sigma_z^2 \ll 1$$

then the second term of the above series for the error function can be neglected and we find

$$\bar{c} \to \frac{Q_l L}{\pi \bar{u} \sigma_y \sigma_z}$$

Comparing this equation with Eq. 10.8 in which y and $H = 0$, we recognize that the finite line source has degenerated to a point source in this limit.

For large values of $L/2\sqrt{2}\sigma_y$ Eq. 10.17 reduces to

$$\bar{c} = \sqrt{\frac{2}{\pi}} \frac{Q_l}{\sigma_z \bar{u}} \tag{10.18}$$

which is the expression for the concentration downwind of an infinite line source normal to the mean wind vector. However, even a line source of finite length may appear to be infinite. From a table of the error function we find $\operatorname{erf} t = 0.95$ when $t = 1.38$, and therefore a line source can be approximated as an infinite line source to an accuracy of 5% or better if $L \geq 3.90\, \sigma_y$. The significance of this result can be appreciated best by a numerical example. At a downwind distance of 1000 m under the neutral dispersion class *D* the horizontal dispersion coefficient is found from Fig. 10.2*a* to be equal to 75 m. Thus the concentration at a point 1000 m directly downwind of the midpoint of a finite line source of crosswind length equal to 293 m is only 5% less than the concentration that would be produced by an infinite line source of equal strength. In this example, the result of which is sensitive to the dispersion class, the angle subtended by the source when viewed from the receptor is only 17° of arc. This illustration shows that the concentration at a receptor is totally controlled by the emissions from that portion of the source which lies within a small angular sector on either side of the upwind direction. This important conclusion can be used to simplify the pollutant concentrations from line sources of an intermediate length.

The line source oriented at an oblique angle to the mean wind vector in principle requires the evaluation of the appropriate integral by numerical

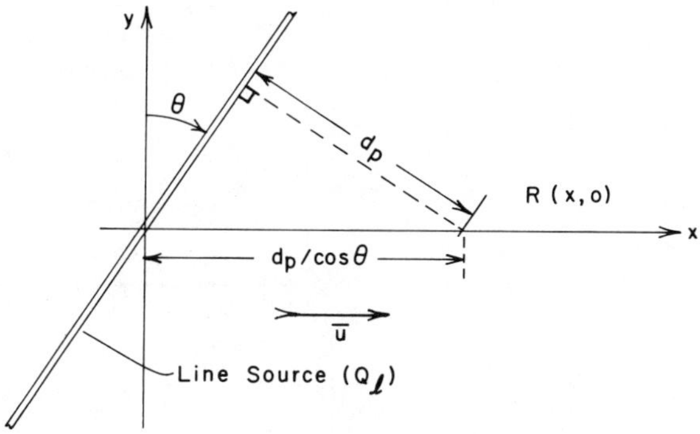

Figure 10.6. The infinite line source at an oblique angle to the wind. The portion of the source near the origin provides the dominant contribution to the concentration at the receptor $R(x, 0)$.

means. However, Calder (Ref. 11) has shown that an approximate solution for the concentration downwind of an infinite line source at ground level is

$$\bar{c} = \sqrt{\frac{2}{\pi}} \frac{Q_l}{\bar{u} \cos\theta \, \sigma_z(d_p/\cos\theta)} \tag{10.19}$$

Here d_p is the perpendicular distance between the receptor and the infinite line source, and θ is the angle between its normal and the wind vector. Note that σ_z is evaluated at the distance $d_p/\cos\theta$ which corresponds to the distance x that a parcel originating at the source travels in reaching the receptor if it is carried by the mean wind. In view of our previous discussion we recognize that the portion of an oblique line source that lies within a small sector directly upwind makes the main contribution to the concentration at the receptor (see Fig. 10.6). The relevant travel distance is therefore $d_p/\cos\theta$, and the apparent source strength is amplified by the factor $1/\cos\theta$ because of the obliquity of the source. Calder found Eq. 10.19 to give agreement with the solution yielded by an exact numerical integration for angles as large as 75° (Ref. 11).

10.4.2. Area Sources

A striplike area source of width x_1 that is normal to the wind vector can be analyzed by an extension of the line source concept. If the crosswind dimension of the source is long enough to be considered of infinite extent, referring to Fig. 10.7a, then it may be divided into a series of infinite line

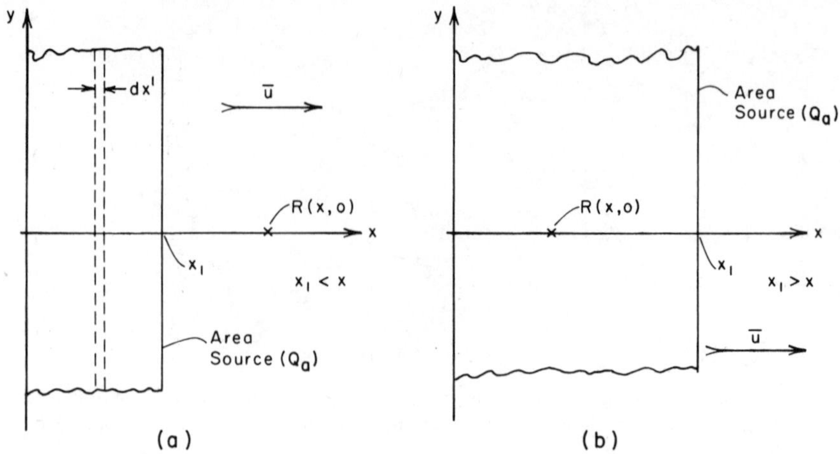

Figure 10.7. An infinite strip area source of width x_1. (*a*) The receptor is exterior to the area source. (*b*) The receptor is interior to the source.

sources of strength $Q_a dx'$ per unit length where Q_a (gm/m^2 s) is the area source strength. The concentration at the receptor located at $(x, 0)$ is expressed by Eq. 10.18 for the infinite line as

$$d\bar{c} = \sqrt{\frac{2}{\pi}} \frac{Q_a dx'}{\bar{u}\sigma_z} \tag{10.20}$$

where σ_z is evaluated at the distance $x - x'$. To integrate this relation we express σ_z in a power law form as

$$\sigma_z = b(x - x')^q$$

where the values of b and q are selected from Table 10.2, according to the dispersion class, to approximate the curved lines shown in Fig. 10.2*b*. The concentration at the receptor therefore is given by

$$\bar{c} = \sqrt{\frac{2}{\pi}} \frac{Q_a}{\bar{u}b} \int_0^{x_1} \frac{dx'}{(x - x')^q} \tag{10.21}$$

This integral is readily evaluated to yield the final result for the infinite strip area source

$$\bar{c} = \sqrt{\frac{2}{\pi}} \frac{Q_a}{\bar{u}b(1 - q)} [x^{1-q} - (x - x_1)^{1-q}] \tag{10.22}$$

For $x_1 = x$ the receptor is located either on the downwind edge or interior to the area source (see Fig. 10.7*b*), and Eq. 10.22 reduces to

$$\bar{c} = \sqrt{\frac{2}{\pi}} \frac{Q_a x}{\bar{u}(1 - q)\sigma_z(x)} \tag{10.23}$$

where

$$\sigma_z(x) = bx^q$$

and x is the upwind distance from the receptor to the leading edge of the area source.

We expect Eq. 10.23 to be a good approximation for a strip that is inclined with respect to the mean wind vector when the receptor is interior to an area source, provided the distance x is measured from the leading edge of the source to the receptor in the windward direction. In this case the area integral is incorrectly evaluated at the leading edge, but this portion of the area source contributes weakly to the concentration at the receptor. This approximation will fail at steep angles of obliquity θ, but the limits of its accuracy have yet to be tested.

Exact closed form solutions are not available for the case when the area source is finite in its crosswind dimension. In this case, the concentration at a receptor at a distance $x - x_1$ from the downwind edge of the finite source (see Fig. 10.8*a*) is given by the integral

$$\bar{c} = \frac{Q_a}{\pi \bar{u}} \int_{y'=-L/2}^{+L/2} \int_{x'=0}^{x_1} \frac{\exp\left[-\dfrac{(y - y')^2}{2\sigma_y^2} - \dfrac{H^2}{2\sigma_z^2}\right]}{\sigma_Y \sigma_Z} dx'dy' \tag{10.24}$$

where $\sigma_y = \sigma_y(x - x')$ and $\sigma_z = \sigma_z(x - x')$. This integral, which may require a considerable numerical effort to evaluate, can usually be avoided by the use of an approximate method given by Calder (Ref. 6). In Section 10.4.1 it was noted that even a relatively short line source normal to the wind appears to be infinite at a downwind receptor. This concept can be applied in the case of an area source to facilitate the calculation of the downwind concentration. Since the portion of the source in a small upwind sector makes the important contribution to the receptor, only a small error is committed by assuming the source to have infinite crosswind extent. In this case we find, from Eq. 10.18 for an infinite line source,

$$d\bar{c} = \sqrt{\frac{2}{\pi}} \frac{Q_a}{\bar{u}\sigma_z} \exp\left(-\frac{H^2}{2\sigma_z^2}\right) dx' \tag{10.25}$$

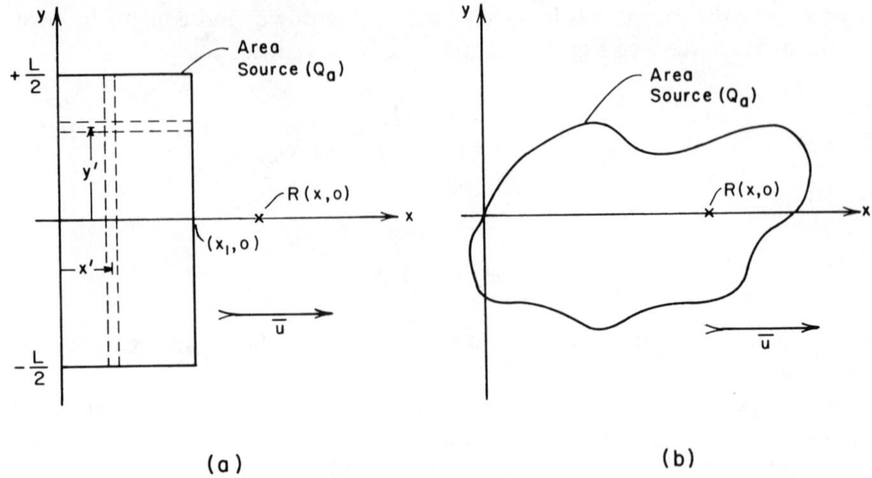

Figure 10.8. Finite area sources. (*a*) A rectangular source with an external receptor and (*b*) an area source of arbitrary shape with an internal receptor.

with $\sigma_z = \sigma_z(x - x')$ as the expression applicable for a differential crosswind increment of an area source that is located at the height H above the ground (see Fig. 10.8*a*). In fact we would expect the use of an infinite crosswind dimension to be even a better approximation for the area source because the heaviest contribution at the receptor will always be received from the area close to its windward side. The concentration caused by the finite area source is then formulated by integrating Eq. 10.25 over x'

$$\bar{c} = \sqrt{\frac{2}{\pi}} \frac{Q_a}{\bar{u}} \int_0^{x_1} \sigma_z^{-1} \exp\left(-\frac{H^2}{2\sigma_z^2}\right) dx' \tag{10.26}$$

with σ_z evaluated during the integration process at the variable downwind distance $x - x'$.

The use of Eq. 10.26 in place of the exact area integral Eq. 10.24, with an enormous saving of computational effort, is referred to as the "narrow plume approximation" (Calder, Ref. 6). While we have described the narrow plume approximation in terms of a rectangular area source normal to the wind, it is also valid for an area source of arbitrary shape with an interior receptor (see Fig. 10.8*b*) where $x_1 = x$ and is interpreted as the distance from the receptor to the upwind edge of the source. Furthermore, the

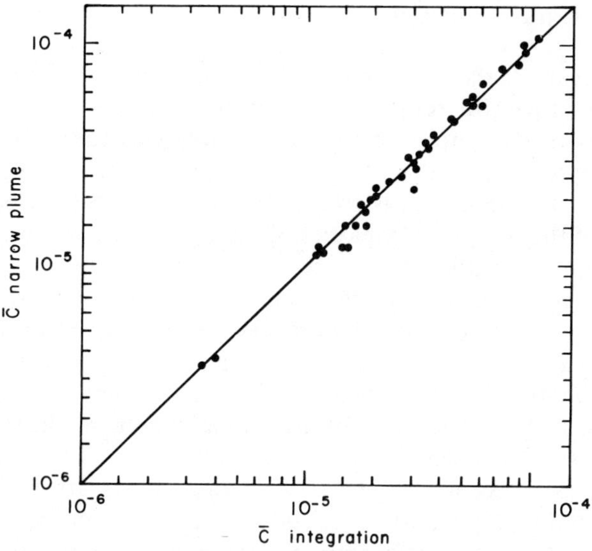

Figure 10.9. A comparison of concentrations predicted by the use of the narrow plume hypothesis (Eq. 10.26) with the exact double integration (Eq. 10.24). The 45° line represents perfect agreement between the two methods. [Based on results obtained by R. C. Kock and S. D. Thayer, and quoted by K. L. Calder, *Atmos. Environ.*, **11**, 408 (1977). Reproduced with permission of Pergamon Press and K. L. Calder.]

upwind edge need not be normal to the wind because the heaviest contribution to the concentration at the receptor is from the portion of the area source that is immediately upwind.

An extensive comparison of the accuracy of the narrow plume approximation is described by Calder (Ref. 6) and is displayed in Fig. 10.9. Here many individual calculations of the concentration downwind of area sources, as determined by the double integration (Eq. 10.24), are compared with the narrow plume approximation (Eq. 10.26). Figure 10.9 shows that the approximate method is of ample accuracy in most cases for the specific geometries that were considered.

Often the height of the area source is ground level, $H = 0$, in which case Eq. 10.26 can be integrated if a power law form for σ_z is employed. The result is, of course, the same as Eqs. 10.22 or 10.23 for the infinite area source. Equation 10.26 may also be evaluated for a series of upwind area sources of various strengths, in which case, the integral is evaluated in a piecewise fashion taking into consideration the local value of the source strength (see Problem 10.4).

10.5. LONG-TERM AVERAGING

The need to evaluate the effectiveness of air pollution control strategies, or to ascertain the impact of projected industrial or residential growth on the compliance with the ambient air quality standards, requires the ability to estimate long-term averages of the concentrations of various pollutants. The Gaussian plume model is frequently used as a basic element in the calculation of the long-term averages of inert or slowly reacting pollutants such as carbon monoxide, particulates, and sulfur dioxide.

A primary requirement for the calculation of the long-term average is the availability of a large accumulation of meteorological data for the locality of interest. This information is available in the United States from the National Climatic Center at Ashville, North Carolina, which maintains records of the weather observations made at all major weather stations. The meteorological factors that enter into the calculation of the concentration of a pollutant at a receptor located among an arbitrary array of sources are the wind direction, the wind speed, and the dispersion class. This information is used to prepare a climatological frequency function which represents the probability of occurrence of a particular set of meteorological variables. The observed wind directions are discretized into 16 angular segments ($k = 1, 2, \ldots, 16$) of 22.5° of arc arranged such that the center of the first segment is in the northerly direction. The observed wind speeds are discretized into six speed classes ($l = 1, 2, \ldots, 6$) which are, typically, 0–3, 4–6, 7–10, 11–16, 17–21, and >21 knots. The dispersion class is similarly represented by integers ($m = 1, 2, \ldots, 6$) corresponding to the letter designations *A*, strong convective activity, to *F* strong outgoing nocturnal radiation with low winds. The climatological data are used to find the number of hourly periods in one year in which the wind direction, speed, and stability category are of specified values. This quantity is then divided by the total number of hourly periods in one year to arrive at the probability of simultaneous occurrences of the specified wind speed, direction, and stability class. The complete specification of all values of ϕ for the allowable combinations of k, l, and m defines the climatological frequency function or matrix $\phi(k, l, m)$.* Clearly, high valves of wind speed will always correspond to the neutral stability class regardless of the radiation boundary condition. Thus, for example, $\phi(k, 6, 6) = 0$ for all wind directions (values of k). The function $\phi(k, l, m)$ is a three-dimensional matrix of dimension $6 \times 6 \times 16$ having 576 elements whose values are positive numbers less than unity and perhaps equal to zero. The sum of all elements of the climatological frequency

* The actual determination of the latter is accomplished by a computer program used by the National Climatic Center entitled STAR. The stability classes used in this program are similar but not identical to the classes defined above by Pasquill.

matrix is, by definition, equal to unity or

$$\sum_{k=1}^{16} \sum_{l=1}^{6} \sum_{m=1}^{6} \phi(k, l, m) = 1.00 \tag{10.27}$$

A climatological frequency matrix for a city on the northeastern coast of the United States reveals that favorable dispersion (classes A, B, and C) occurs 12.5% of the time, mostly as class C; neutral dispersion (calss D) persists for 62% of the time; and adverse nocturnal dispersion (classes E and F) is present 25.5% of the time. These results are specific to one location but are probably representative of eastern coastal areas.

In general the calculation of the long-term average must give consideration to the correlation between the emission strengths and the meteorological variables. For example, the seasonal variation of the pollutant emissions due to the combustion of fossil fuels used for heating of homes, offices, and industrial buildings will be correlated with stability classes and wind speeds that are associated with the winter months. Such a correlation will complicate the averaging process and will require appropriate weighting methods. If the emission strengths are uncorrelated with the meteorological variables then the long-term concentration $\langle c \rangle$ can be calculated by

$$\langle c \rangle = \sum_{k} \sum_{l} \sum_{m} \bar{c}(k, l, m)\phi(k, l, m) \tag{10.28}$$

where $\bar{c}(k, l, m)$ represents the concentration resulting from all point, line, and area sources for the values of wind direction k, the midrange wind speed corresponding to l, and the stability class m. The influence of the finite height of the mixed layer is considered in the calculation of $\bar{c}(k, l, m)$, and is thereby included in the calculation of the long-term concentration. The contribution by area sources can be calculated by means of the narrow plume approximation. Line sources are normally treated as a series of line segments that are each considered to be a point source. A more general discussion of the calculation of long-term averages is given by Calder (Ref. 6) who describes economical methods which circumvent the use of a discretized wind direction variable.

10.6. COMPUTER PROGRAMS EMPLOYING THE GAUSSIAN PLUME METHOD

The Gaussian plume method has been used as the basis for a number of computer programs which are available for selected special purposes. These programs, which are usually written in Fortran, are summarized below and are described in more detail in the references cited.

HIWAY is a computer program consisting of essentially four parallel line sources of finite length that are inclined at an arbitrary angle with respect to the wind vector (Ref. 12). The line sources can be used to represent a highway consisting of two pairs of traffic lanes separated by a median strip. The width of the highway and the median can be specified, and the downwind receptor location is arbitrary. The pollutant is a nonreactive species such as carbon monoxide. The six dispersion classes conform to Pasquill's definitions, although the methods of finding σ_x and σ_y are modified to be more suitable for this particular application. Automotive emission factors are available from the literature of the Environmental Protection Agency (Ref. 13). The Fortran listing of the program is given (Ref. 12).

The ability of the HIWAY model to predict pollutant concentrations very near to a four lane highway has been subjected to controlled, large scale, experimental tests at the proving grounds of the General Motors Corporation (Refs. 14 to 16). This study leads to a number of interesting conclusions—including the tendency of the HIWAY model to overpredict ground level concentrations at the highway's edge under stable atmospheric conditions. Traffic generated turbulence and the buoyancy caused by the heat of the exhaust effluent were believed to contribute to overestimation of surface level pollutant concentrations. The study suggests a revision of the stability classes and revised estimates of the dispersion coefficients as well for this type of problem. The practical limitations of the Gaussian plume method are illustrated by this study, which was concerned with pollutant concentrations at distances less than 100 m from the highway.

RAM is a computer program described by Novak and Turner (Ref. 17) for point and area sources for average concentrations up to 24 hr in duration. Plume rise from point sources is considered. Meteorological input data are wind direction and speed, stability class, and mixing height. The program uses the Gaussian plume model with a finite mixing height, and area sources are treated by a simplified method which is based upon a numerical integration along an upwind ray originating at the receptor.

APRAC is a program designed to give hourly average carbon monoxide concentrations in an urban location (Ref. 18). The program attempts to incorporate the influence of multistory buildings that form urban canyons which impede removal of pollutants.

CDM, the climatological dispersion model, provides estimates of either seasonal or annual pollutant concentrations at a ground level receptor (Ref. 19). An extensive source inventory and the appropriate climatological frequency matrix are required inputs. The model will give concentration isopleths resulting from an extensive array of point or area sources. The report describing the CDM gives an interesting discussion of the application of six different dispersion models to predict the long-term average SO_2 and suspended particulate material in the New York region.

10.7. THE GAUSSIAN PLUME MODEL COMPARED WITH OBSERVATIONS

The quality of the results of the application of the Gaussian plume model is strongly dependent upon the availability of accurate meteorological data in the vicinity of the source-receptor area. Even with this requirement fulfilled the model is subject to moderate uncertainties. For example, the vertical gradient of wind, $d\bar{u}/dz$, which the Gaussian plume does not take into account, is an important factor that influences the concentrations downwind of ground level sources. Ground level concentrations from elevated sources are sensitive to the value of the vertical dispersion coefficient which appears in the exponent and also in the denominator of the pre-exponential term as well. Thus Gifford (Ref. 3) quotes Pasquill as concluding that the concentrations of pollutants estimated by the Gaussian plume model "... may be accurate to within 20% for long-term averages, given good quality emissions and meterological data, but may exceed a factor of 2 for short-term values." A moderate literature is now available that permits a comparison between the predictions of the Gaussian plume model and the measured pollutant concentrations.

Short-term calculated average concentrations of SO_2 emitted primarily from large point sources were compared with observed values by Turner (Ref. 20), Smith and Singer (Ref. 21), Shieh et al. (Ref. 9), and Gibson and Peters (Ref. 22). Since SO_2 is converted in the atmosphere to sulfates, an exponential decay of this compound based on travel time from the point of release is often assumed. This is accomplished by multiplying the calculated concentration from each individual source by a chemical attenuation factor,

$$\exp\left(-\frac{0.693x}{\bar{u}\tau_{ch}}\right)$$

where τ_{ch} is the chemical half life of the compound and x is the downwind distance of the receptor as previously defined. A value of τ_{ch} for SO_2 at 25°C is approximately 2 hr under "urban" conditions and depends upon the extent to which other pollutants are present (Ref. 23). These studies are not easily summarized, but some individual results are worth noting. The analysis of Smith and Singer shows that about 67% of the predicted levels of SO_2 are to within a factor of 2 of the observed values. On the other hand, a few predictions are in error by a factor of 10, either too high or too low. Turner also found that 70% of the predictions were within a factor of 2 of the observed values under the proviso that "zero readings" were excluded from consideration. Shieh et al. calculated 6 and 24 hr averages of SO_2 at New York, and their results show excellent agreement with observations

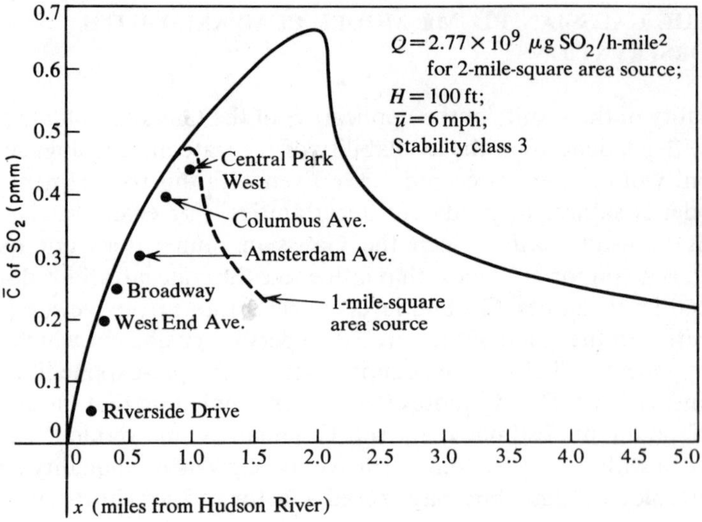

Figure 10.10. A comparison of SO_2 concentrations predicted by the area source double integration (Eq. 10.24) and observed values. The location is 79th Street, New York, on March 9, 1966 between 0615 and 0825 hr EST. [Reprinted with permission from L. J. Shieh and *I.B.M. J. Res. Develop.*, **16**, 164 (March 1972); copyright 1972 by International Business Machines Corporation.]

for the small number of cases that were reported. Their results for a square area source are shown in Fig. 10.10. The height of this area source is placed at 30.5 m above the ground which presumably represents the height of the releases from apartments, and so forth. Gibson and Peters' results for 342 separate comparisons of calculations and observations in Louisville are probably realistic of short-term predictions. They eliminated 32% of the data points on the basis of the questionable reliability of the air quality measurements; the remaining pairs of observed and predicted values show a correlation coefficient of 0.72. Gibson and Peters conclude that the necessary input data for the prediction are difficult to obtain and that wind direction is the most critical meteorological input. They conclude that a more sophisticated model is not warranted without improved quality of the input data. Short-term average values of CO in an urban topography were discussed by Dabberdt et al. (Ref. 18).

Long-term average predictions using the Gaussian plume model were discussed by Busse and Zimmerman (Ref. 19). Their predictions for the St. Louis area using the climatological dispersion model (CDM) strongly overpredict the SO_2 concentrations at 40 receptor locations in St. Louis.

This result is in disagreement with Pasquill's anticipated 20% accuracy for long-term averages. Three defects were suggested as possible contributors to the discrepancy. The emissions inventory is estimated and assumes constant emission rates. For example, daytime industrial emissions were projected to occur over a 24 hr period, which includes the nighttime periods of poor dispersion, rather than during the daytime periods of favorable dispersion characteristics. A second factor cited was the use of climatological data from a nearby airport rather than from the city itself. Finally, the method of estimating the height of the mixed layer was considered crude. This quantity was considered especially critical for clear nighttime conditions which represented 25% of the total observations, but contributed more to the calculated concentration than all other atmospheric conditions. The CDM program was used more successfully to describe SO_2 and particulate pollutants in New York City. This calculation is compared with other methods, most of them derived from the basic Gaussian plume model, in the Appendix of Ref. 19.

10.8 THE GAUSSIAN PUFF MODEL FOR INSTANTANEOUS RELEASES

The discussion throughout this chapter has been devoted to the description of the concentration downwind of a source emitting at a constant rate into a quasi-steady atmospheric flow. A description of an instantaneous or puff release is readily formulated if it is assumed that the centroid of the released mass is advected by the mean wind and that the concentration distribution is Gaussian. If the release occurs at an effective height H above an impenetrable ground plane, the concentration at the surface ($z = 0$) is given by

$$\bar{c}(x, y, 0) = \frac{2Q_{pf}}{(2\pi)^{3/2}\sigma_x\sigma_y\sigma_z} \exp\left[-\frac{(x - \bar{u}t)^2}{2\sigma_x^2} - \frac{y^2}{2\sigma_y^2} - \frac{H^2}{2\sigma_z^2}\right] \quad (10.29)$$

where Q_{pf} (gm) is the mass of material released; and σ_x, σ_y, and σ_z are the windward, crosswind, and vertical dispersion coefficients, respectively. In this equation, which is of identical form to Eq. 9.21, the source is located at the height H above the ground plane and an image source is placed at $z = -H$. Equation 9.21 was given as a solution for the hypothetical problem of anisotropic molecular diffusion in a laminar flow. Molecular diffusion, which results from the random motion of the gas molecules, is isotropic and the solution is actually useful in that context with $\sigma_x = \sigma_y = \sigma_z = 2\mathscr{D}_{AB}t$, where $\mathscr{D}_{AB}$ is the binary diffusion coefficient. Equation 10.29 applies to turbulent dispersion if the concentration distribution is Gaussian and the dispersion is anisotropic.

The values of $\sigma_y(x)$ and $\sigma_z(x)$ derived from steady source flow experiments (e.g., Fig. 10.2 or Table 10.2) are used in the puff solution. No specific information is available for the value of the windward dispersion coefficient σ_x and it is set equal to σ_y. The centroid position of the puff in steady wind-oriented coordinates is given by ($\bar{u}t$, 0, H), and all dispersion coefficients are calculated at the local value of travel distance $x = \bar{u}t$. The puff solution given above was used as the basis of an urban model for Chicago (Ref. 24) which incorporates the unsteadiness of the controlling meteorological features.

REFERENCES

1. F. A. Gifford, Jr., "Use of Routine Meteorological Observation for Estimating Atmospheric Dispersion," *Nuclear Safety*, **2**, 47–51 (1961).
2. F. Pasquill, *Atmospheric Diffusion* (Halsted Press of John Wiley & Sons, New York, 1974). Second Edition.
3. F. A. Gifford, "Turbulent Diffusion Typing Schemes: A Review," *Nuclear Safety*, **17**, 68–86 (1976).
4. J. Tadmor and Y. Gur, "Analytical Expressions for the Vertical and Lateral Dispersion Coefficients in Atmospheric Diffusion," *Atmos. Environ.*, **3**, 688–689 (1969).
5. D. B. Turner, "Workbook of Atmospheric Dispersion Estimates," Public Health Service Publication No. 999-AP-26 (N.T.I.S. PB-191482), Revised 1970.
6. K. L. Calder, "Multiple-Source Plume Models of Urban Air Pollution—The General Structure," *Atmos. Environ.*, **11**, 403–414 (1977).
7. R. J. Yamartino, Jr., "A New Method for Computing Pollutant Concentrations in The Presence of Limited Vertical Mixing," *J. Air Pollut. Control Assoc.*, **27**, 467–468 (1977).
8. M. E. Smith, and I. A. Singer, "An Improved Method of Estimating Concentration and Related Phenomena from a Point Source Emission," *J. Appl. Meteorol.*, **5**, 631–639 (1966).
9. L. J. Shieh et al., "Air Quality Diffusion Model; Application to New York City," *I.B.M. J. Res. and Dev.*, **16**, 162–170 (1972).
10. G. T. Csanady, "Crosswind Shear Effects on Atmospheric Diffusion," *Atmos. Environ.*, **6**, 221–232 (1972).
11. K. L. Calder, "On Estimating Air Pollution Concentrations from a Highway in an Oblique Wind," *Atmos. Environ.*, **7**, 863–868 (1973).
12. J. R. Zimmerman and R. S. Thompson, "User's Guide for HIWAY, A Highway Air Pollution Model," E.P.A. 650/4-74-008, U.S. Environmental Protection Agency, February 1975.
13. U.S.E.P.A., "Compilation of Air Pollutant Emission Factors," 2nd Edition, U.S. Environmental Protection Agency, Publication No. AP-42, April 1973.

14. S. H. Cadle et al., "General Motors Sulfate Dispersion Experiment: Experimental Procedures and Results," *J. Air Pollut. Control Assoc.*, **27**, 33–38 (1977).
15. D. P. Chock, "General Motors Sulfate Dispersion Experiment: Assessment of EPA HIWAY Model," *ibid.*, **27**, 46–51 (1977).
16. W. E. Wilson, "General Motors Sulfate Dispersion Experiments: Summary of E.P.A. Measurements," *ibid.*, **27**, 46–51 (1977).
17. J. H. Novak and D. B. Turner, "An Efficient Gaussian-Plume Multiple-Source Air Quality Algorithm," *J. Air Pollut. Control Assoc.*, **26**, 570–576 (1976).
18. W. F. Dabberdt, F. L. Ludwig, and W. B. Johnson, Jr., "Validation and Application of an Urban Diffusion Model for Vehicular Pollutants," *Atmos. Environ.*, **7**, 603–618 (1973).
19. A. D. Busse and J. R. Zimmerman, "User's Guide for the Climatological Dispersion Model," EPA-RH-73-024, U.S.E.P.A., Research Triangle Park, N. C., December 1973.
20. D. B. Turner, "A Diffusion Model for an Urban Area," *J. Applied Meteorol.*, **3**, 83–91 (1964).
21. M. E. Smith and I. A. Singer, "Application of a Multi-source System to an Urban Study," in *Proceedings of the Second International Clean Air Congress*, H. M. England and W. T. Beery, Editors (Academic Press, New York, 1971).
22. L. V. Gibson, Jr., and L. K. Peters, "A Short Term Air Quality Model for Sulfur Dioxide in Louisville Kentucky," *J. Air Pollut. Control Assoc.*, **27**, 218–223 (1977).
23. E. Mézáros, D. J. Moore, and J. P. Lodge, Jr., "Sulfur Dioxide-Sulfate Relationships in Budapest," *Atmos. Environ.*, **11**, 345–349 (1977).
24. J. J. Roberts, E. J. Croke, and A. S. Kennedy, "An Urban Dispersion Model," in Proceedings of Symposium on Multiple-Source Urban Diffusion Models, U.S.E.P.A., A.P.C.A. Pub. No. AP-86, Research Triangle Park, N. C. (1970).

PROBLEMS

10.1. Show that the total mass flow of a pollutant that is far downwind of a source, and which is confined by the ground plane and an elevated inversion, is equal to the source strength. That is, show that

$$\bar{u}\int_{z=0}^{H}\int_{y=-\infty}^{+\infty}\bar{c}(x, y, z)dydz = Q_p$$

for all values of x when $\bar{c}(x, y, z)$ is given by Eq. 10.15.

10.2. Use the approximation suggested by Yamartino (see text following Eq. 10.14) to show that Eq. 10.13 reduces to Eq. 10.15 for large values of downwind distance.

10.3. Derive an expression for the concentration at a receptor directly downwind of a line source of length x_1 which is perfectly aligned with the mean wind vector (see sketch). Both the source and the receptor are located at the ground level and the source strength is Q_l(gm/m s).

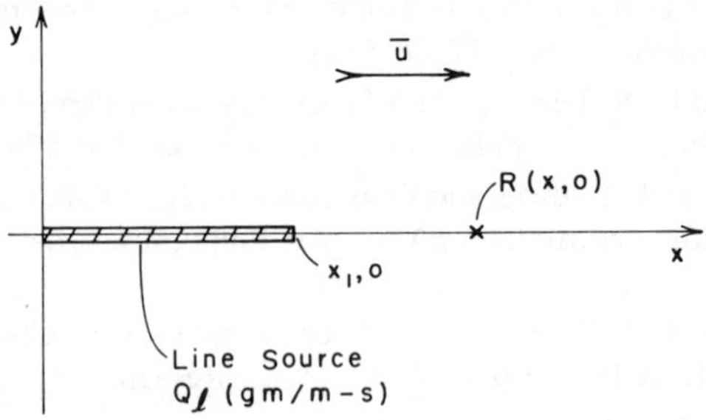

10.4. Derive an equation for the concentration at a receptor $R(x, 0)$ downwind of a series of three adjoining infinite strip sources of varying strengths that are normal to the mean wind (see sketch). The receptor is on the downwind edge of the third source.

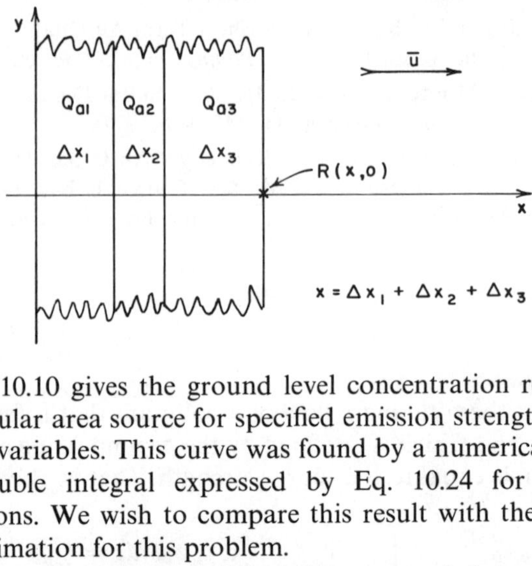

10.5. Figure 10.10 gives the ground level concentration resulting from a rectangular area source for specified emission strength and meteorological variables. This curve was found by a numerical evaluation of the double integral expressed by Eq. 10.24 for the particular conditions. We wish to compare this result with the narrow plume approximation for this problem.

1. What equations apply when the receptor is internal and external to the rectangular source in the narrow plume approximation?
2. Perform the numerical integration necessary to compare $\bar{c}$ at $x = 2.0$ miles with the results given in Fig. 10.10. To permit a comparison of the results, it is suggested that the numerical integration be carried out by dividing x into five equal increments, and

that all quantities be first converted to metric units. Assume that dispersion class 3 is the same as Pasquill class C.

10.6. The effective stack height of an electric generating station is 300 m and the unit emits 2800 g/s of SO_2 when coal containing 2.3% sulfur is used as a fuel.

1. Find the ground level concentration of SO_2 at a distance 10.0 km downwind when the stability class is D corresponding to nighttime conditions with partial cloudiness. The wind speed is 8.5 m/s.
2. Find the ground level concentration 10.0 km downwind when the stability class is A corresponding to strong convective conditions and when an elevated inversion is present at a height of 600 m. The wind speed is now 1.95 m/s. At this downwind distance there is perfect mixing in the vertical, $0 < z < 600$ m, and turbulent transport occurs in the crosswind direction.

10.7. A long highway, which runs in a north-south direction, has a traffic volume (TV) of 1000 vehicles/hr traveling in both directions. The CO emission factor (EF) for the 1975 mix of traffic is 60 gm/vehicle-mile. The line source emission strength is then given by

$$Q_l = 0.174\,(EF)(TV) = 10{,}440\ \mu\text{g/m s}$$

1. Calculate the CO concentration 1 km east of the highway if the wind speed is 5 m/s from the west when the sky is overcast.
2. Calculate the CO concentration 1 km east of the highway if the wind speed is 5 m/s from the northwest under similar atmospheric conditions.

11

DISPERSION OF POLLUTANTS—REACTING COMPONENTS AND UNSTEADY FLOWS

The Gaussian plume model discussed in Chapter 10 requires that the atmospheric flow field be independent of time and that the pollutant be essentially nonreactive. Clearly the method will not be applicable if the time-dependent evolution of the mixed layer in the morning hours influences pollutant concentrations or if the pollutant is of secondary origin. In this chapter we discuss methods that have the potential of incorporating the full detail of the time-dependent meteorological features, possible unsteady source emission rates, and also the chemical interactions that create or consume the pollutant in question. The application of these methods in the most general cases will require a large amount of input data. The methods can, however, be applied to more restricted problems in order to illustrate their properties or to facilitate comparisons of their predictions with one another.

11.1. THE ATMOSPHERIC DISPERSION EQUATION (K THEORY)

In Chapter 9 we derived the basic conservation equation, Eq. 9.28, for the ith chemical species in a reacting turbulent mixture as

$$\frac{\partial \bar{c}_i}{\partial t} + \bar{u}\frac{\partial \bar{c}_i}{\partial x} + \bar{v}\frac{\partial \bar{c}_i}{\partial y} + \bar{w}\frac{\partial \bar{c}_i}{\partial z} = \frac{\partial(-\overline{u'c_i'})}{\partial x} + \frac{\partial(-\overline{v'c_i'})}{\partial y} + \frac{\partial(-\overline{w'c_i'})}{\partial z} + \bar{R}_i + S_i \tag{11.1}$$

This equation was derived by time-averaging the diffusion equation that applies to the ith component of a mixture of pollutants in atmospheric air.

We have added in Eq. 11.1 a source term, $S_i = S_i(x, y, z, t)$, which represents the strength of the various point, line, and area sources. On some occasions the sources are included in the boundary conditions rather than through a term in the equation itself. Molecular diffusion represented by terms such as $\mathscr{D}_{im}\, \partial \bar{c}_i/\partial x$ has been neglected in comparison with mass transport by the turbulent Reynolds dispersion terms, $-\overline{u'c'_i}$, and so on. The chemical production rate $\bar{R}_i$ is dependent upon the concentrations of the various components that produce or destroy the ith species, the reaction rate constants, and other quantities that affect the net production rates. Even if a full prescription of boundary conditions, the mean wind components, and the reaction rate terms were to be provided, Eq. 11.1 cannot be solved in its present form because it contains the unknown concentration $\bar{c}_i$ and also the three Reynolds dispersion terms, $\overline{u'c'_i}$, $\overline{v'c'_i}$, and $\overline{w'c'_i}$, which are also unknown functions of position and time. The presence of the unknown Reynolds dispersion terms in the equation for the species conservation provides another illustration of the closure problem of turbulence. In the following subsections we discuss a widely used method of closure and some solutions that result from its adoption.

11.1.1. The Gradient-Transport Hypothesis and the Atmospheric Dispersion Equation

The Reynolds equations for the flow in the atmospheric boundary layer (Section 6.3.2) provided our first encounter with the closure problem. In that case, the turbulent shear stress was postulated to be proportional to the gradient of the mean velocity. This concept—the proportionality of the turbulent flux of a quantity being transferred to the gradient of the time-averaged value of the property that serves as the driving force—yields a set of commonly used rudimentary closure relationships for the transport of momentum, heat, and mass in a turbulent flow. These relationships are suggested by corresponding expressions in laminar flow where momentum, heat, and mass will be transferred by molecular processes if gradients of velocity, enthalpy, or mixture ratio, respectively, are present. The constants of proportionality in these cases have been defined by Eqs. 6.56, 6.57, and 9.2 and they are summarized in Table 11.1. The kinematic viscosity, the thermal diffusity, and the binary diffusion coefficient, which may be thought of as the molecular diffusivities for the transfer of momentum, heat, and mass, can be determined for any gas and are known to depend on its pressure and temperature. Just as we have written the gradient-transport hypotheses in Chapter 6 for turbulent momentum and heat transfer, we may also write a

Table 11.1. Analogs in Momentum, Heat, and Mass Transfer by Molecular Diffusion and Turbulent Dispersion

Molecular Transfer Process	*Constitutive Relationship* (one-dimensional process)
Momentum (kinematic viscosity, ν)	$-\tau_{zx} = -\rho\nu \dfrac{du}{dz}$ (Eq. 6.56)
Heat (thermal diffusivity, α_h)	$q_z = -\rho\alpha_h \dfrac{d(c_p T)}{dz}$ (Eq. 6.57)
Mass (binary diffusion coefficient, $\mathscr{D}_{im}$)	$j_i = -\rho\mathscr{D}_{im} \dfrac{d}{dz}\left(\dfrac{c_i}{\rho}\right)$ (Eq. 9.2)

$\dfrac{\nu}{\alpha_h} \equiv$ Prandtl No., Pr. For air at STP, $Pr = 0.72$

$\dfrac{\nu}{\mathscr{D}_{im}} \equiv$ Schmidt No., Sc. For O_2 in air at STP, $Sc = 0.81$

Turbulent Transfer Process	*Gradient-Transport Hypothesis*
Momentum (eddy kinematic viscosity, K_m)	$-\tau^T_{zx} = -\rho K_m \dfrac{d\bar{u}}{dz}$ (Eq. 6.58)
Heat (eddy diffusivity of heat, K_h)	$q^T_z = -\rho K_h \dfrac{d(c_p \bar{T})}{dz}$ (Eq. 6.59)
Mass (turbulent diffusivity, K_{zz})	$j^T_{iz} = -\rho K_{zz} \dfrac{d}{dz}\left(\dfrac{\bar{c}_i}{\rho}\right)$ (Eq. 11.2)

$\dfrac{K_m}{K_h} \equiv$ Turbulent Prandtl No., Pr^T $Pr^T \simeq 0.7$ (experiment)

$\dfrac{K_m}{K_{zz}} =$ Turbulent Schmidt No., Sc^T $Sc^T \simeq 1.0$

corresponding one-dimensional relationship for turbulent mass transfer of species i as

$$j^T_{iz} = +\overline{w'c_i} = -\rho K_{zz} \frac{d}{dz}\left(\frac{\overline{c_i}}{\rho}\right) \tag{11.2}$$

where j^T_{iz} is the average flow in the z direction of the ith component per unit area owing to the turbulent flow field, and K_{zz} is called the turbulent diffusivity.

Thus, the quantities K_m, K_h, and K_{zz} (Table 11.1) may be thought of as the diffusivities for turbulent transfer of momentum, heat, and mass. They are properties of the fluid and also the particular flow field as well and, unlike their laminar flow analogs, their values cannot be derived from any theory of the microscopic structure of the fluid substance.

We will assume that the density of air is constant in the atmospheric boundary layer where Eq. 11.1 is to be applied. The gradient-transport hypotheses for the three Reynolds dispersion terms are then expressed, as suggested by Eq. 11.2 with ρ = constant, by

$$-\overline{u'c_i'} = +K_{xx}\frac{\partial \bar{c}}{\partial x} \qquad -\overline{v'c_i'} = +K_{yy}\frac{\partial \bar{c}}{\partial y} \qquad -\overline{w'c_i'} = +K_{zz}\frac{\partial \bar{c}}{\partial z} \tag{11.3}$$

The closure of the equation for species conservation will be accomplished only if the three components of turbulent diffusivity, K_{xx}, K_{yy}, and K_{zz}, can be expressed as prescribed functions of space or of mean flow properties.

The above relations imply that the turbulent diffusivity has three components, but the more careful formulation given by Calder (Ref. 1) demonstrates that it is a second order tensor with nine components. Thus, for example, the x component of the Reynolds dispersion terms is more rigorously given by

$$-\overline{u'c_i'} = +K_{xx}\frac{\partial \overline{c_i}}{\partial x} + K_{xy}\frac{\partial \overline{c_i}}{\partial y} + K_{xz}\frac{\partial \overline{c_i}}{\partial z}$$

However, no information on the off-diagonal terms such as K_{xy} and K_{xz} is available, and it is customary to assume that they are zero. Furthermore, we usually assume the horizontal eddy diffusivities are equal, $K_{xx} = K_{yy}$. The equation for turbulent dispersion then becomes

$$\frac{\partial \overline{c_i}}{\partial t} + \bar{u}\frac{\partial \overline{c_i}}{\partial x} + \bar{v}\frac{\partial \overline{c_i}}{\partial y} + \overline{w}\frac{\partial \overline{c_i}}{\partial z} = \frac{\partial}{\partial x}\left(K_y\frac{\partial \overline{c_i}}{\partial x}\right) + \frac{\partial}{\partial y}\left(K_y\frac{\partial \overline{c_i}}{\partial y}\right) + \frac{\partial}{\partial z}\left(K_z\frac{\partial \overline{c_i}}{\partial z}\right) + \overline{R_i} + S_i \tag{11.4}$$

where K_y and K_z designate the vertical and horizontal components of the turbulent diffusivity. This equation, or its minor variants, are often referred to as the atmospheric dispersion equation, or K theory. In view of the hypothetical nature of the closure relationship and the other approximations involved, Eq. 11.4 must be regarded as an approximate, semi-empirical expression. The appropriateness of the atmospheric dispersion equation is to be judged by its ability to yield predictions that agree either with observations under a variety of conditions, or with solutions to a series of problems obtained by another method of known veracity. A detailed description of the conditions that are necessary for the validity of Eq. 11.4 is given by Reynolds et al. (Ref. 2).

The mean wind components $\bar{u}$, $\bar{v}$, $\bar{w}$ in Eq. 11.4 are prescribed functions of space and time (x, y, z, t). Their values are based upon observations if pollution concentrations that existed in the past are sought. Predicted wind components and predicted source strengths must be used if future pollutant concentrations are being estimated. The initial concentration $\bar{c}_i(x, y, z, 0)$ is a required initial condition and even at sunrise it usually will be finite as a result of advection of pollutants generated during the previous day. Finally, certain boundary conditions must be prescribed at the surfaces that enclose the volume within which the above equation is applied. From Eq. 11.2 with density constant we find the average flux of component i resulting from turbulent dispersion is given by $j_{iz}^T = -K_z(\partial\bar{c}_i/\partial z)$. Therefore an impenetrable surface, such as the ground or a strong elevated inversion, will constitute a plane along which $\partial\bar{c}_i/\partial z = 0$. An area source of strength Q_a gives rise to a boundary condition of the type $Q_a = -K(\partial\bar{c}_i/\partial z)$ along a specified surface. Far from the source the concentration will vanish, $\bar{c}_i \to 0$, at a particular limiting value of a coordinate. The application of some of these boundary conditions is illustrated by several specific solutions given below.

Equation 11.4 can in principle be used to calculate the concentration of the ith species of a mixture of pollutants if the velocity field is known or has been predicted. For photochemical pollutants these solutions will involve a number of species and a corresponding system of simultaneous equations which are linked to one another through the chemical reaction terms, and numerical methods will be required for their solution. A particular example of a calculation of this type for photochemical air pollution is provided in Refs. 2, 3, and 4. One difficulty encountered in representing Eq. 11.4 in finite difference form is that the approximations involved in this process can create an apparent mixing to occur even when the diffusive terms are set equal to zero. This effect is termed numerical pseudodiffusion, and it can cause apparent diffusion that is as large as turbulent dispersion itself. The suppression of numerical pseudodiffusion is one of several important considerations that influence the choice of the specific numerical procedure. A comparison of several numerical methods commonly used for the solution of the atmospheric dispersion equation is given in Refs. 5 and 6. Further reference to the numerical methods is made in Section 11.1.5 and in the references cited therein.

The components of the turbulent diffusivity can be related, as shown by Csanady (Ref. 7) and also in Problem 9.1, to the mean-square particle displacements by

$$K_y = \frac{1}{2}\frac{d(\sigma_y^2)}{dT} \qquad \text{and} \qquad K_z = \frac{1}{2}\frac{d(\sigma_z^2)}{dT} \tag{11.5}$$

where T is the particle travel time. These relationships have been used to derive values of the turbulent diffusivity from observations of the motion of a finite number of neutrally buoyant particles in a turbulent flow field. The earlier discussion of the mean-square displacement of particles in a turbulent flow (Section 9.4) enables us to relate K_y and K_z to the properties of the turbulent flow which is homogeneous and isotropic. Thus, if the travel time of the particle is long compared with the Lagrangian time scale ($T \gg \tau_L$), then we find

$$K_y = \overline{v'^2}\tau_{LY} \qquad K_z = \overline{w'^2}\tau_{LZ}$$

These relationships do not apply within the atmospheric boundary layer where the turbulence is anisotropic and inhomogeneous. The determination of the turbulent diffusivities for use in atmospheric flows is discussed below.

11.1.2. The Turbulent Diffusivity

Useful information on the nature of the diffusivities for transport processes is gained by considering their ratios which are also summarized in Table 11.1. For example, the ratio of kinematic viscosity to the diffusivity of heat by molecular processes, ν/α_h, is defined to be the Prandtl number, and its value for air at standard conditions is well known by theory and experiments to be near 0.71. The ratio $\nu/\mathscr{D}_{im}$ is the Schmidt number which depends upon the composition of the binary mixture. For oxygen diffusing through air, a case in which the molecular weights of the two components are nearly equal, the Schmidt number is 0.81.

The corresponding quantities in turbulent flow are the turbulent Prandtl number, $Pr^T = K_m/K_h$ and the turbulent Schmidt number, $Sc^T = K_m/K_z$ (see Table 11.1). These quantities, like the turbulent diffusivities, cannot be derived from a theory of the microscopic structure of matter. Experiments do show that, although the magnitude of the diffusivities does vary widely, their ratios are relatively constant. For example, experiments conducted in the laboratory and also in the atmosphere (Plate, Ref. 8) give $Pr^T = 0.74$ for neutral stability. The turbulent Schmidt number is customarily assumed to be unity under neutral conditions. Through this assumption we are able to derive the turbulent diffusivity from our knowledge of the velocity profile and its derivatives.

A layer of constant shear stress, such as the earth's surface layer under neutral stability, displays a logarithmic velocity profile (see Section 6.2.3),

$$\frac{\bar{u}}{u_*} = \frac{1}{\kappa}\ln\frac{z}{z_0} \tag{11.6}$$

where u_*^2 is the surface level shear stress divided by the density, κ is the von Karman constant usually taken to be 0.4, and z_0 is the aerodynamic roughness. Evaluating $d\bar{u}/dz$ from the logarithmic profile we find from Eq. 6.58 (see Table 11.1) that K_m varies linearly with height and is given by

$$K_m = \kappa u_* z \tag{11.7}$$

The presumption that the turbulent Schmidt number is unity therefore yields the vertical turbulent diffusivity as

$$K_z = \kappa u_* z \tag{11.8}$$

This relationship must be restricted to a limited range of height. The flow field and the value of K_z will depend on the specific geometry of the topographic features of the surface when z is smaller than the height of the roughness elements. At higher levels the wind vector turns with increasing altitude as a result of the combined influence of the Coriolis, pressure, and frictional forces. Equation 11.8 is applicable in the intermediate stratum between these two layers and is often also applied as a fictitious description of the turbulent diffusivity within the roughness elements.

The friction velocity u_* is determined from the Rossby number similarity theory for the neutral atmospheric boundary layer (Section 6.3.4). The results of this theory can be approximated by a simple arithmetic relationship for the geostrophic drag coefficient u_*/q_G of the type suggested by Lettau (Ref. 9),

$$\frac{u_*}{q_G} = \frac{0.16}{\log_{10} \text{Ro} - 1.8} \tag{11.9}$$

where the surface Rossby number is given by

$$\text{Ro} = \frac{q_G}{z_0 f}$$

and q_G is the geostrophic wind and f is the Coriolis parameter. The aerodynamic roughness is estimated to range from 0.01 to 0.1 m in open fields, 0.1 to 1.0 m in forested country, and 1.0 to 10 m in urban areas (see results quoted by Plate, Ref. 8). The geostrophic drag coefficient has often been observed in meteorological experiments. A favorable comparison between the various observations reported by Blackadar (Ref. 10) and Eq. 11.9 is displayed in Fig. 11.1.

The above description of the turbulent diffusivity in the surface layer where the shear stress is constant under neutral conditions is well documented and widely accepted. Unfortunately our knowledge of the behavior of

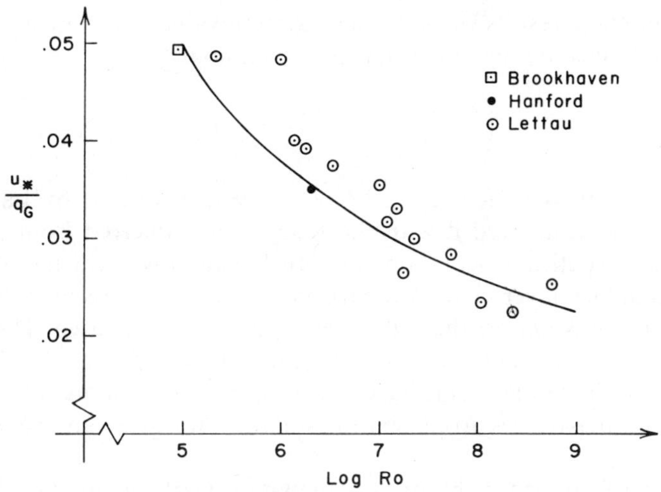

Figure 11.1. The geostrophic drag coefficient u_*/q_G versus the surface Rossby number Ro. The solid line is Eq. 11.9. [Observed data points from A. K. Blackadar, *J. Geophys. Res.*, **67**, 3098 (1962), copyrighted by the American Geophysical Union. Reproduced with permission of the latter and A. K. Blackadar.]

this quantity in the remainder of the atmospheric boundary layer is quite limited. The generally accepted view is that above the surface layer, K_z must (a) agree asymptotically with Eq. 11.8, (b) have a maximum at some level above the surface layer, and (c) decrease to a value appropriate for the atmosphere at the level of the geostrophic wind. A relationship that fulfills these requirements proposed by Shir and Shieh (Ref. 11) is

$$K_z = \kappa u_* z \exp\left(-\frac{z}{h_{\max}}\right) \tag{11.10}$$

where $h_{\max}$ is the level at which K_z has its maximum value. Shir and Shieh set $h_{\max} = H_b/4$.* The dimensionless height of the neutral boundary layer H_b is expressed by Clarke (Ref. 12), based on observations made over flat terrain in Australia, as

$$\frac{H_b f}{u_*} = 0.20 \tag{11.11}$$

* Many of the numerical values that are cited in this section are those which commonly appear in the literature at the time of publication of this text. Their values are expected to be refined as further research improves our knowledge of the properties of the atmospheric boundary layer.

Combining these results we find the maximum value of turbulent diffusivity at $z = H_b/4$ given by the dimensionless expression

$$\left(\frac{K_z f}{u_*^2}\right)_{z=h\ \max} = 7.36 \times 10^{-3} \tag{11.12}$$

The above expression for $K_z(z)$, Eq. 11.10, was evaluated by Lamb et al. (Ref. 13), who compared it with the $K_z(z)$ profile inferred from a detailed numerical theoretical calculation of the turbulent flow in the neutral atmospheric boundary layer. This comparison showed fair agreement if a larger value of the constant on the right side of Eq. 11.11 was used. The detailed form of $K_z(z)$ may undergo some adjustment as additional information becomes available. The currently accepted methods of modifying K_z to account for departures from neutral stability are given in the references cited above.

The above formulae enable one to calculate $K_z(z)$ throughout the boundary layer once the geostrophic wind q_G and the aerodynamic surface roughness z_0 are prescribed. For example, if q_G is the representative value of 15 m/s and z_0 is 1.0 m typical of the urban topography, then Ro is 150,000 (for $f = 10^{-4}\text{s}^{-1}$). From the above relationships, we find, with $\kappa = 0.40$, that $K_z = 0.284\,z$ within the surface layer (MKS units) and $K_{z,\max} = 37.2\text{m}^2/\text{s}$. On the other hand, if $q_G = 5.0$ m/s corresponding to a stagnation condition, then $K_{z,\max}$ is reduced to 5.61 m^2/s (see Table 11.2 below). Clearly the geostrophic wind, which controls u_* and H_m as well as K_z, exerts a dominant role in the removal of pollutants emitted near the surface.

11.1.3. Solutions to the Simplified Atmospheric Dispersion Equation

Equation 11.4 can be simplified for certain cases of practical interest. For steady unidirectional flow over a flat terrain in combination with steady pollutant emission rates, we choose the x coordinate to correspond to the direction of the wind and set $\bar{v}$, $\bar{w}$, and $\partial\bar{c}/\partial t$ equal to zero. Often the transport in the x direction by the wind is dominant over turbulent dispersion in the x direction so that the $\partial\bar{c}/\partial x$ term can be ignored. We restrict the pollutants of interest to chemically inert substances and include the source terms in the boundary conditions. The atmospheric dispersion equation reduces to

$$\bar{u}\frac{\partial\bar{c}}{\partial x} = \frac{\partial}{\partial y}\left(K_y\frac{\partial\bar{c}}{\partial y}\right) + \frac{\partial}{\partial z}\left(K_z\frac{\partial\bar{c}}{\partial z}\right) \tag{11.13}$$

Some solutions are available for the case in which $K_y = K_z =$ constant throughout space. However, the variation of K_z with the height above the ground is essential to the realistic description of turbulent dispersion near the surface of the earth. The vertical structure of the atmosphere can be approximated by power law expressions for the vertical profiles of velocity and turbulent diffusivity,

$$\bar{u}(z) = c_u z^m \qquad \text{and} \qquad K_z = c_{kz} z^n \tag{11.14}$$

These equations may be specialized to the practical forms

$$\bar{u}(z) = \bar{u}(z_1)\left(\frac{z}{z_1}\right)^m \qquad \text{and} \qquad K_z = \kappa u_* z \tag{11.15}$$

where m is typically in the range of 0.2 to 0.6 and n has been set equal to unity as appropriate for the constant shear stress layer. Comparing Eqs. 11.14 and 11.15 we find

$$c_u = \frac{\bar{u}(z_1)}{z_1^m} \qquad \text{and} \qquad c_{kz} = \kappa u_* \tag{11.16}$$

Thus for the light wind illustrative example cited above ($q_G = 5$ m/s, $z_0 = 1$ m, and $f = 10^{-4}\text{s}^{-1}$), we find, if we choose $m = 1/3$ (see Table 11.3 below) and determine u_1 at $z_1 = 40$ m from the logarithmic velocity profile Eq. 11.6, $c_u = 0.744\, m^{2/3}/\text{s}$ and $c_{kz} = 0.110$ m/s with $n = 1.0$. The properties of the neutral atmospheric boundary layer for the two illustrative cases cited above are summarized in Table 11.2.

A further simplification of Eq. 11.13 is necessary to afford closed-form solutions from K theory that incorporate the vertical structure of the atmosphere as described by Eqs. 11.14. This is achieved by restricting the source geometry to be of infinite extent in the crosswind direction. In this case the crosswind diffusion term $\partial\bar{c}/\partial y$ in Eq. 11.13 also vanishes, and the equation reduces to simply a balance between advection in the x direction and vertically directed turbulent dispersion,

$$\bar{u}\,\frac{\partial \bar{c}}{\partial x} = \frac{\partial}{\partial z}\left(K_z\,\frac{\partial \bar{c}}{\partial z}\right) \tag{11.17}$$

In this case the mean concentration is a function of two independent variables, $\bar{c} = \bar{c}(x, y)$. Several potentially useful solutions for this restricted form of the atmospheric dispersion equation are given in the following subsections. Solutions for sources of infinite crosswind geometry are often applicable to finite sources because of the narrow plume hypothesis that was discussed in Chapter 10.

Table 11.2. Typical Values of Properties of a Neutral Atmospheric Boundary Layer over Urban Topography ($z_0 = 1.0$ m, $f = 10^{-4}\ s^{-1}$)

	Low Wind Condition	Average Wind Condition
q_G (m/s)	5.0	15.0
Ro	50,000	150,000
u_* (m/s)	0.276	0.711
H_b (m)	552	1420
τ_{z0} (N/m^2)	0.093	0.619
$\bar{u}_1$ ($z_1 = 40$ m) (m/s)	2.54	6.56
$\bar{u}(z)$ (m/s)	$0.744z^{1/3}$	$1.92z^{1/3}$
K_z in surface layer (m^2/s)	$0.110z$	$0.284z$
$K_{z,\max}$ (m^2/s)	5.61	37.2
$h_{\max}$ (m)	138	355

Notes: 1. The exponent for the power law velocity profile is $m = \frac{1}{3}$.
2. A typical roughness height is about $30z_0$. The velocity $\bar{u}_1$ was calculated at $z_1 = 40$ m, about 10 m above the roughness elements, from the logarithmic law, Eq. 11.6.
3. The density was assumed to be 1.225 Kg/m^3 for purposes of calculating the shear stress at the surface $\tau_{z0} = \rho u_*^2$.
4. The above values are quoted to 3 significant figures for numerical consistency although this accuracy is not justified for quantities that are dependent upon a measured wind speed.

11.1.3.1. The K *Theory Solutions for an Infinite Line Source*

An unpublished exact solution for an infinite steady line source at the surface was given by Roberts (see Calder, Ref. 14). The boundary conditions are

$$
\begin{aligned}
\bar{c}(x, z) &\to 0 \qquad \text{as} \qquad x, z \to \infty \\
\bar{c}(x, z) &\to \infty \qquad \text{at} \qquad x = z = 0 \\
K_z \frac{\partial \bar{c}}{\partial z} &= 0 \qquad \text{for} \qquad z = 0, x > 0 \qquad (11.18) \\
\int_0^\infty \bar{c}(x, z)\, &\bar{u}(z) dz = Q_l
\end{aligned}
$$

where Q_l is the line source strength (gm/m s). The third statement asserts that no transport of pollutant occurs through the ground plane. The last boundary condition expresses the conservation of the component at all downwind locations. The solution for $r = m - n + 2 > 0$ is

$$\bar{c}(x, z) = \frac{Q_l r}{\Gamma(s) c_u} \left(\frac{c_u}{c_{kz} r^2 x} \right)^s \exp\left(- \frac{c_u z^r}{c_{kz} r^2 x} \right) \tag{11.19}$$

where $s = (m + 1)/r$ and $\Gamma(s)$ is the Gamma function. The latter is a widely tabulated function that is defined by

$$\Gamma(s) = \int_0^\infty e^{-x} x^{s-1} dx$$

and possesses the property that $\Gamma(s + 1) = s!$ if s is a positive integer.

For the typical values of practical interest, $m = 1/3$ and $n = 1$, Eq. 11.19 reduces to

$$\bar{c}(x, z) = \frac{3Q_l}{4c_{kz} x} \exp\left(- \frac{9c_u z^{4/3}}{16 c_{kz} x} \right) \tag{11.20}$$

In this case we find that K theory predicts that the surface concentration $\bar{c}(x, 0)$ varies inversely as the downwind distance from the source.

A solution given by Smith (Ref. 15) for the infinite line source at height h_l above the ground, for m arbitrary but n restricted to the value $1 - m$, is

$$\bar{c}(x, z) = \frac{Q_l (z h_l)^{m/2}}{t c_{kz} x} \exp\left[- \frac{c_u (z^t + h_l^t)}{t^2 c_{kz} x} \right] I_{-p} \left[\frac{2 c_u (z h_l)^{t/2}}{t^2 c_{kz} x} \right] \tag{11.21}$$

where $t = 2m + 1$, $p = m/t$ and I_{-p} is the modified Bessel function of the first kind of order $-p$. The latter is tabulated for certain values of p and has the asymptotic properties

$$I_{-p}(x) \sim \frac{\left(\frac{2}{x}\right)^p}{\Gamma(1 - p)} \qquad \text{for} \qquad x \to 0$$

and

$$I_{-p}(x) \sim \frac{e^{+x}}{\sqrt{2\pi x}} \qquad \text{for} \qquad x \to \infty \tag{11.22}$$

Unfortunately the important illustrative case of $m = 1/3$, $n = 1$ cannot be recovered from Eq. 11.21. On the other hand, under stable stratification the value of m becomes small, typically 0.1 to 0.15 for smaller roughness heights. In this event the value $m = 0$ (hence $n = 1$) is of interest and we find, from Eq. 11.21,

$$\bar{c}(x, z) = \frac{Q_l}{c_{kz} x} \exp\left[- \frac{c_u (z + h_l)}{c_{kz} x} \right] I_0 \left[\frac{2 c_u (z h_l)^{1/2}}{c_{kz} x} \right] \tag{11.23}$$

The latter expression reduces, when $h_l \to 0$, to the corresponding specialized form of Eq. 11.19.

11.1.3.2. The K Theory Solution for the Semi-Infinite Area Source—Similarity Solution

A solution to the atmospheric dispersion equation in the form of Eq. 11.17 is available for an area source of strength Q_a (gm/m^2 s) consisting of a semi-infinite plane surface at $z = 0$ for $x > 0$. Thus the leading edge of the source is the y axis which is normal to the wind. The boundary conditions in this case are

$$\bar{c}(x, z) = 0 \qquad \text{for } x = 0, z > 0$$
$$\bar{c}(x, z) = 0 \qquad \text{for } z \to \infty \tag{11.24}$$

and

$$-K_z \frac{\partial \bar{c}}{\partial z} = +Q_a \qquad \text{at } z = 0 \qquad \text{for } x > 0$$

The wind speed and eddy diffusivity are given by the power law expressions Eqs. 11.14. Lebedeff and Hameed (Ref. 16) reduced the partial differential Eq. 11.17 to an ordinary differential equation by the adoption of a similarity variable containing both dependent variables x and z.

The solution for the ground level concentration is, for $n < 1$,

$$\bar{c}(x, 0) = \frac{Q_a}{c_{kz}} \frac{r^{2\nu - 1}}{\nu\Gamma(1 - \nu)} \left(\frac{c_{kz}x}{c_u}\right)^{\nu} \tag{11.25}$$

with $\nu = (1 - n)/r$ and r is defined as previously. The concentration at the surface tends to infinity if $n = 1$ and $\nu = 0$ because the area source is then located at the surface where the turbulent diffusivity vanishes. Nevertheless, the solution may be of interest for a case when ν can be considered less than unity.

11.1.3.3 The Semi-Infinite Area Source—Integral Method of Solution

A solution for the semi-infinite area source was also given by Lebedeff and Hameed (Ref. 17) using an integral solution. This method of solution has been widely exploited in the analysis of heat transfer by conduction and convection to give approximate solutions to many problems that cannot be solved by more vigorous methods. The expression for the turbulent diffusivity adopted by Lebedeff and Hameed for the integral solution is

$$K_z = K_{z0} + c_{k0}z^n \tag{11.26}$$

and they used the conventional power law expression for the wind profile.

In the integral method of analysis the algebraic form of the vertical dependence of the concentration is prescribed somewhat arbitrarily. Thus Lebedeff and Hameed (Ref. 17) assumed that the concentration profile is of the form

$$\bar{c}(x, z) = \bar{c}(x, 0)\left(1 - \frac{z}{H_p}\right)^3 \tag{11.27}$$

where H_p is a function of x and represents the height of the polluted layer resulting from the area source. In terms of a Gaussian plume model, H_p represents the height at which the pollutant concentration is reduced to a small fraction, say 1%, of its ground level value. In the integral method the pollutant is assumed to be confined totally within the layer $0 < z \leq H_p$ for the purpose of calculating the surface distribution $\bar{c}(x, 0)$. At the upwind edge of an area source, the height of the pollutant layer is normally substantially less than the height of the mixed layer. In view of Eq. 11.27 and the last boundary condition in Eqs. 11.24 we find

$$\bar{c}(x, 0) = \frac{H_p(x)Q_a}{3K_{z0}} \tag{11.28}$$

The surface pollutant concentration is then specified once $H_p(x)$ is determined. A peculiarity of this solution revealed by Eq. 11.28 is that $\bar{c}(x, 0)$ depends only on K_{z0} and is independent of c_{K0} and n that appear in Eq. 11.26.

By twice integrating both sides of the atmospheric dispersion equation in the form of Eq. 11.17 and using the last boundary condition given in Eq. 11.24, we find the integral form of the conservation equation as (see Problems 11.1 and 11.2)

$$\int_0^{H_p} \bar{u}(z)\bar{c}(x, z)dz = \int_0^x Q_a dx \tag{11.29}$$

At this point, Q_a is assumed to be constant and Eq. 11.28 and the velocity profile are inserted into the left side of Eq. 11.29 which is integrated to yield the height of the polluted layer as

$$H_p(x) = \left[\frac{3K_{z0}x}{B(m + 1, 4)c_u}\right]^{1/(m+2)} \tag{11.30}$$

Here, $B(p, q)$ represents the Beta function, an integral which for $p, q > 0$ is defined by

$$B(p, q) = \int_0^1 t^{p-1}(1 - t)^{q-1}dt$$

The Beta function is evaluated readily by the identity relating it to the Gamma function

$$B(p, q) = \frac{\Gamma(p)\Gamma(q)}{\Gamma(p + q)}$$

Since $\Gamma(n + 1) = n\Gamma(n)$ for $n > 0$ and $\Gamma(n + 1) = n!$, if n is a positive integer, the constant term within the bracket of Eq. 11.30 is found to be

$$\frac{3}{B(m + 1, 4)} = \frac{(m + 1)(m + 2)(m + 3)(m + 4)}{2}$$

Thus for a typical value of $m = 1/3$ the constant in Eq. 11.30 is

$$\frac{3}{B(4/3, 4)} = 22.5$$

We find the surface concentration by combining Eqs. 11.28 and 11.30 to obtain

$$\bar{c}(x, 0) = \frac{Q_a}{3K_{z0}} \left[\frac{3K_{z0}x}{B(m + 1, 4)c_u} \right]^{1/(m+2)} \tag{11.31}$$

The solution for $\bar{c}(x, z)$ in this case does not satisfy the governing partial differential Eq. 11.17, but it does comply with an integrated form of the conservation relationship. Even though the form of the concentration profile was imposed somewhat arbitarily, the integral method often provides satisfactory accuracy because the quantity of principal interest is weakly dependent upon this particular assumption. Thus Lebedeff and Hameed (Ref. 17) show that Eq. 11.31 is only in error by 2% as compared with an exact solution that is available when $m = n = 0$ (see Problem 11.3).

Pollutants that are sometimes considered to be emitted by area sources are, in fact, released above the surface level. For example the carbon monoxide emitted by automobiles in a densely populated residential area is mixed by its buoyancy and by the turbulent wakes so that its effective height of release is several meters above the ground. Equation 11.31 can be used to find the concentration due to an area source at a height h_a above the ground by using the turbulent diffusivity applicable at this level for K_{z0}. In this case it is probably reasonable to assume that the concentration is uniform in the underlying volume $0 < z < h_a$ provided $h_a \ll H_p$.

An integral solution for an infinite line source was given by Sharma and Myrup (Ref. 18). These authors give a useful graphical presentation of the velocity exponent m as a function of the roughness height and the Monin-Obukhov length scale (see Eq. 6.46, p.115) as a stability parameter. For large values of the latter corresponding to neutral stability they find the values of m versus z_0 that are given in Table 11.3.

Table 11.3. The Relationship Between the Velocity Power Law m and Surface Roughness z_0 in Neutral Stability (from Ref. 18)

z_0 (m)	m
0.3	0.23
0.5	0.27
1.0	0.33
1.5	0.38
2.0	0.45
3.0	0.59

11.1.4. Comparisons of Predictions of the Gaussian Plume Model and the Atmospheric Dispersion Equation

Several comparisons have been made of the predictions of the Gaussian plume model with those of the atmospheric dispersion equation. Using a numerical solution to the latter, Ragland and Dennis (Ref. 19) performed a study of this type for a nonbuoyant effluent released from a point source 85 m above ground in an atmospheric boundary layer of height equal to 275 m. Their numerical solution used a fully implicit matrix inversion which was found to be economical, stable, and accurate. Figure 11.2 shows their favorable comparison of the two methods when $\bar{u}$ and K_z are independent of z. In this case the dispersion coefficients σ_y and σ_z for the Gaussian plume model are the Pasquill-Gifford values (Fig. 10.2, p. 225) for class D stability. The values of K_y and K_z were found from the relations

$$\sigma_y^2 = 2K_y\left(\frac{x}{\bar{u}}\right) \quad \text{and} \quad \sigma_z^2 = 2K_z\left(\frac{x}{\bar{u}}\right) \tag{11.32}$$

which follow from Eqs. 11.5 when K_y and K_z are assumed independent of particle travel time T which is replaced by $x/\bar{u}$. The favorable comparison between these methods appears to attest to the applicability of the atmospheric dispersion equation, notwithstanding the approximations employed in its derivation and, further, to indicate the numerical procedures were also reliable. Excellent agreement was also reported for the lateral variation of concentration $\bar{c}(x, y)$ as calculated by these two methods under the conditions given above.

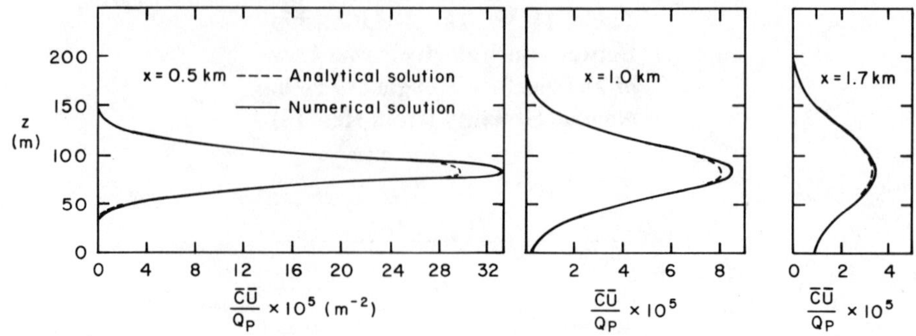

Figure 11.2. Comparison of a numerical solution of the atmospheric dispersion equation with a solution by the Gaussian plume method. [From K. W. Ragland and R. L. Dennis, *Atmos. Environ.*, **9**, 182 (1975). Reprinted with permission of Pergamon Press and K. W. Ragland.]

The influence of the vertical variation of $\bar{u}(z)$ and $K_z(z)$ was tested by Ragland and Dennis for neutral stability by systematically varying z_0. The surface and boundary layers were held constant even though several values of roughness z_0 were used. The vertical component of turbulent diffusivity K_z was taken from Eq. 11.8 for $0 < z < 85$ m, and u_* was found from Eq. 11.9 which includes the dependence of z_0 through the surface Rossby number, Ro. The value of K_z was held constant in the interval $85 \text{ m} < z < 275$ m, and K_y was found from

$$K_y = K_z \left(\frac{\sigma_y}{\sigma_z}\right)^2 \tag{11.33}$$

which follows from Eq. 11.32. The roughness height z_0 influences K_z through its effect on the surface Rossby number and friction velocity u_*. The effect of z_0 on the concentration profile $\bar{c}(x, z)$ is shown in Fig. 11.3 panels *b*, *c*, and *d*, while the Gaussian plume model, with the Pasquill-Gifford dispersion coefficients (applicable for nonbuoyant sources over a rural terrain), gives the results depicted in panel *a*. The sensitivity of $\bar{c}(x, z)$ to z_0 is well demonstrated by this study. On the other hand, the change of surface roughness by a factor of 100 suggests that an adjustment of dispersion coefficients appropriate to the changing terrain is also in order. Thus the study also demonstrates the necessity for careful selection of the dispersion coefficients that are used in the Gaussian plume model.

A second quite different type of comparison of the atmospheric dispersion equation (K theory) with the Gaussian plume solutions was performed by Lamb et al. (Ref. 13). These investigators used a detailed numerical model for the flow in the turbulent atmospheric boundary layer to trace the

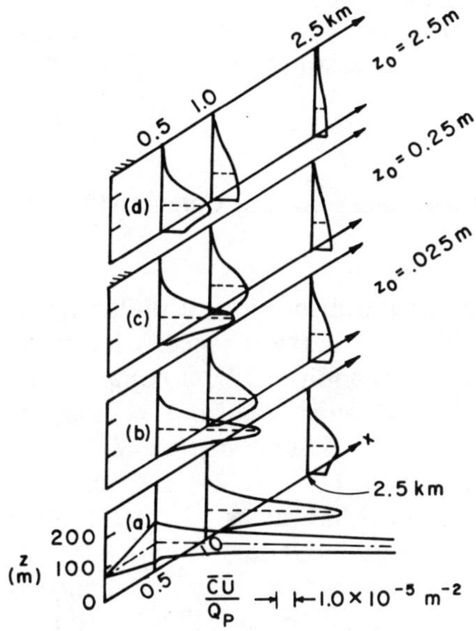

Figure 11.3. A comparison of the vertical concentration profiles downwind of a point source calculated by two methods. (*a*) Gaussian plume model with Pasquill–Gifford dispersion coefficients; (*b*), (*c*), and (*d*) *K* theory numerical solutions for three roughness heights z_0. [From K. W. Ragland and R. L. Dennis, *Atmos. Environ.*, **9**, 183 (1975). Reprinted with permission of Pergamon Press and K. W. Ragland.]

motion of 800 particles for several cases of stability. The computed particle trajectories provided the downwind concentrations. The values of σ_z were calculated directly from the particle positions after a specified travel time. A turbulent diffusivity profile was calculated which was an optimal fit, as described in Ref. 13, for the observed concentration distribution and the derived $K_z(z)$ was used in a numerical solution of the atmospheric dispersion equation for comparison with the predictions of the turbulence model. The investigators concluded that the solutions yielded by the atmospheric dispersion equation were far superior to those obtained by the Gaussian plume model. The concentration profiles predicted by the numerical model of turbulence were not of the Gaussian form, and the Pasquill-Gifford dispersion parameters were found "... not applicable to emissions from elevated sources unless some allowance is made for wind shear effects." For the particular conditions for which their study was conducted, $\text{Ro} = u_*/fz_0 = 1.5 \times 10^{+7}$, they found the best fit value of σ_z under neutral stability to be given by

$$\frac{\sigma_z f}{u_*} = 0.032 + 0.0044 \left(\frac{xf}{u_*}\right) \tag{11.34}$$

for

$$0.23 < \frac{xf}{u_*} < 4.0$$

Thus the study of Lamb et al. (Ref. 13) suggests that improved values of dispersion coefficients are required for the Gaussian plume model and that the K theory is a promising method.

11.1.5. Solutions to the Atmospheric Dispersion Equation Compared to Observed Concentrations

The most important test of any theory of the dispersion of pollutants is its ability to predict the concentrations that are observed at specific locations in a particular urban airshed under a variety of conditions. Such comparisons are frequently hampered by the lack of adequate input detail and accuracy for both meteorological factors and also for the emissions inventory. In this section we discuss several comparisons (Refs. 4, 11, 20 to 24) of theory and the observations of concentrations made in urban areas where extensive programs were established to collect the relevant input data.

Randerson (Ref. 20) performed one of the earliest numerical analyses using the atmospheric dispersion equation, Eq. 11.4, to study SO_2 concentrations in Nashville. He used a polynomial $K_z(z)$ profile and a constant value of K_y in a model that was applied to the heights restricted to $0 < z < 60$ m. Exponential decay of SO_2 with time after emission using a 4.5 hr half-life was employed to account for chemical depletion. Significant discrepancies between theory and observations were evident, and they were attributed to the lack of accurate input data. This study did demonstrate the capability of the method in incorporating the meteorological data and source inventory in a complex urban airshed.

Eschenroeder and Martinez (Ref. 21) used Eq. 11.4 together with a photochemical reaction sequence to study the production of O_3 concentrations in Los Angeles. Their two-dimensional model used a "moving air mass" or semi-Lagrangian formulation in which the particular air parcels are analyzed during their motion through the airshed, and vertical gradients owing to dispersion are considered. Their $K_z(z)$ profile increased linearly near the surface, was constant at an intermediate level, and decreased linearly to the inversion level. The study was the first ambitious attempt to incorporate a detailed chemical reaction sequence into the dispersion analysis. Extreme sensitivity to the initial conditions was found and correction factors for emissions and chemical rate constants were used to give agreement with observations. The study showed the potential of the numerical solutions of K theory to accommodate a complex chemical reaction sequence.

Reynolds et al. (Ref. 4) performed a detailed analysis of CO and photochemical pollution in the Los Angeles airshed. They used a version of Eq. 11.4 which was transformed by a coordinate system to account for the regional topographical features that exert important influences on the advection and confinement of pollutants. The $K_z(z)$ function was similar to that used in Ref. 21 and K_y was held constant. Their model was first tested against CO observations at four locations within Los Angeles County. The agreement was generally favorable, although no correlation coefficient was reported. The photochemical modeling was not as successful; nevertheless, the results were considered highly encouraging. A sensitivity analysis to determine which of the many details of the complex model exert the more important influences was recommended.

Sklarew et al. (Ref. 22) studied the time-dependent CO and photochemical concentrations in Los Angeles using the particle-in-cell method to solve Eq. 11.4. The daily averaged CO concentrations agreed well with observations yielding a correlation coefficient of 0.73 at 12 locations in Los Angeles. The photochemical simulation was less successful, apparently because of inadequacies of the chemical kinetic model which caused NO_2 to be formed from NO too rapidly, and predicted excessive concentrations of O_3.

Shir and Shieh (Ref. 11) studied SO_2 concentrations in greater St. Louis during 25 consecutive days of February 1965. The time-dependent concentrations in a three-dimensional space were calculated taking into consideration large point sources and area sources. Extensive meteorological input detail was used. The $K_z(z)$ profile was given by Eq. 11.10 and K_y was assumed to be a constant. Chemical decay of SO_2 was included. Excellent comparison of theory and observations were reported, and correlation coefficients of 0.873 and 0.899 were found for 24 h average and 2h average concentrations, respectively. The consistent performance of the model during both strong and light wind conditions was considered significant. Their comparison of theoretical predictions with observations for the 24 h averages is given in Fig. 11.4. This study presents the most extended analysis of the pollution concentrations that is available.

Liu and Goodin (Ref. 23) observed that carbon monoxide in the Los Angeles airshed is well mixed in the marine layer, which is typically 200 to 400 m deep above the ground level. On this basis they employed a version of Eq. 11.4 that applied to the vertically averaged properties within a cellular subdivision of the airshed and neglected the vertical diffusion term. Numerical solutions yielded under these conditions showed that the pollutant concentration histories were controlled by advection and that horizontal diffusion could be ignored except at extremely low wind speeds. These results are presumably specific to the Los Angeles area, but they may have application in other areas where vertical gradients below an inversion layer are known to be small.

Perhaps the most comprehensive theoretical analysis of the dispersion of reacting air pollutants is the LIRAQ model described by MacCracken

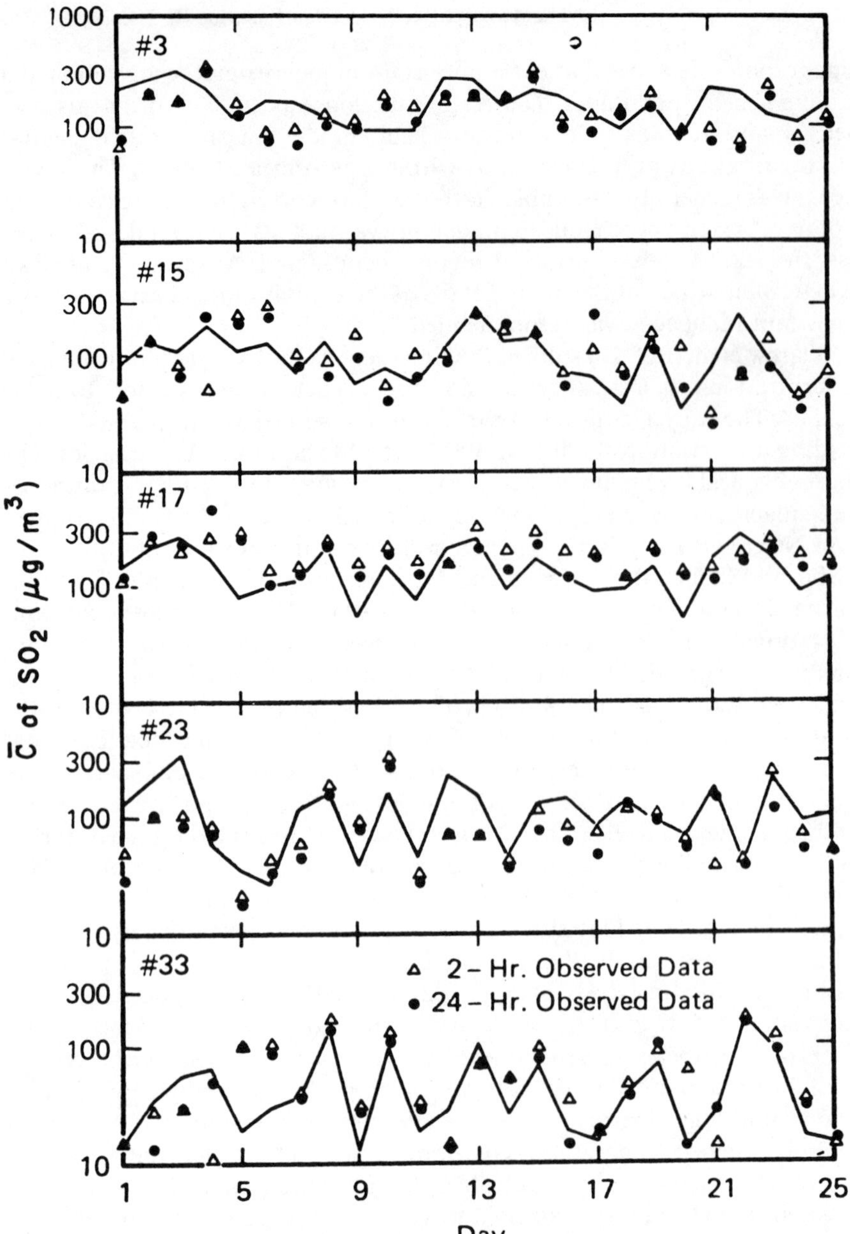

Figure 11.4. Observed and calculated concentrations of SO_2 during February 1 to 25, 1965, at ten locations in St. Louis. The calculated concentrations are numerical solutions using K

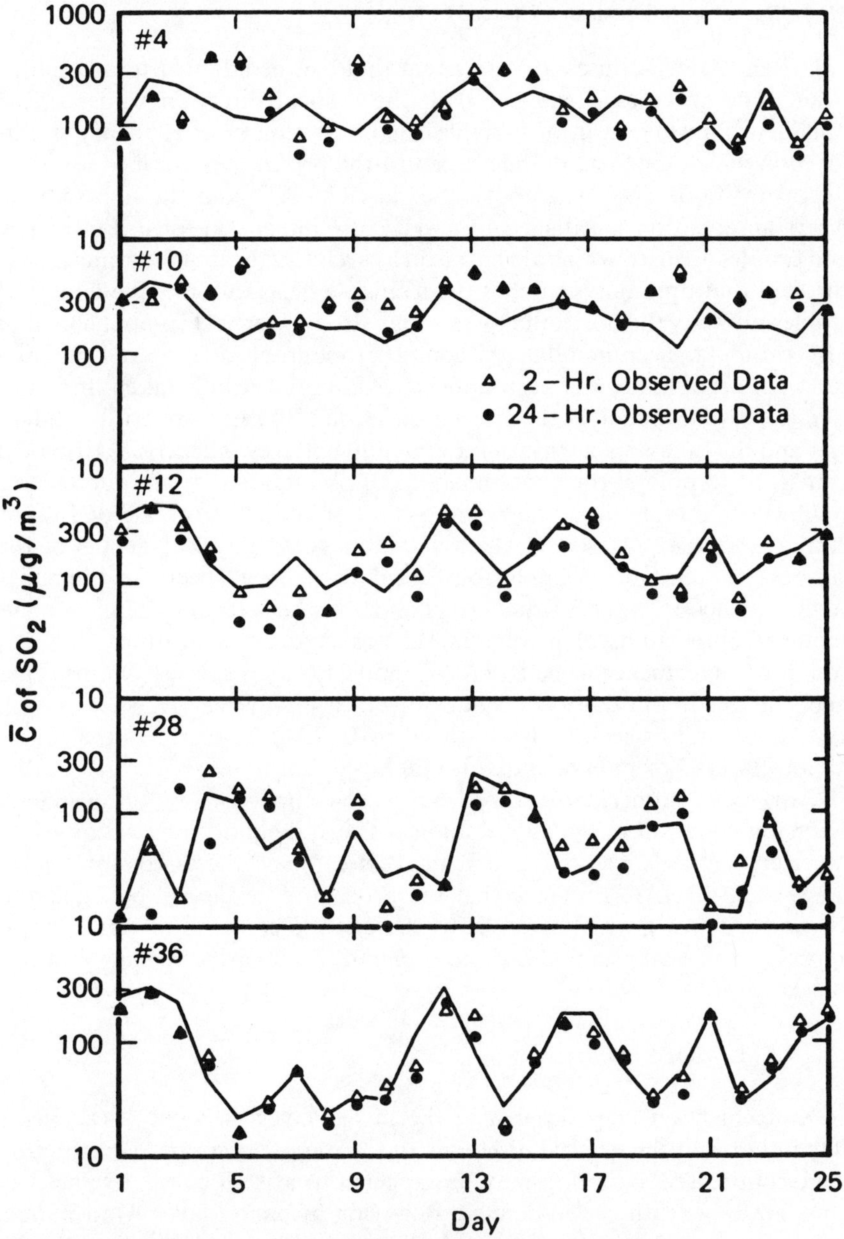

theory. [From C. C. Shir and L. J. Shieh, *J. Appl. Meteorol.*, **13**, 193 (1974). Reproduced with permission of the American Meteorological Society and L. J. Shieh.]

et al. (Ref. 24) which uses a grid system that can be adapted to a complex terrain. The grid has a single cell in the vertical direction whose upper boundary is an inversion of variable height. A version of K theory is employed in which the form of the concentration profile is prescribed as in the integral methods discussed above (Section 11.1.3.3), and the atmospheric dispersion equation is satisfied in a vertically integrated form. Power law wind profiles are used which are based on a careful analysis of the atmospheric flow field that optimally matches the available observations. The flow field includes not only the horizontal winds, but also the vertical components that must consider to accommodate the complex topography of the San Francisco area where the model has been applied. Additional details taken into consideration by the model are the diurnal variation of the atmospheric boundary layer and the time-dependent source emission patterns. The LIRAQ-1 model is intended to predict the concentrations of pollutants that are nonreactive or subject to simple decay and employs a horizontal grid scale of 1, 2, or 5 km. The LIRAQ-2 model, which sacrifices some physical details of the transport of pollutants, is applicable for photochemical species and includes the finite reaction rates for numerous reactions between 19 chemical compounds. Comparisons of predicted and observed concentrations (Ref. 25) show good agreement for the LIRAQ-1 model for inert species. On the other hand, the maximum daily concentrations of a highly reactive species such as ozone are not as accurately predicted by LIRAQ-2 because of the course temporal grid (3 hr intervals) that is employed.

A variety of numerical methods have been applied to the atmospheric dispersion equation in the above studies. The systematic comparison of six numerical methods by Long and Pepper (Ref. 5) and the discussion by Liu and Seinfeld (Ref. 26) indicate that the procedure developed by Egan and Mahoney (Ref. 27), which is explicitly designed to conserve material and to suppress numerical pseudodiffusion, is one of several attractive methods.

11.2. CELL MODELS

The concentration of pollutants in the mixed layer on some occasions is quite uniform in the vertical direction and transport is controlled primarily by advection. This situation may be expected to arise if convective activity is moderate beneath a well-defined, low-lying inversion layer. Under these circumstances pollution concentrations can be analyzed by dividing the airshed into an array of communicating cells. Turbulent mixing is considered to be efficient so that each cell represents a well-stirred reactor for the purposes of calculating chemical production or depletion of pollutants. If the airshed is divided into many cells then the model can include a large amount of detail on the regional meteorological features, a chemical reaction sequence, or the source inventory. In this section we discuss several cell models used to calculate atmospheric dispersion.

11.2.1. The Dynamic Airshed Model

A flexible multicell formulation of considerable generality is the dynamic airshed model formulated by Kyan and Seinfeld (Ref. 28). According to this model the material balance for the ith species of a mixture of pollutants in the kth cell is given by

$$\frac{d}{dt}(\bar{c}_{ik}\mathscr{V}_k) = +\sum \mathscr{Q}_{jk}\bar{c}_{ij} - \bar{c}_{ik}\sum \mathscr{Q}_{kl} + E_{ik} - d_i + r_i \tag{11.35}$$

with

$$\bar{c}_{ik}(t_0) = \bar{c}_{ik}(0)$$

The term on the left side of Eq. 11.35 represents the rate of change of mass of the ith species within the kth cell whose momentary volume is $\mathscr{V}_k$. The latter may change if, for example, the cell extends from the surface to the top of the mixed layer whose height varies with time. The subscript j designates those cells from which transport occurs into the kth cell, while l signifies the cells which receive material from the kth cell. The volume flows into and out of the kth cell are given by $\mathscr{Q}_{jk}$ and $\mathscr{Q}_{kl}$, respectively. The aggregate source strength for all sources within the kth cell of the ith component is E_{ik}, the chemical production rate is given by r_i, and d_i is the deposition rate. The latter applies to particulate materials or to gaseous species that react on the ground by surface reactions. Since Eq. 11.35 is a first order differential equation a single boundary condition, the initial value of $\bar{c}_{ik}$, is required. The cells are normally of constant rectangular area a_k in the horizontal plane and therefore

$$\frac{d\mathscr{V}_k}{dt} = a_k\frac{dH_k}{dt} \tag{11.36}$$

If a single cell is used between the ground plane and an elevated inversion, then H_k is equal to H_m.

The above Eq. 11.35 can be expressed and solved using matrix algebra as discussed in Ref. 28 where this formulation is applied to a reacting mixture to study pollution control strategies. Graedel et al. (Ref. 29) used a multicell model of northern New Jersey to analyze a complex photochemical reaction sequence involving 76 species that chemically interact through 143 reactions.

11.2.2. The Single Cell Dynamic Model

Interesting results are yielded by a single cell model proposed by Tennekes (Ref. 30) which is derivable from Eq. 11.35 for a chemically inert pollutant. The cell includes an entire urban region of dimensions $D \times D \times H_c$ enclosing an area source of strength Q_a(gm/m^2 s). The wind speed $\bar{u}$ in this case is considered to be constant in time and space and normal to two opposite sides of the cell (see Fig. 11.5). The concentration within the cell $\bar{c}$ is assumed uniform, while the concentration above the cell is c^*. The height of the cell H_c is a time dependent quantity which is either the height to which pollutants are dispersed by turbulent dispersion H_p (discussed in Section 11.1.3) or, if wind speed is very low, the height of the mixed layer H_m (discussed in Section 6.4). Turbulent dispersion from the sides of the box is ignored. The flow of pollutant out the downwind side is $\bar{c}\bar{u}H_cD$, and the pollutant inflow on the upwind side need not be considered since it gives rise merely an additive constant. The first three terms of Eq. 11.35 and the source term then yield

$$\frac{d}{dt}(\bar{c}D^2H_c) = c_* D^2 \frac{dH_c}{dt} - c\bar{u}H_cD + Q_aD^2 \tag{11.37}$$

The above equation simply states that the mass rate of change of the pollutant within the cell is increased by the entrainment from aloft as a result of the increasing height of the mixed layer, is reduced by the effiux on the downwind side, and is increased by the surface emissions which constitute the principal source. Dividing Eq. 11.37 by D^2 and rearranging, we find the

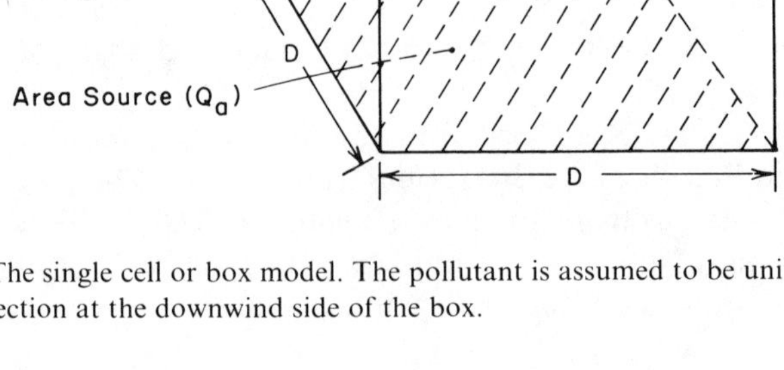

Figure 11.5. The single cell or box model. The pollutant is assumed to be uniformly mixed in the vertical direction at the downwind side of the box.

basic equation given by Tennekes as

$$H_c \frac{d\bar{c}}{dt} = Q_a - \frac{\bar{u} H_c \bar{c}}{D} - (\bar{c} - c_*) \frac{dH_c}{dt} \tag{11.38}$$

with $\bar{c} = \bar{c}_0$ at $t = 0$.

Several special cases of this equation are of interest. The simplest of these is the steady state case in which both time derivatives vanish in Eq. 11.38; the equivalence of the source and the efflux terms yields the "box model" which gives the concentration as

$$\bar{c} = \frac{Q_a D}{\bar{u} H_c} \tag{11.39}$$

The height H_c in Eq. 11.39 will usually be replaced by the height to which significant turbulent transport is limited, H_p, which on some occasions coincides with the height of the mixed layer H_m.

If the dynamic behavior of the mixed layer can be ignored, $dH_c/dt = 0$, then Eq. 11.38 reduces to

$$\frac{d\bar{c}}{dt} + \frac{\bar{u}}{D} \bar{c} = \frac{Q_a}{H_c}$$

By integrating the above equation subject to the initial condition $\bar{c} = 0$ at $t = 0$, we find that the buildup of the pollutant concentration is given by

$$\bar{c} = \frac{Q_a D}{\bar{u} H_c} \left[1 - \exp\left(-\frac{\bar{u} t}{D} \right) \right] \tag{11.40}$$

Thus the pollutant concentration reaches 63% of its final value in the time interval $D/\bar{u}$ which is called the ventilation time of the airshed.

If C_* is zero because advection has removed the previous day's accumulated pollutants, Eq. 11.38 is integrated to yield

$$\bar{c} H_c - \bar{c}_r H_{cr} = \left(\frac{Q_a D}{\bar{u}} - \bar{c}_r H_{cr} \right) \left[1 - \exp\left(-\frac{\bar{u} t}{D} \right) \right]$$

where time t is measured from sunrise and the subscript r designates the sunrise values. The concentration at sunset $\bar{c}_s$ is found by setting $t = T/2$ where, if days and nights are taken to be of equal duration and $T = 24$ hr, the result is

$$\bar{c}_s H_c - \bar{c}_r H_{cr} = \left(\frac{Q_a D}{\bar{u}} - c_r H_{cr} \right) \left[1 - \exp\left(-\frac{\bar{u} T}{2D} \right) \right] \tag{11.41}$$

The influence of the ventilation time is further illustrated by this result. If the ventilation time is small, $D/\bar{u} \ll T/2$, the bracket of Eq. 11.41 tends to unity, the transient behavior decays rapidly, and Eq. 11.39 is recovered. In the other extreme, namely, under light wind conditions when the ventilation time is large compared with 12 hr, $D/\bar{u} \gg T/2$, accumulation of pollutants occurs over several days, and adverse dispersion conditions are persistent.

An important feature of Tennekes' box model is that it can be linked to the dynamics of the convectively, driven mixed layer, and therefore has the potential of allowing the calculation of the concentration history to properly consider the dynamics of the convective atmospheric boundary layer.

11.2.3. The Climatological Inverse Wind Correlation

The Gaussian plume model for an area source, Eq. 10.23, and the box model predict the concentration at the downwind edge of an area source to be directly proportional to the area source strength, the lateral dimension of the source, and inversely proportional to wind speed and the height of the polluted layer. Gifford and Hanna (Ref. 31) argue that the meteorological factors controlling the height of the pollutant layer are essentially the same for all cities in the contiguous United States. They conclude that the long-term average concentrations $\bar{c}_{av}$ are distinctive from one city to the next, primarily as a result of differences of long-term average wind speeds $\bar{u}_{av}$ and area source strength Q_{av}. Accordingly, they suggest that a dimensional relationship of the form

$$\bar{c}_{av} = k_i \frac{Q_{av}}{u_{av}} \tag{11.42}$$

is an adequate expression for the long-term average pollutant concentrations. The constant of proportionality k_i is a dimensional quantity which is determined on a statistical basis from observations of a large number of cities. A value of $k_i = 225$ is recommended for particulates when Q_{av} is in units of $\mu g/m^2$ s, $\bar{u}_{av}$ is in m/s, and $\bar{c}_{av}$ is expressed as $\mu g/m^3$. This value of k_i is an average that is established by examining the observations for a large number of cities. These authors report that the individual values of k_i for two-thirds of the cities that they studied fall within 50% of the average value quoted above. Hanna and Gifford contend that Eq. 11.42 has a higher accuracy for long-term prediction than the detailed numerical dispersion models. The value of k_i found for SO_2 was substantially lower than the above value, apparently because of the reactive nature of this pollutant and the different character of its sources. The inverse wind correlation thus appears to have its greatest utility for chemically inert pollutants emitted by effective area sources. A more detailed examination of this model is the subject of problem 11.8 at the end of this chapter.

11.3. OTHER METHODS OF ANALYSIS

Pollutant concentrations can be analyzed by several other theoretical methods (see the references at the end of Chapter 10), but only one of these has the potential of including information on the chemical interactions of reacting species in a turbulent flow. This is the method that is based on the use of second-order closure techniques and which is being used to attack a variety of problems in turbulent flows. The procedure has been applied to atmospheric dispersion by Lewellen and Teske (Ref. 32). In this method the Reynolds dispersion terms, $\overline{u'c'}$, $\overline{v'c'}$, and $\overline{w'c'}$, are retained in species conservation equations and are developed by using approximate equations for their local values. The resulting "model" equations involve numerical constants that are presumed to be universal and can be determined by reference to existing experimental data. The method is clearly a more fundamental closure model, and it involves the numerical solution of a system of simultaneous partial differential equations. Extensive numerical analysis is therefore required. A comparison of some results of this method with laboratory experiments of Deardorff and Willis (Ref. 33) is displayed in Fig. 11.6 where concentration isopleths are shown downwind of a point source near ground level that is emitting a neutrally buoyant effluent. The boundary layer is convectively driven and capped by an elevated inversion. The more interesting point of the comparison is that both experiments and this theory show that the altitude of maximum concentration increases with downwind distance. Deardorff and Willis have pointed out that neither the Gaussian plume model nor the K theory predict the maximum concentration

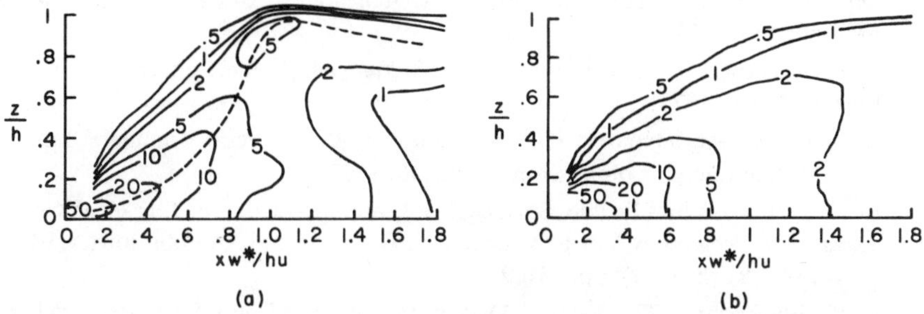

Figure 11.6. Concentration isopleths in the x-z plane downwind of a point-source release at $0.05z/h$ in a convective boundary layer under an elevated inversion: (a) one experiment by Deardorff and Willis and (b) a theoretical prediction of the ensemble average by Lewellen and Teste using second-order closure modeling. [From W. S. Lewellen and M. E. Teske, *Boundary Layer Meteorol.*, **10**, 86 (1976). Reprinted with permission of Reidel Publishing Company, Dordrecht, Holland, and W. S. Lewellen.]

will occur at higher altitudes with increasing downwind distances. Thus the turbulence modeling by the second-order closure technique is capable of describing the physical reality in greater detail than do other available methods. Lewellen and Teske noted that Gaussian-shaped concentration profiles were predicted by the second order closure method under neutral conditions. However, departures from Gaussian profiles were predicted for both stable and convective conditions.

REFERENCES

1. K. L. Calder, "On the equation of atmospheric diffusion," *Quart. J. Roy. Meteorol. Soc.*, **91**, 514–517 (1965).
2. S. D. Reynolds, P. M. Roth, and J. H. Seinfeld, "Mathematical Modeling of Photochemical Air Pollution. I. Formulation of the Model," *Atmos. Environ.*, **7**, 1033–1061 (1973).
3. P. M. Roth et al., "Mathematical Modeling of Photochemical Air Pollution. II. A Model and Inventory of Pollutant Emissions," *Atmos. Environ.*, **8**, 97–130 (1974).
4. S. D. Reynolds et al., "Mathematical Modeling of Photochemical Pollution. III. Evaluation of the Model," *Atmos. Environ.*, **8**, 563–596 (1974).
5. P. E. Long and D. W. Pepper, "A Comparison of Six Numerical Schemes for Calculating the Advection of Atmospheric Pollution," in *Third Symposium on Atmospheric Turbulence, Diffusion and Air Quality* (American Meteorological Society, Boston, Mass., October 1976), pp. 181–187.
6. W. R. Goodin, G. J. McRae, and J. H. Seinfeld, "Validity and Accuracy of Atmospheric Air Quality Models," in *Third Symposium on Atmospheric Turbulence, Diffusion and Air Quality* (American Meteorological Society, Boston, Mass., 1976), pp. 366–373.
7. G. T. Csanady, *Turbulent Diffusion in the Environment* (Reidel Publishing Company, Boston, Mass., 1973).
8. E. J. Plate, "Aerodynamic Characteristics of Atmospheric Boundary Layers," U.S. Atomic Energy Commission TID-25465 (1971).
9. H. H. Lettau, "Wind Profile, Surface Stress, and Geostrophic Drag Coefficients," *Advances in Geophysics*, Vol. 6, Atmospheric Diffusion and Air Pollution (Academic Press, New York, 1959), pp. 241–256.
10. A. K. Blackadar, "The Vertical Distribution of Wind and Turbulent Exchange in a Neutral Atmosphere," *J. Geophys. Res.*, **67**, 3095–3102 (1962).
11. C. C. Shir and L. J. Shieh, "A Generalized Urban Air Pollution Model and Its Application to the Study of SO_2 Distributions in the St. Louis Metropolitan Area," *J. Appl. Meteorol.*, **13**, 185–204 (1974).
12. R. H. Clarke, "Observational Studies in the Atmospheric Boundary Layer," *Quart, J. Roy. Meteorol. Soc.*, **96**, 91–114 (1970).

13. R. G. Lamb, W. H. Chen, and J. H. Seinfeld, "Numerico-Empirical Analyses of Atmospheric Diffusion Theories," *J. Atmos. Sci.*, **32**, 1794 (1975).

14. K. L. Calder, "Eddy Diffusion and Evaporation in Flow Over Aerodynamically Smooth and Rough Surfaces," *Quart. J. Mech. Appl. Math.*, **2**, 153–176 (1949).

15. F. B. Smith, "The Diffusion of Smoke from a Continuous Elevated Point Source in Turbulent Atmosphere," *J. Fluid Mech.*, **2**, 49–76 (1957).

16. S. A. Lebedeff and S. Hameed, "Steady-State Solution of the Semi-Empirical Diffusion Equation for Area Sources," *J. Appl. Meteorol.*, **14**, 546–549 (1975).

17. S. A. Lebedeff and S. Hameed, "Study of Atmospheric Transport Over Area Sources by an Integral Method," *Atmos. Environ.*, **9**, 333–338 (1975).

18. V. Sharma and L. O. Myrup, "Diffusion From a Line Source in an Urban Atmosphere," *Atmos. Environ.*, **9**, 907–922 (1975).

19. K. W. Ragland and R. L. Dennis, "Point Source Atmospheric Diffusion Model with Variable Wind and Diffusivity Models," *Atmos. Environ.*, **9**, 175–189 (1975).

20. D. Randerson, "A Numerical Experiment in Simulating the Transport of Sulfur Dioxide Through the Atmosphere," *Atmos. Environ.*, **4**, 615–632 (1970).

21. A. Q. Eschenroeder and J. R. Martinez, "Concepts and Applications of Photochemical Smog Models," in *Photochemical Smog and Ozone Reactions*, *Advances in Chemistry*, Series 113, American Chemical Society, 1972.

22. R. C. Sklarew, A. J. Fabrick, and J. E. Prager, "Mathematical Modeling of Photochemical Smog Using the PICK Method," *J. Air Pollut. Control Assoc.*, **22**, 865–869 (1972).

23. C. Y. Liu and W. R. Goodin, "A Two-Dimensional Model for the Transport of Pollutants in an Urban Basin," *Atmos. Environ.*, **10**, 513–526 (1976).

24. M. C. MacCracken, "The Livermore Regional Air Quality Model: I. Concept and Development," *J. Appl. Meteorol.*, **17**, 254–272 (1978).

25. W. H. Duewer, M. C. MacCracken, and J. J. Walton, "The Livermore Regional Air Quality Model: II. Verification and Sample Applications in the San Francisco Bay Area," *J. Appl. Meteorol.*, **17**, 273–311 (1978).

26. M. K. Liu and J. H. Seinfeld, "On the Validity of Grid and Trajectory Models of Urban Air Pollution," *Atmos. Environ.*, **9**, 555–574 (1975).

27. B. A. Egan and J. R. Mahoney, "Numerical Modeling of Advection and Diffusion of Urban Area Source Pollutants," *J. Appl. Meteorol.*, **11**, 312–322 (1972).

28. C. P. Kyan and J. H. Seinfeld, "Real-Time Control of Air Pollution," *J. A.I.Ch.E.*, **19**, 579–589 (1973).

29. T. E. Graedel, L. A. Farrow, and T. E. Weber, "Kinetic Studies of the Photochemistry of the Urban Troposphere," *Atmos. Environ.*, **10**, 1095–1116 (1976).

30. H. Tennekes, "Observations on the Dynamics and Statistics of Simple Box Models with a Variable Inversion Ltd," in *Third Symposium on Atmospheric Turbulence, Diffusion and Air Quality* (American Meteorological Society, Boston, Mass., 1976), pp. 397–402.

31. F. A. Gifford and S. R. Hanna, "Modelling Urban Air Pollution," *Atmos. Environ.*, **7**, 131–136 (1973).

32. W. S. Lewellen and M. E. Teske, "Second-Order Closure Modeling of Diffusion in the Atmospheric Boundary Layer," *Boundary Layer Meteorol.*, **10**, 69–90 (1976).
33. J. W. Deardorff and G. E. Willis, "Physical Modelling of Diffusion in the Mixed Layer," in *Symposium on Atmospheric Diffusion and Air Pollution*, Santa Barbara, Ca. (American Meteorological Society, Boston, Mass., September 1974).

PROBLEMS

11.1. Integrate Eq. 11.17 first with respect to z over the interval 0 to H_p, and then with respect to x over the interval 0 to x, to derive Eq. 11.29,

$$\int_0^{H_p} \bar{u}(z)\bar{c}(x, z)dz = \int_0^x Q_a(x)dx$$

where the area source $Q_a(x)$ is an arbitrary function of x.

11.2. Use an elementary box model to derive Eq. 11.29.

11.3. The exact solution for the problem of an area source of uniform strength when $K_z = K_{z0}$ and $\bar{u}$ = constant is given in Ref. 17 as

$$\bar{c}(x, 0) = \frac{2}{\sqrt{\pi}} \frac{Q_a x^{1/2}}{(\bar{u}K_{z0})^{1/2}}$$

Find the corresponding solution by the approximate integral method in this particular case by specializing Eq. 11.31. Find the error in the approximate method by comparing it with the exact solution.

11.4. Compare the ground level concentration profiles predicted by Smith's K theory solution for the elevated infinite line source with an appropriate Gaussian plume solution when the line source is 5.0 m above the ground. Assume the wind speed is 5 m/s and is independent of altitude. In both cases calculate and compare the quantity $\bar{c}\bar{u}/Q_l$ at downwind distances of 1.0, 3.0, and 10 km. Assume the friction velocity is 0.5 m/s and the stability is neutral. Why does K theory give answers which are low in this case?

11.5. Repeat Problem 11.4 for a line source at ground level with $\bar{u}$ = 5.0 m/s and K_z = 2.0 m^2/s, both independent of altitude.

11.6. The following problem refers to the calculation of the annual average particulate concentration for a coastal city where the average particulate area source strength Q_{av} is 4.26 $\mu g/m^2$ s, the annual mean transport wind is 8.3 m/s, and the average height of the mixed layer is 700 m. Assume the city may be considered to be a rectangle whose sides are 12.4 km long.

1. Find the annual average particulate concentration using the climatological inverse wind correlation. Recall that this method gives the average particulate concentration as

$$\bar{c}_{av} = 225 \frac{Q_{av}}{\bar{u}_{av}}$$

2. Find the concentration using the box model with the height of the box equal to the average height of the mixed layer. Therefore

$$\bar{c}_{av} = \frac{Q_{av} x}{\bar{u}_{av} H_m}$$

3. Find the concentration using the Gaussian plume model for an infinite strip of width x, for which

$$\bar{c} = \sqrt{\frac{2}{\pi}} \frac{Q_a x}{\bar{u}\sigma_z(x)(1 - q)}$$

where $\sigma_z = bx^q$. Use values of p and q that are recommended for neutral stability which is the most probable dispersion class. Here we will assume Q_a and $\bar{u}$ are equal to their average values and that $\bar{c}$ at the downwind edge of the strip is representative of the annual average.

4. The observed average particulate concentration in 1969 was 113 μg/m^3. How do you account for the large discrepancy between the observed value and that value calculated in part 2? Why is it possible to use the infinite strip approximation for a city of rectangular shape?

11.7. Use the K theory solution by the integral method, Eq. 11.34, to determine the particulate concentration at the downwind edge of a city 12.4 km in width when the area source strength is 4.26 μg/m^2 s and the wind speed is 8.3 m/s at a height that is 10 m above the roughness elements. Frequently the information necessary to find u_* is not available, and it has been suggested (Ref. 32) that an estimate is one-tenth the wind speed that is measured 10 m above the ground. Note that the data given in Table 11.2 are in agreement with this estimate. On this basis we choose $u_* = 0.83$ m/s. For purposes of calculating K_{z0} assume that the effective release height of the particulate material is $h_a = 5.0$ m. Use a value of the velocity exponent $m = 1/3$ which is applicable for neutral stability and moderate roughness (see Table 11.3). Compare this answer with the results of Problem 11.6.

11.8. The following problem is on the subject of the analysis of annual average concentrations of particulate air pollutants and involves the use of the Data for 44 United States cities published by Gifford and Hanna (Ref. 31). The problem is conveniently performed with a programmable calculator or digital computer with a routine designed to find the straight line which is the best fit by the method of least squares to an arbitrary number of pairs of data points.

1. Use the above data to investigate the suitability of a model for the annual-average particulate concentration which is of the form

$$\bar{c}_{av} = C_1 + C_2 Q_{av}$$

where Q_{av} is the annual average area source strength for particulates given in the above reference. The most probable values of C_1 and C_2 are given by the method of least squares,

$$C_1 = \frac{(\Sigma Y_i)(\Sigma X_i^2) - (\Sigma X_i Y_i)(\Sigma X_i)}{N\Sigma X_i^2 - (\Sigma X_i)^2}$$

$$C_2 = \frac{N\Sigma X_i Y_i - (\Sigma X_i)(\Sigma_i Y_i)}{N\Sigma X_i^2 - (\Sigma X_i)^2}$$

where $X_i = Q^i_{av}$ and $Y_i = \bar{c}^i_{av}$ for 44 ($=N$) cities. Calculate the values of C_1 and C_2.

2. Use the linear correlation coefficient defined by

$$r = \frac{N\Sigma X_i Y_i - \Sigma X_i \Sigma Y_i}{[N\Sigma X_i^2 - (\Sigma X_i)^2]^{1/2}[N\Sigma Y_i^2 - (\Sigma Y_i)^2]^{1/2}}$$

to evaluate the effectiveness of the above model ($X_i = C_1 + C_2 Q^i_{av}$ and $Y_i = \bar{c}^i_{av}$. Determine a linear correlation coefficient for the model proposed by Gifford and Hanna ($X_i = 225\, Q^i_{av}/\bar{u}$ and $Y_i = \bar{c}^i_{av}$

3. Discuss the comparison of the correlation coefficients found in part 2 in terms of the implications with regard to a physical model for suspended particulates in urban areas.

For a discussion of the statistical analysis of data see, for example, H. D. Young, *Statistical Treatment of Experimental Data*, McGraw-Hill, New York, 1962.

12

AIR POLLUTION, THE METEOROLOGICAL RECORD, AND LONG-DISTANCE TRANSPORT OF POLLUTANTS

An extensive network of meteorological stations has been in existence in the United States for many years. The information obtained by this system is stored in a central location and constitutes a vast history of the atmospheric behavior. Systematic observations of the prevalent air pollution concentrations have been gathered by various state and federal agencies in more recent years. These two sets of records have been used to analyze the circumstances under which sustained periods of high air pollution concentration may occur. Such studies show the seasonal and geographic distribution of periods of adverse atmospheric dispersion and serve as a guide to future likelihood of similar episodes. Finally, a number of studies have been performed of the transport of pollutants from their sources to geographical locations far downwind. The results of some of these studies are also reviewed in this chapter.

12.1. THE ATMOSPHERIC MONITORING NETWORK IN THE CONTIGUOUS UNITED STATES

The U.S. National Weather Service has established a network of over 60 upper air stations within the contiguous United States from which rawinsondes are launched to probe the atmosphere. The locations of these stations during the early 1970s is shown in Fig. 12.1. The rawinsondes are launched twice daily, at noon and midnight Greenwich Mean Time (GMT), by international agreement in order to give a worldwide picture of the atmosphere for weather prediction and commercial purposes. The noon GMT

Figure 12.1. The upper air monitoring network in the contiguous United States circa 1970. (Courtesy of G. C. Holzworth.)

launch time is at the hour 0700 Eastern Standard Time and the corresponding earlier hours in the other United States time zones. Measurements of temperature, pressure, and humidity are telemetered from the rawinsonde directly to the ground station for subsequent data reduction. The hypsometric equation is used to convert these observations to functions of altitude. The wind speed and direction are inferred by differentiating the tracked position of the rawinsonde. The reduced data are conveyed to the National Meteorological Center at Suitland, Maryland, where it is used as part of the input data for numerical weather forecasts, and is also sent to the National Climatic Center at Asheville, North Carolina, for storage. The studies that are discussed below are based in part on the data acquired by the upper air stations in recent years.

The upper air stations do not constitute an optimum data gathering network for recording the meteorological variables of direct concern in air pollution. These stations are frequently located at airports that are removed from the cities where numerous pollution sources and high population densities are coexistent. The launch times of the upper air rawinsondes are not the most suitable. Finally, their flight profile has been designed to probe the entire troposphere, and thus they provide a very limited amount of information about the atmospheric boundary layer which is of immediate interest for air pollution studies. A limited program was established about 1970 to provide, in certain locations of the United States, a modified type of rawinsonde station which would function independently of the upper air network. These stations (Ref. 1), called environmental meteorological support units (EMSU), were established in about a dozen cities where high air pollution concentrations were relatively frequent. These rawinsondes were launched twice daily at local sunrise and near local noontime. This launch schedule provides information on the state of the boundary layer prior to the morning hour of heavy traffic activity and again after the mixed layer under clear skies is well established. The EMSU rawinsondes ascend at the slow rate of 200 m/min to provide a higher vertical resolution and thereby to secure more numerous observations of such important properties as wind speed, wind direction, and temperature as a function of altitude within the atmospheric boundary layer. Thus the requirements of air pollution meteorology were fulfilled by this program. The data obtained by the EMSU network are used mainly by local air pollution meteorologists for prognostic purposes.

12.2. CLIMATOLOGICAL STUDIES OF THE ATMOSPHERIC BOUNDARY LAYER

The records of the U.S. National Weather Service have been used to analyze the dispersive properties of the atmosphere. An early study of the frequency of nocturnal radiation inversions below 150 m in height was conducted by Hosler (Ref. 2). Radiation inversions are likely when light winds and clear nighttime skies prevail, and they can strongly influence early morning pollution concentrations. In general, it was noted that rural areas tend to have more frequent radiation inversions than urban areas where mechanical turbulence induced by high buildings and the urban heat effects are present. The regional-seasonal distributions of radiation inversions are summarized in Table 12.1 which shows estimates of the frequency of inversions in all areas except along the West and Northwest Pacific coastal region. Studies of this type establish the average behavior of those properties of the atmosphere that pertain to its dispersive characteristics and constitute climatological studies of boundary layer meteorology.

Table 12.1. Approximate Frequency of Occurrence of Radiation Inversions below 150 m in the Contiguous United States (from Ref. 2)

Region	Frequency Spring–Summer	Frequency Fall–Winter
Appalachian Mountains	30–45%	30–45%
Great Plains	20–30%	30–45%
Great Lakes Area	20–30%	20–30%
Gulf-Atlantic Coast, South of Long Island	10–35%	10–35%
Pacific Northwest Coast	Infrequent	Infrequent
Rocky Mountains	30–40%	40–55%[a]
West Coast	Less prominent[b]	More prominent[b]

[a] More frequent occurrence of radiation inversions in these regions is attributed to frequent anticyclones in the fall and to the long nights during the winter.

[b] A subsidence inversion is more dominant in the southern portion of this region because of the semi-permanent anticyclone frequently located off the California coast. Radiation and subsidence inversions often merge and cooperate with unfavorable wind and topographic features to create an adverse climate for air pollution in valley and basin areas.

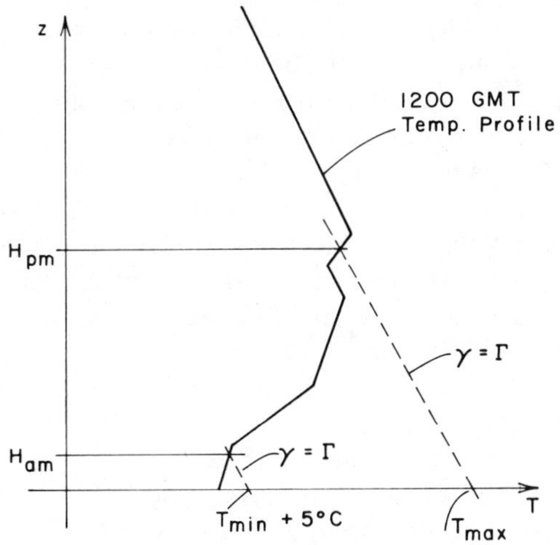

Figure 12.2. Definitions of morning (H_{am}) and afternoon (H_{pm}) mixing heights. T_{min} is the minimum surface temperature observed between 0200 and 0600 local time; T_{max} is the maximum surface temperature observed between 1200 and 1600 local time.

The properties of the atmospheric boundary layer have been the subject of a series of retrospective studies by Holzworth (Refs. 3 to 5). These studies have examined in detail the quantities defined by Holzworth as the morning mixing depth, the afternoon mixing depth, and the mixing layer wind speeds.

The morning mixing depth H_{am} is intended to be an estimate of the height of the mixed layer at the hour of peak morning traffic activity, and it is calculated with the knowledge of the minimum surface temperature T_{min} observed between 0200 and 0600 local standard time and the profile of temperature versus altitude for the 1200 GMT sounding. The intersection of the latter curve with an adiabatic temperature profile drawn from a surface temperature which is 5°C greater than T_{min} (see Fig. 12.2) defines the morning mixing height. This procedure is based on the assumptions that (a) all but the lower portion of the 1200 GMT profile will be applicable at the hour of maximum traffic activity, (b) that T_{min} + 5°C represents the true urban surface air temperature at this hour, and (c) that an equilibrium convective mixed layer is rapidly established. The adjustment of 5°C to T_{min} is intended "... to allow for urban-rural differences in morning surface temperature and for some solar heating of the surface after sunrise." Various objections can be lodged against this procedure, which is advanced as a rough estimate, such as the varying local time of the 1200 GMT sounding in the several time zones of the United States, the seasonal variation of the relationship between sunrise and the hour of maximum traffic, and the imposition of a single incremental adjustment to the minimum temperature for all locations and seasons. Unfortunately the detailed information necessary to refine this procedure is not available.

The afternoon mixing depth H_{pm} is defined by Holzworth as the altitude at which the $T(z)$ profile determined by the 1200 GMT sounding is intersected by an adiabatic profile drawn from the maximum air temperature observed near the ground between 1200 and 1600 local time. The basis for estimating the afternoon mixing depth appears to be more reliable, and the definition probably corresponds rather closely to the height of the mixed layer H_m as discussed previously.

Finally, the mixing layer wind speed was calculated by Holzworth by averaging the winds observed at the surface and aloft to an altitude equal to the mixing depth. This definition coincides with the transport wind speed q_T (Eq. 6.90), provided H_{am} and H_{pm} represent valid determinations of the mixed layer heights. Figure 12.3 shows interesting monthly averages of the mean mixing depths at various cities. The three interior cities display nearly identical values of H_{am} which is nearly constant throughout the year. The value of H_{pm} increases dramatically during the summer months owing to the greater isolation. A lower value of H_{am} during the winter months caused by stronger radiative cooling is a reasonable expectation that is found in Hosler's observations (Ref. 2), but it is not displayed in the data shown in Fig. 12.3*a*. The value of H_{pm} at Salt Lake City shows a sevenfold seasonal variation caused by the annual variation of insolation. The data for New York display higher morning mixing depths which are tentatively explained to be caused by the proximity of water masses and possibly by the urban heat island. Holzworth (Ref. 3) also shows transport wind speeds which, for many cities, are typically highest (7 to 9 m/s) during March and April and lowest (4 to 5 m/s) during July and August. A detailed report on the regional variation of H_{am} and H_{pm} and the corresponding values of transport winds during the various seasons of the year is also provided (Ref. 4).

The significance of these results on the incidence of high air pollution concentrations also has been examined by Holzworth (Refs. 3 to 5). A set of criteria for the definition of limited atmospheric dispersion similar to those used for forecast purposes was formulated. These requirements were (a) values of H_{am} and $H_{pm} \leq 1500$ m, (b) $q_T \leq 4.0$ m/s (hence the ventilation factor is $\leq 6000\ \mathrm{m^2/s}$), (c) the absence of significant precipitation, and (d) the persistence of the above conditions for at least 48 hr. The fulfillment of these criteria is considered to generate one episode lasting at least two days. The total number of episode-days for all episodes lasting at least two days for the contiguous United States was calculated with the data from the upper

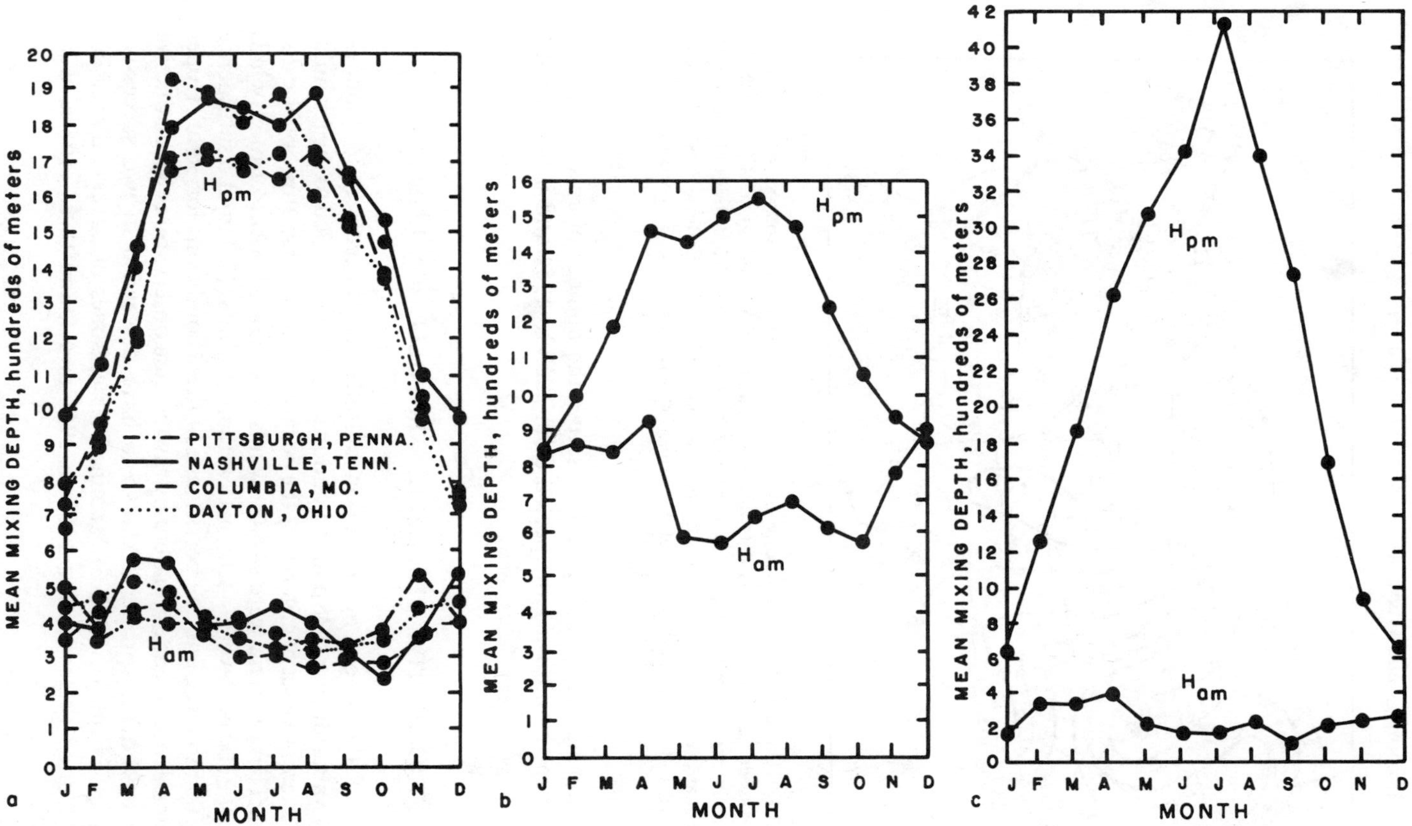

Figure 12.3. The monthly mean mixing depths H_{am} and H_{pm} at (*a*) four interior cities, (*b*) New York, and (*c*) Salt Lake City. [From G. C. Holzworth, *J. Appl. Meteorol.*, **6**, 1040, 1041 (1967). Reprinted with permission from the American Meteorological Society and G. C. Holzworth.]

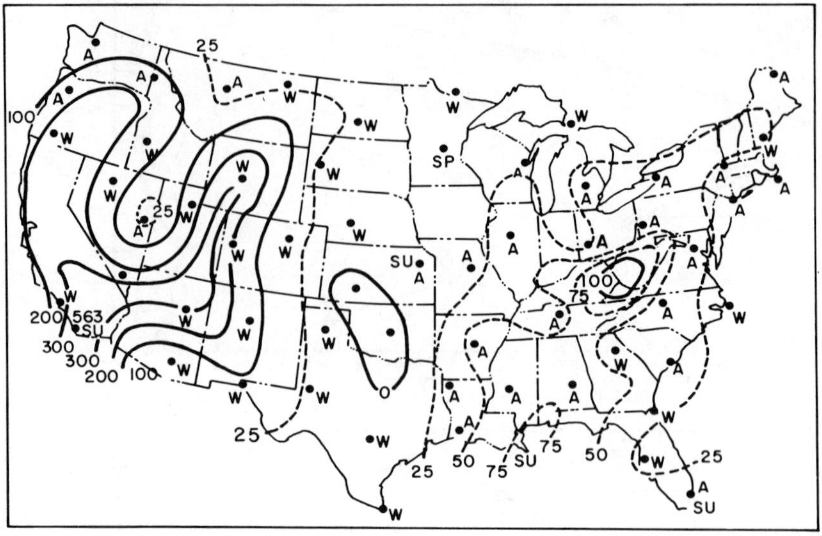

Figure 12.4. The number of episode days of limited dispersion during 1960 to 1964. The season of most episode days is indicated as *W* (winter), *SU* (summer), *SP* (spring), or *A* (autumn). See text for definition of limited dispersion. (From G. C. Holzworth, Proceedings of the Second International Clean Air Congress, H. M. Englund and W. T. Beery, Editors, Paper ME-20-C, p. 1080. Reproduced with permission of Academic Press Inc. and G. C. Holzworth.)

air monitoring network. Figure 12.4 shows the number of episode-days of limited dispersion during the period from 1960 to 1964 and it is discussed below.

12.3. RETROSPECTIVE STUDIES OF AIR POLLUTION EPISODES

Studies of the severe air pollution episodes that occurred in Donora, Pennsylvania, during 1948 and in London during 1952 had led to the formulation of a tentative set of criteria for the occurrence of adverse atmsopheric dispersion. These criteria were surface winds less than 8 knots (4.1 m/s), winds aloft to the 600 mb level ($\sim$4200 m) less than 25 knots (12.9 m/s), a subsidence at the 600 mb level, and the persistence of these conditions for 36 hr. Upper level subsidence is made apparent by rawinsonde observations if a distinctive feature in the $T(z)$ profile is shown to have descended in two successive soundings. Niemeyer (Ref. 6) examined six episodes of adverse dispersion in the United States during 1957 to 1958 and found these criteria were fulfilled.

Two major air pollution episodes in the eastern United States in the 1960s were the subject of detailed reports. During late 1962 a severe air pollution episode persisted over a large portion of the eastern United States for a period of seven days (Ref. 7). Twenty-three major cities and 87 million people were exposed to high air pollution concentrations during this period. A second similar episode (Ref. 8) occurred in November 1966 when an anticyclone, whose center stalled over New York, expanded to cover the eastern region from Maine to Mississippi. An increase in the daily death rate in New York amounting to approximately 24/day was observed. The report of this study first suggested a value of the ventilation factor, the product of the transport wind times the mixing layer height, equal to 6000 m^2/s, or an average low level wind speed of 4 m/s, were critical values below which adverse pollution concentrations in urban areas were likely. Other criteria for adverse dispersion added during this time period were the absence of significant precipitation and the persistence of all earlier criteria over a moderate geographical area. These conditions were noted to accompany slow-moving anticyclonic wind-pressure systems.

A study of stagnating anticyclones during the 30 year period from 1936 to 1965 was conducted by Korshover (Ref. 9). He found the seasonal variation of these events in the eastern United States that is displayed in Fig. 12.5 where the months of highest frequency are found to be May and October.

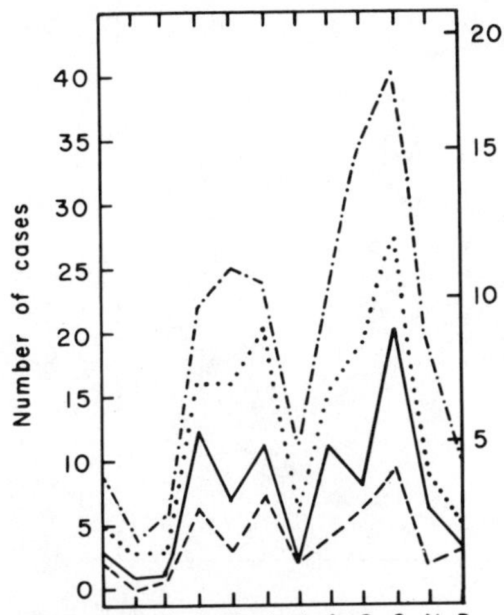

Figure 12.5. Monthly frequency of stagnation cases of four or more days duration east of the Rocky Mountains, 1936 to 1965. (From J. C. Fensterstock and R. K. Fankhauser, "Thanksgiving 1966 Air Pollution Episode in the Eastern United States," U.S. Dept. of H.E.W. Publication No. AP-45, 1968.)

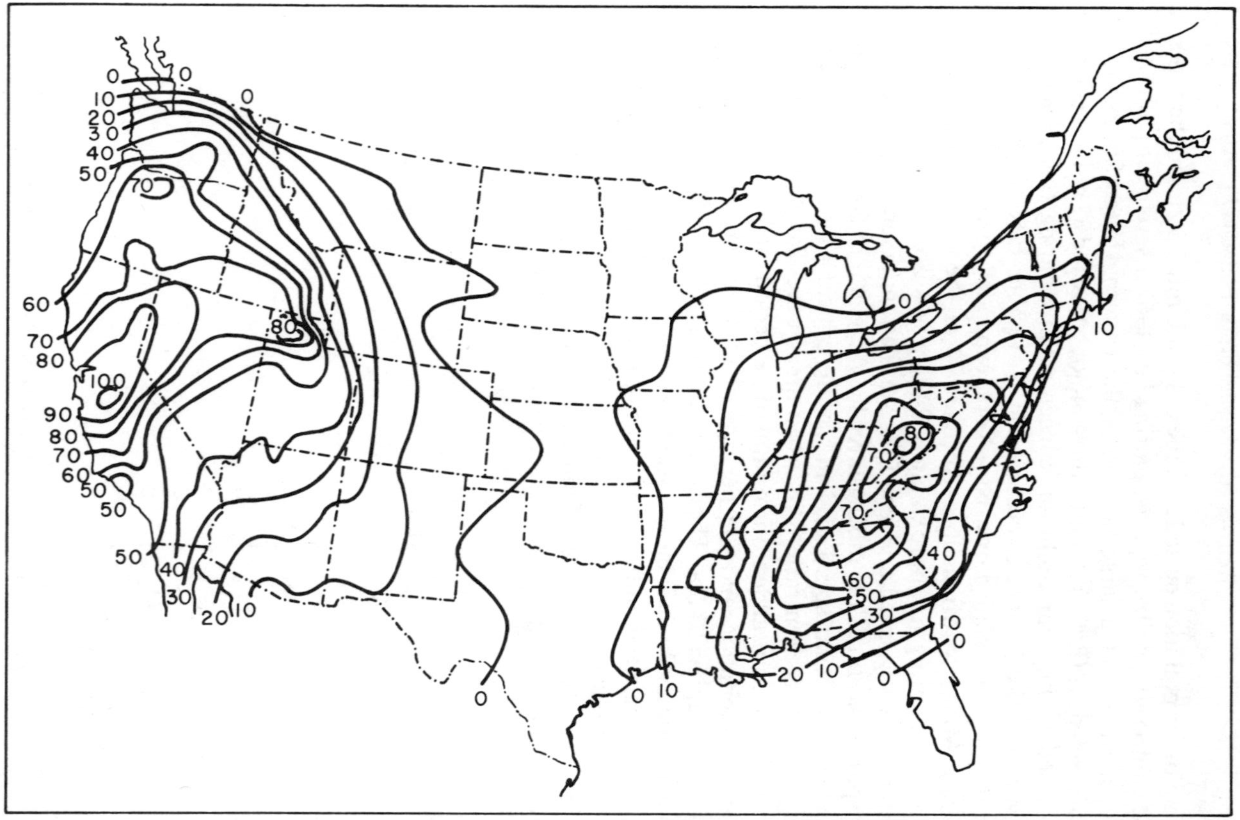

Figure 12.6. The number of days of potentially high concentrations of air pollution. The isopleths are based on forecasts involving 75 episodes in eastern United States (August 1960 to April 1970) and 39 episodes in western United States (October 1963 to April 1970). (From E. Gross, "The National Air Pollution Potential Forecast Program," U.S. Dept. of Commerce, WBTM NMC47, 1970.)

The above studies led to the formulation by the U.S. National Weather Service of a forecast procedure that is based largely on the above criteria as discussed in Section 6.5.1 and in detail in Ref. 10. The forecasts were first implemented in 1960 for the eastern United States and in 1963 for all of the contiguous states. The number of days for which adverse atmospheric dispersion was forecast is shown in Fig. 12.6. The two regions of the map shown in Fig. 12.6 where adverse dispersion was more frequently predicted, the greater Appalachian region and coastal California, both have large urban centers with strong pollutant sources and high population densities. This map can be compared with Fig. 12.4 with which it is closely related. Both figures suggest that the most frequent cases of adverse atmospheric dispersion are to be expected over the central region of the Appalachian Mountains and over central and southern California. Both Figs. 12.4 and 12.6 show infrequent occurrences of adverse dispersion in the Great Plains region of the United States where transport wind speeds are higher than in most other regions. The minor differences between the two figures can be attributed to the fact that slightly different criteria were used in their derivation, and that forecasts were used to arrive at Fig. 12.4, while the measured conditions that actually developed were the basis for Fig. 12.6.

12.4. SYNOPTIC SCALE TRANSPORT OF POLLUTANTS

In the early 1970s it was widely assumed that high air pollution concentrations were restricted to urban areas of high source density and to their more immediate downwind localities. This viewpoint has now been discarded as a result of the subsequent observations. Urban pollution is now known to be transported for long distances by the large-scale weather patterns. The accumulative effects of the pollution generated throughout the continental United States have the greatest impact on the densely populated eastern seaboard as a result of the eastward motion of the midlatitude, large-scale circulation. In the following sections we discuss the evidence and some of the consequences of the long distance transport of pollution.

12.4.1. Acid Precipitation and Sulfate Transport

The serious effect of the acidification of rainfall in certain locations in North America and in Northern Europe has attracted widespread attention. Pure water in equilibrium with atmospheric concentrations of CO_2 establishes a pH level of 5.6, somewhat on the acid side of the neutral value of 7.0.

Rain or snow possessing a pH less than 5.6 is described as being acid precipitation. Likens (Ref. 11) reports the earliest measurement in the United States during a single rainstorm, made at Brooklin, Maine in August 1939 by H. G. Houghton, as yielding a pH equal to 5.9. The first nationwide survey of pH in the United States was conducted by the National Center for Atmospheric Research between 1964 and 1966. These measurements disclosed pH levels of greater than 7 in the west which decreased to less than 4 in New England.

The trend of acidity of the precipitation in the eastern United States and in the Scandinavian peninsula over about two decades is shown in Fig. 12.7. The increased acidity of precipitation is apparent in both locations over the time span of one of two decades. Sulfuric and nitric acids are considered to be the major sources of acidity in precipitation although their relative amounts vary in time and location. In northeastern United States sulfuric acid has been the dominant component, but the nitric acid content has increased markedly during the past two decades. Sulfuric acid is considered to be the dominant cause of acid precipitation in Sweden. Likens (Ref. 11) discusses various compounds that contribute to a lower pH in more detail. A current focus of attention is the generation of sulfur and nitrogen compounds and the transport of these materials over long distances.

Estimates on the worldwide emissions of sulfur and nitrogen by anthropogenic and natural sources vary considerably and do not permit the formulation of detailed global budgets of these materials. The anthropogenic sources of sulfur are concentrated in industrial regions and their total output is estimated to be $50 \times 10^{+9}$ to $65 \times 10^{+9}$ kg/yr. Despite the uncertainties, Likens concludes that anthropogenic sources appear to be the major contributors to acidic precipitation in the northeastern United States and in northern Europe because of the following factors:

1. The increased acidity correlates with the increased anthropogenic emissions of sulfur dioxide and oxides of nitrogen over the past 20 years.
2. The onset of widespread acidity of precipitation in both locations coincides with the more widespread use of higher smokestacks which disperse the pollutants more widely.
3. With respect to the prevailing wind direction, the major anthropogenic sources are upwind, while the major natural sources are downwind at both of the above locations.
4. Studies of air mass trajectories show a high correlation between the passage of storms over areas of high anthropogenic emissions and the subsequent episodes of acid precipitation.

5. Likens' own estimates are that the sulfur fallout in the United States during 1955 to 1956 was only one-third of the nation's anthropogenic sulfur emissions. Thus the latter sources have ample capacity to provide the amount of sulfur deposited.
6. The deposition through rainfall of other clearly anthropogenic pollutants, namely heavy metals, has increased in unison with acid precipitation at both locations.

Thus the widely accepted viewpoint is that the acidification of rain in these regions is caused by sulfuric and nitric acids whose precursors originate at anthropogenic sources located hundreds of miles away. The transport of these pollutants from the sources to the distant receptors is presumed to result from the large scale atmospheric motion.

The ecological effects of the acidic precipitation depend upon many factors. Certain lakes have low buffering capacity and their fish populations have suffered severely. Large losses of trout and salmon have occurred in the lakes of Norway and Sweden. Ninety percent of the lakes above 600 m in altitude in the Adirondack Mountains of New York were devoid of fish during 1976, while only 4 % of these lakes were without fish life during the 1930s (Ref. 11). It has been suggested that aluminum released by the increased acidity causes damage to the gills of fish. Forest growth rates are also believed to be reduced by acidic precipitation whose long range influence on vegetation is viewed as a critical unknown factor by Likens.

The need for a greater knowledge of the chemical content of rain throughout North America has lead to the formation of a coordinated network of stations to analyze the composition of precipitation on a systematic basis. In the United States a group of 80 to 100 stations is expected by 1981 to 1982 (Ref. 12) to constitute the National Atmospheric Deposition Network. This program is expected to permit analysis of ecological impact, meteorological dispersal and deposition, and atmospheric chemical transformations.

The transport of sulfates was studied by Wilson and colleagues (Ref. 13) in the downwind wake or urban plume of St. Louis. These authors list seven forms (see Table 12.2) of sulfates known to be present in the atmosphere. The primary SO_2 emitted by sources is converted by atmospheric oxidation into SO_3 which readily combines with water vapor to form sulfuric acid, H_2SO_4. The latter can combine with natural or anthropogenic ammonia gas NH_3 to produce the three forms of ammoniated sulfates that are given in Table 12.2. Examination of this table shows that H_2SO_4 and the various ammoniated sulfates are all present in the 0.1 to 1.0 μm diameter range. It is this size range that constitutes respirable dust and which also most directly affects the deterioration of atmospheric visibility. The remaining sulfates

Acidity of precipitation has increased markedly in the eastern U.S. . . .

Average pH of annual precipitation

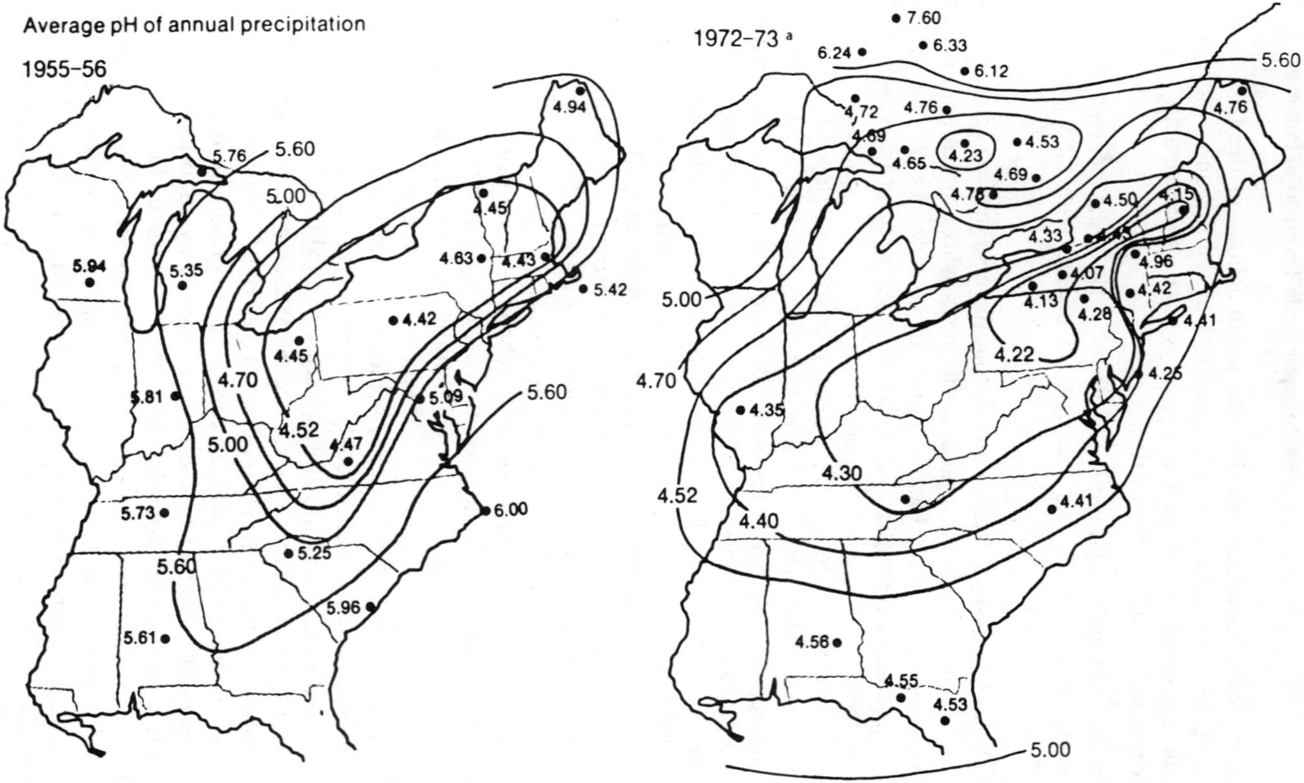

Source: C. V. Cogbill and G. E. Likens

a Data from Oak Ridge, Tenn., for 1973–74; data from Tallahassee, Fla., for 1974–75; and from Gainesville, Fla., for 1976. **Source:** Various, including C. V. Cogbill, Thomas Burton, Patrick Brezonik, and Gray Henderson

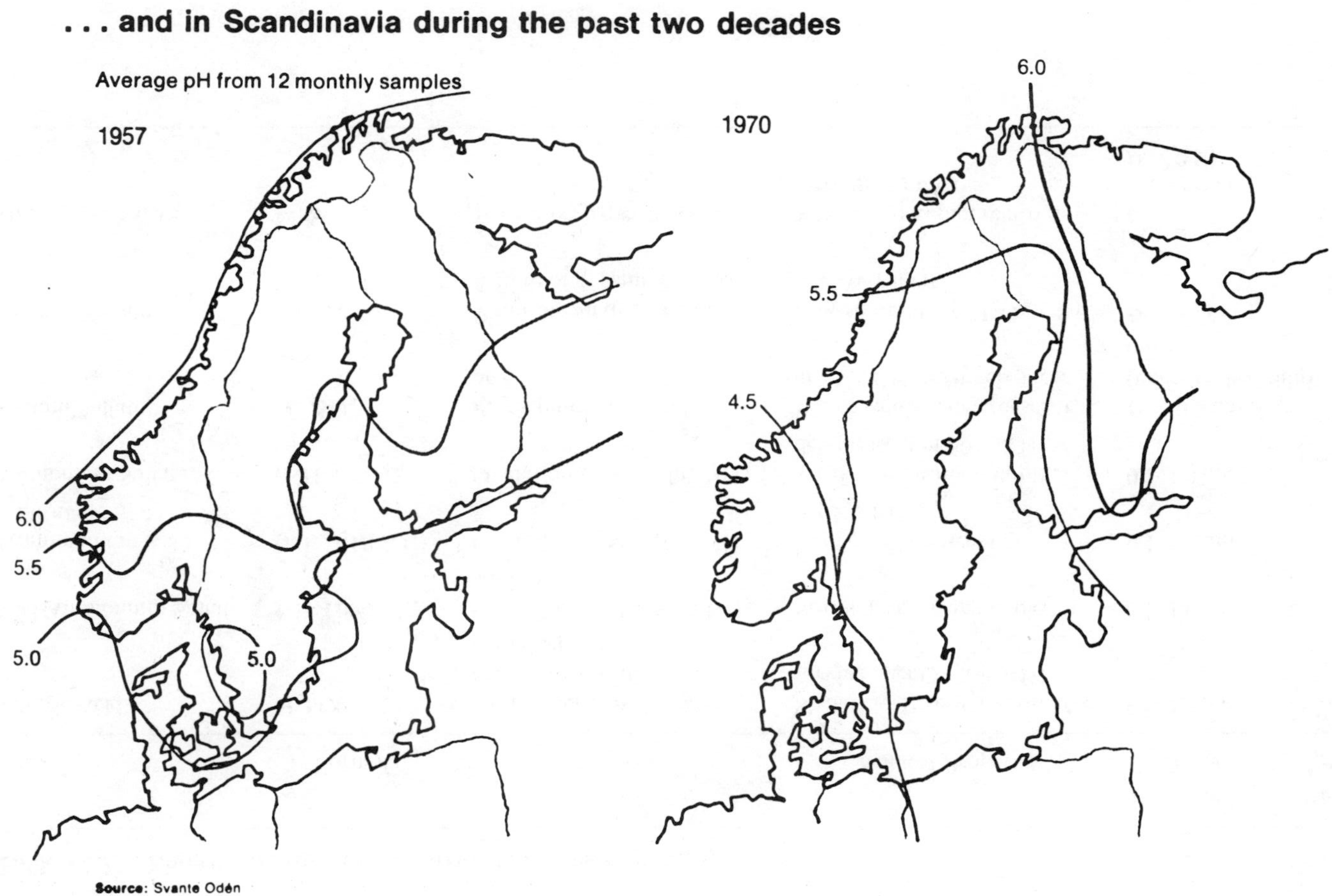

Figure 12.7. The trend of acidity in precipitation over eastern United States and Scandinavia over two decades. [From G. E. Likens, *Chem. Eng. News*, **54**, 31 (Nov. 22, 1976). Reproduced with permission of the American Chemical Society and G. E. Likens.]

Table 12.2. Known Atmospheric Sulfates (from Wilson et al., Ref. 13)

Name	Formula	Sources	Chemical Properties	Probable Size Class Diameter (D)
Sulfuric Acid	H_2SO_4	Atmospheric oxidation of SO_2; direct from manufacturing	Strong acid; very hygroscopic (drying agent at low RH)	0.1–1.0 μm
Acid Ammonium Sulfate	NH_4HSO_4	Oxidation of SO_2 with NH_3 addition	Strong acid; hygroscopic	0.1–1.0 μm
Triammonium Acid Disulfate	$(NH_4)_3H(SO_4)_2$	Oxidation of SO_2 plus NH_3	Acidic; deliquescent at ~65% RH	0.1–1.0 μm
Ammonium Sulfate	$(NH_4)_2SO_4$	Oxidation of SO_2 plus NH_3	Weak acid; water-soluble; deliquescent at 80% RH	0.1–1.0 μm
Sodium Sulfate	Na_2SO_4	Paper pulping by kraft process	Water-soluble; deliquescent at 84% RH; relatively inert	$D > 0.5$ μm (uncontrolled pulp mill)
Calcium Sulfate	$CaSO_4$	Wind-blown dust; manufacture of gypsum products	Low solubility in H_2O; relatively inert	$D > 1$ μm
Magnesium Sulfate	$MgSO_4$	Sea spray; paper pulping	Very hygroscopic; relatively inert and nontoxic	$D > 1$ μm (sea spray) $D > 0.5$ μm

of Na, Ca, and Mg present in the atmosphere generally appear as larger particles (> 0.5 or 1.0 μm) and may be either of natural or anthropogenic origin.

The first five sulfates given in Table 12.2 are secondary pollutants and the remaining three are primary. The compounds listed in Table 12.2 possess varying degrees of hygroscopicity or deliquescence, and many of them will adsorb sufficient moisture to form particles in the 0.5 μm size range that contribute heavily to light scattering and atmospheric haze.

Wilson et al. (Ref. 13) examined the sulfur budgets downwind of St. Louis where the sources of SO_2 were ground level emissions and low level stacks of height less than 110 m. In this urban plume the total flow rate of sulfur in both gaseous and particulate form was found to decrease. A characteristic distance for a $1/e$ decay of the initial concentration was found to be 90 km, a distance achieved in 5 hr under the prevailing wind conditions. The formation of sulfate aerosol was in evidence after 1 to 2 hr had elapsed. On the other hand, power plant plumes located in rural areas showed no detectable loss of sulfur within the region of 50 to 100 km from the source. In one test the aerosol fraction of the total sulfur was 6% at 10 km and 13% at 40 km.

These investigators conclude that SO_2 emitted in the urban plume relatively near the ground can be depleted by reactions that occur on ground level surfaces. Such reactions are described as dry deposition (as opposed to wet deposition through rainfall). In contrast to this situation, the sulfur dioxide emitted from an isolated high stack is converted to particulate forms after dilution occurs, and the small particles formed have a long residence time. These particles may be held aloft by an underlying but elevated inversion, and thus dry deposition of either gaseous or particulate sulfur compounds is prevented. The resulting long residence times provide ample opportunity for conversion to sulfate particulates and dispersal over a relatively wide geographic area. Thus tall stacks reduce ground level SO_2 concentrations, ground level SO_2 deposition, and increase the conversion of SO_2 to sulfate. Wilson and colleagues (Ref. 13) also sampled urban plumes at downwind distances of 250 km and power plant plumes 60 km from their sources and concluded that air pollution episodes could be caused at these distances.

Husar et al. (Ref. 14) have compared the light scattering coefficient b_{sca} measured by a nephelometer, a device that measures the light scattered by the particulate phase of the aerosol, with the extinction coefficient b_{ext} calculated from the routine observations of visibility by the U.S. Weather Service. They concluded that a high correlation exists between these two quantities, and that the haze is associated with the presence of sulfates. This conclusion is important because of the extensive network of some 300 stations in the contiguous United States where visual range is observed and forwarded to the National Climatic Center for storage and possible future use. Husar and colleagues then used $b_{ext} = 3.91/L_v$ (see discussion in Section 8.5.2), where L_v is the visual range determined by the nationwide weather network, to be a quantity that can be used to track the large scale motion of hazy air masses that contain high concentrations of pollutants. Six hazy air masses were tracked during the summer of 1975 and were found to have an average residence time of eight days during their meandering passage from the midwest to the Atlantic seaboard.

A survey article by Altshuller (Ref. 15) contains a discussion of the trends and relationships between concentrations of sulfur dioxide and sulfate aerosols in urban and nonurban sites throughout the United States between 1963 and 1972. Although SO_2 concentrations have decreased substantially in the eastern United States, the sulfate concentration has decreased only slightly. Altshuller concludes that a reduction of sulfate aerosol in the eastern United States would necessitate the substantial reduction of SO_2 emissions throughout a large geographic region that is not yet defined. He cites the topics for continuing study to be the rates of conversion of SO_2 to sulfate within plumes of various origins; the relative contributions to sulfate aerosols from various types of SO_2 sources; the role of precipitation and other removal mechanisms on transport and deposition of sulfates; their impact on visibility, materials, water quality, soil chemistry; and the improvement of instrumentation and experimental procedures.

12.4.2. Long-Distance Transport of Ozone

The formation of ozone from anthropogenic nitric oxide NO and hydrocarbons HC has been observed in the Los Angeles basin since the early studies by A. J. Haagen-Smit in the late 1940s. The current understanding of the formation and transport of this pollutant in the eastern United States has developed since the establishment of air quality monitoring stations some 30 years later. The measurements of ozone conducted in the eastern United States during the early 1970s showed the surprising result that the National Ambient Air Quality Standard was frequently exceeded in rural areas far from the sources of NO and HC that were considered to be the necessary precursor species. Some progress has been made in understanding the origin of ozone as a pollutant in the eastern United States since that time.

The urban and rural ozone measurements obtained during the summer of 1973 in New York State were studied by Stasiuk and Coffey (Ref. 16). In this study the ozone concentrations were monitored at four widely separated cities and at two mountain-top locations as indicated on the map shown in Fig. 12.8*a*. At Mt. Utsayantha the O_3 monitoring device was located on a

(a)

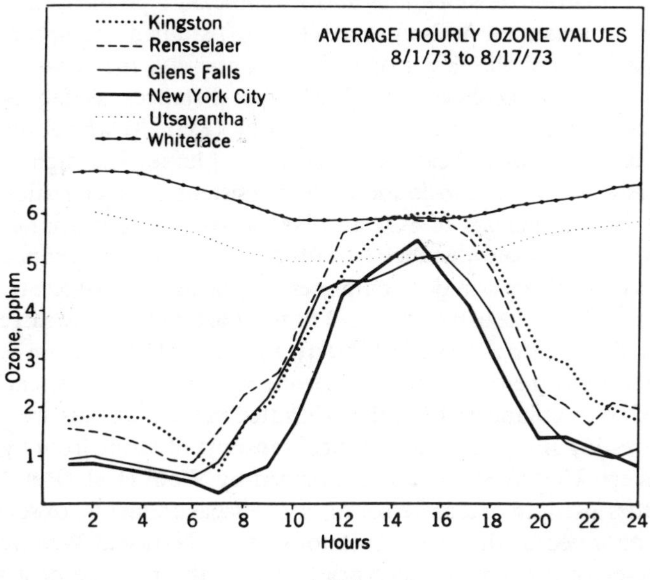

(b)

Figure 12.8. The hourly average ozone concentrations during 17 consecutive days of August 1973 at six locations in the state of New York. (*a*) The locations of the monitoring sites and (*b*) the time-dependent ozone concentrations. [From W. N. Stasiuk and P. E. Coffey, *J. Air Pollut. Control Assoc.*, **24**, 565, 567 (1974). Reproduced with permission of the publisher and W. N. Stasiuk.]

12 m tower above the summit which is 400 m above the surrounding rural terrain. At Whiteface Mountain the monitoring station was located at the summit, above the treeline of a coniferous forest, and some 800 m above the remote forested area. Although these locations are separated by 260 km, both stations showed ozone concentrations that were nearly equal and which displayed similar time dependence. The ozone concentrations in urban locales of New York State often reached the rural values at their daily maxima during the early afternoon but otherwise remained substantially lower. These striking results, shown in Fig. 12.8*b*, lead Stasiuk and Coffey to suggest that the origin of the observed ozone was external to the state of New York.

Subsequent studies by Cleveland et al. (Ref. 17) showed that ozone production does occur in the urban plume of the New York City metropolitan area. The time development of ozone generation could be tracked by the monitoring stations across Connecticut to northeastern Massachusetts for downwind distances of 300 km. Thus it appears from the conclusions of Refs. 16 and 17 that eastern cities are subject to high background concentrations of ozone to which they add a significant increment that is evident at downwind localities. Wilson et al. (Ref. 13) studied the urban plume of St. Louis and found the flow rate of both ozone and the aerosol scattering coefficient b_{sca} to increase with downwind distances as far as 135 km. Similar results were reported by White et al. (Ref. 18) who found O_3 and b_{sca} were conspicuous indicators of the urban plume. The high readings of both O_3 and b_{sca} were also found to be persistent characteristics and contrasted with the behavior of SO_2 and NO whose concentrations decreased rapidly with downwind distance from the source. Sulfate compounds were reported to constitute a major component of the aerosol associated with the high value of b_{sca} in the urban plume. The total ozone flow rate increased with downwind distance to a level of 125 metric tons/hr at 150 km from St. Louis. It is estimated that one to two parts of ozone were produced for each part (as carbon) of nonmethane hydrocarbons emitted in the St. Louis area.

An interesting study of the long range transport of ozone from the midwest to the eastern United States was performed by Wolff et al. (Ref. 19). These investigators used an extensive set of ozone measurements, observations of visibility obtained at the ground stations of the National Weather Service, and analyses of air parcel trajectories to plot the passage of a particular high pressure system that carried high concentrations of ozone. They found that ozone in amounts exceeding the federal standard could be transported some 1200 km and that a half-life of O_3 of 30 hr in the atmosphere is implied. This long-distance transport provides a high background level which can be increased downwind of eastern cities by their additional effluvia of anthropogenic precursors. The extent to which ozone production is enhanced by natural hydrocarbon emissions and stratospheric injection is not clear.

Subsequent studies in the northeastern United States have shown that the diurnal variation displayed in Fig. 12.8*b* in urban areas does not apply during overcast weather. Under these circumstances the daily ozone maximum may occur during hours of darkness. This observation strengthens the hypothesis that the ozone is a result of advection rather than local generation. Hence the low ozone concentrations observed during the clear nocturnal hours may result from its depletion at the surface either by the chemical recombination with NO through the dark reaction (see Section 8.4.1) or by dry deposition at the surface. Since the sampling ports of ozone monitors are located among the ground level roughness elements, the air that they sample may be locally depleted of ozone. Concentrations of O_3 at levels of 50 to 100 m above the ground may be significantly higher than those indicated by the monitor if a nocturnal inversion is present. Thus the midday peak ozone readings shown in Fig. 12.8*b* are probably the result of fumigation of ozone from above the surface layer during the breakup of the nocturnal inversion by the morning solar radiation as it establishes the convective atmospheric boundary layer. A photolytic contribution to the peak ozone readings is to be expected mainly at monitors located downwind of large urban areas.

A discussion of the important topics relating to the nature of ozone and photochemical pollutants is given in the article edited by Dimitriades and Altshuller (Ref. 20). This summary presents the viewpoints of a group of specialists in various aspects of photochemical pollutants. The topics examined include the importance of the relative reactivity of hydrocarbons, the possible role of naturally emitted hydrocarbons, stratospheric ozone injection, the importance of long-distance transport, the value and limitations of theoretical analysis and laboratory simulation, the difficulties of quantitative measurements, and the optimum control strategy. The consensus of the specialists was that elevated concentrations of ozone can be transported for distances up to 1000 km or more. The interaction of transported pollutants with various combinations of locally emitted anthropogenic pollutants and natural hydrocarbons was considered to be unclear. The implications of ozone formation and transport on the control strategies was discussed (based on the then current knowledge), and the majority of the specialists concluded that the reduction of ozone and photochemical oxidants does require control of both hydrocarbons and NO emissions.

REFERENCES

1. B. H. Kirschner, "Environmental Meteorological Support Unit: A New Weather Bureau Program Supporting Urban Air Quality," in *Proceedings of the Second International Clean Air Congress*, H. M. Englund and W. T. Beery, Editors (Academic Press, New York, 1971).

2. C. R. Hosler, "Low-level Inversion Frequency in the Contiguous United States," *Monthly Weather Review*, **89**, 319–339 (1961).
3. G. C. Holzworth, "Mixing Depths, Wind Speeds and Air Pollution Potential for Selected Locations in the United States," *J. Appl. Meteorol.*, **6**, 1039–1044 (1967).
4. G. C. Holzworth, "Mixing Heights, Wind Speeds, and Potential for Urban Air Pollution Throughout the Contiguous United States," Office of Air Programs Pub. No. AP-101, Environmental Protection Agency, Research Triangle Park, N. C. (January 1972).
5. G. C. Holzworth, "Meteorological Potential for Urban Air Pollution in the Contiguous United States," in *Proceedings of the Second International Clean Air Congress*, H. M. Englund and W. T. Beery, Editors (Academic Press, New York, 1971).
6. L. E. Niemeyer, "Forecasting Air Pollution Potential," *Monthly Weather Review*, **88**, 88–96 (1960).
7. D. A. Lynn, B. J. Steigerwald, and J. H. Ludwig, "The November–December 1962 Air Pollution Episode in the Eastern United States," U.S. Dept. of H.E.W., P.H.S. Pub. No. 999-AP-7 (September 1964).
8. J. C. Fensterstock and R. K. Fankhauser, "Thanksgiving 1966 Air Pollution Episode in the Eastern U.S.," U.S. Dept. of H.E.W., N.A.P.C.A. Pub. No. AP-45 (July 1968).
9. J. Korshover, "Climatology of Stagnating Anticyclones East of the Rocky Mountains, 1936–1965," U.S. Dept. of H.E.W., P.H.S. Pub. No. 999-AP-34 (1967).
10. E. Gross, "The National Air Pollution Potential Forecast Program," E.S.S.A. Tech. Memorandum NMC 47, National Meteorological Center, Washington, D.C. (May 1970).
11. G. E. Likens, "Acid Precipitation," *Chem. Eng. News*, **54**, 29–44 (1976).
12. J. N. Galloway and E. B. Cowling, "The Effects of Precipitation on Aquatic and Terrestrial Ecosystems: A Proposed Precipitation Chemistry Network," *J. Air Pollut. Control Assoc.*, **28**, 229–235 (1978).
13. W. E. Wilson et al., "Sulfates in the Atmosphere," 69th Annual Meeting of the Air Pollution Control Association, Portland, Oregon (1976), and EPA Report 600/7-77-021, March 1977.
14. R. B. Husar, N. V. Gillani, and J. D. Husar, "Long Range Transport of Pollutants Observed Through Visibility Contour Maps, Weather Maps and Trajectory Analysis," *Third Symposium on Atmospheric Turbulence, Diffusion and Air Quality* (American Meteorological Society, Boston, Mass., 1976), pp. 344–347.
15. A. P. Altshuller, "Regional Transport and Transformation of Sulfur Dioxide to Sulfates in the U.S.," *J. Air Pollut. Control Assoc.*, **26**, 318–324 (1976).
16. W. N. Stasiuk and P. E. Coffey, "Rural and Urban Ozone Relationships in New York State," *J. Air Pollut. Control Assoc.*, **24**, 564–568 (1974).

17. W. S. Cleveland et al., "Photochemical Air Pollution: Transport from the New York City Area into Connecticut and Massachusetts," *Science*, **191**, 179–181 (1976).
18. W. H. White et al., "Formation and Transport of Secondary Air Pollutants: Ozone and Aerosols in the St. Louis Urban Plume," *Science*, **194**, 187–189 (1976).
19. G. T. Wolff et al., "An Investigation of Long-Range Transport of Ozone Across Midwestern and Eastern U.S.," *Atmos. Environ.*, **11**, 797–802 (1977).
20. B. Dimitriades and A. P. Altshuller, "International Conference on Oxidant Problems: Analysis of the Evidence/Viewpoints Presented. Part II. Evidence/Viewpoints on Key Issues," *J. Air Pollut. Control Assoc.*, **28**, 207–212 (1978).

PROBLEMS

12.1. Use the equations given in Chapter 11 to derive an expression for the dependence of the ventilation factor on the geostrophic wind under neutral conditions. Assume that the transport wind is approximately proportional to the geostrophic wind.

12.2. Using the information given in Table 11.2, p. 260, calculate the ventilation factors if $z_0 = 1.0$ m (a) when the geostrophic wind q_G is a typical value of 15 m/s, and (b) when $q_G = 5$ m/s corresponding to stagnation conditions. To calculate $u(z)$, use the power law expression from $z = z_0$ to the intermediate level z_i at which $\bar{u}$ is predicted to equal q_G. The value of $\bar{u}$ is then set equal to q_G for $z_i < z < H_b$. Calculate the wind speeds at 100 m heights for the higher wind condition and at 50 m heights for the low wind condition.

APPENDIX I

VECTOR IDENTITIES, VECTOR OPERATORS, AND THE PARTICLE DERIVATIVE IN SEVERAL COORDINATE SYSTEMS

1. A PARTIAL LIST OF VECTOR IDENTITIES

$$\nabla \times (\nabla \Phi) = 0$$

$$\nabla \cdot (\nabla \times \mathbf{A}) = 0$$

$$\nabla \times (\nabla \times \mathbf{A}) = \nabla(\nabla \cdot \mathbf{A}) - \nabla^2 \mathbf{A}$$

where

$$\nabla^2 \mathbf{A} = \nabla^2 A_x \mathbf{i} + \nabla^2 A_y \mathbf{j} + \nabla^2 A_z \mathbf{k}$$

$$\mathbf{A} \times (\mathbf{B} \times \mathbf{C}) = (\mathbf{A} \cdot \mathbf{C})\mathbf{B} - (\mathbf{A} \cdot \mathbf{B})\mathbf{C}$$

$$\nabla \times (\mathbf{A} \times \mathbf{B}) = (\mathbf{B} \cdot \nabla)\mathbf{A} - \mathbf{B}(\nabla \cdot \mathbf{A}) - (\mathbf{A} \cdot \nabla)\mathbf{B} + \mathbf{A}(\nabla \cdot \mathbf{B})$$

$$\nabla \times (\Phi \mathbf{A}) = \Phi(\nabla \times \mathbf{A}) + \nabla \Phi \times \mathbf{A}$$

2. THE GRADIENT OF A SCALAR

Cartesian coordinates:

$$\nabla \Phi = \frac{\partial \Phi}{\partial x}\mathbf{i} + \frac{\partial \Phi}{\partial y}\mathbf{j} + \frac{\partial \Phi}{\partial z}\mathbf{k}$$

Cylindrical coordinates:

$$\nabla\Phi = \frac{\partial\Phi}{\partial r}\,\mathbf{e}_r + \frac{1}{r}\frac{\partial\Phi}{\partial\theta}\,\mathbf{e}_\theta + \frac{\partial\Phi}{\partial z}\,\mathbf{e}_z$$

Spherical coordinates:

$$\nabla\Phi = \frac{\partial\Phi}{\partial r}\,\mathbf{e}_r + \frac{1}{r}\frac{\partial\Phi}{\partial\theta}\,\mathbf{e}_\theta + \frac{1}{r\sin\theta}\frac{\partial\Phi}{\partial\phi}\,\mathbf{e}_\phi$$

3. THE DIVERGENCE OF A VECTOR A

Cartesian coordinates:

$$\nabla\cdot\mathbf{A} = \frac{\partial A_x}{\partial x} + \frac{\partial A_y}{\partial y} + \frac{\partial A_z}{\partial z}$$

Cylindrical coordinates:

$$\nabla\cdot\mathbf{A} = \frac{1}{r}\frac{\partial(rA_r)}{\partial r} + \frac{1}{r}\frac{\partial A_\theta}{\partial\theta} + \frac{\partial A_z}{\partial z}$$

4. THE CURL OF A VECTOR A

Cartesian coordinates:

$$\nabla\times\mathbf{A} = \left(\frac{\partial A_z}{\partial y} - \frac{\partial A_y}{\partial z}\right)\mathbf{i} + \left(\frac{\partial A_x}{\partial z} - \frac{\partial A_z}{\partial x}\right)\mathbf{j} + \left(\frac{\partial A_y}{\partial x} - \frac{\partial A_x}{\partial y}\right)\mathbf{k}$$

Cylindrical coordinates:

$$\nabla\times\mathbf{A} = \left(\frac{1}{r}\frac{\partial A_z}{\partial\theta} - \frac{\partial A_\theta}{\partial z}\right)\mathbf{e}_r + \left(\frac{\partial A_r}{\partial z} - \frac{\partial A_z}{\partial r}\right)\mathbf{e}_\theta + \left(\frac{1}{r}\frac{\partial rA_\theta}{\partial r} - \frac{1}{r}\frac{\partial A_r}{\partial\theta}\right)\mathbf{e}_z$$

5. THE PARTICLE DERIVATIVE OF THE VELOCITY VECTOR

Cartesian coordinates:

$$\frac{D\mathbf{q}}{Dt} = \frac{\partial\mathbf{q}}{\partial t} + u\frac{\partial\mathbf{q}}{\partial x} + v\frac{\partial\mathbf{q}}{\partial y} + w\frac{\partial\mathbf{q}}{\partial z}$$

where

$$\mathbf{q} = u\mathbf{i} + v\mathbf{j} + w\mathbf{k}$$

hence

$$\frac{D\mathbf{q}}{Dt} = \left(\frac{\partial u}{\partial t} + u\frac{\partial u}{\partial x} + v\frac{\partial u}{\partial y} + w\frac{\partial u}{\partial z}\right)\mathbf{i} + \left(\frac{\partial v}{\partial t} + u\frac{\partial v}{\partial x} + v\frac{\partial v}{\partial y} + w\frac{\partial v}{\partial z}\right)\mathbf{j} + \left(\frac{\partial w}{\partial t} + u\frac{\partial w}{\partial x} + v\frac{\partial w}{\partial y} + w\frac{\partial w}{\partial z}\right)\mathbf{k}$$

Cylindrical coordinates:

$$\frac{D\mathbf{q}}{Dt} = \frac{\partial \mathbf{q}}{\partial t} + u\frac{\partial \mathbf{q}}{\partial r} + \frac{v}{r}\frac{\partial \mathbf{q}}{\partial \theta} + w\frac{\partial \mathbf{q}}{\partial z}$$

where

$$\mathbf{q} = u\mathbf{e}_r + v\mathbf{e}_\theta + w\mathbf{e}_z$$

from which it follows that

$$\frac{D\mathbf{q}}{Dt} = \left(\frac{\partial u}{\partial t} + u\frac{\partial u}{\partial r} + \frac{v}{r}\frac{\partial u}{\partial \theta} - \frac{v^2}{r} + w\frac{\partial u}{\partial z}\right)\mathbf{e}_r + \left(\frac{\partial v}{\partial t} + u\frac{\partial v}{\partial r} + \frac{v}{r}\frac{\partial v}{\partial \theta} + \frac{uv}{r} + w\frac{\partial v}{\partial z}\right)\mathbf{e}_\theta + \left(\frac{\partial w}{\partial t} + u\frac{\partial w}{\partial r} + \frac{v}{r}\frac{\partial w}{\partial \theta} + w\frac{\partial w}{\partial z}\right)\mathbf{e}_z$$

6. THE LAPLACIAN OPERATOR (THE DIVERGENCE OF THE GRADIENT OF A SCALAR)

Cartesian coordinates:

$$\nabla^2\Phi = \frac{\partial^2\Phi}{\partial x^2} + \frac{\partial^2\Phi}{\partial y^2} + \frac{\partial^2\Phi}{\partial z^2}$$

Cylindrical coordinates:

$$\nabla^2\phi = \frac{1}{r}\left[\frac{\partial}{\partial r}\left(r\frac{\partial\Phi}{\partial r}\right) + \frac{\partial}{\partial \theta}\left(\frac{1}{r}\frac{\partial\Phi}{\partial \theta}\right) + \frac{\partial}{\partial z}\left(r\frac{\partial\Phi}{\partial z}\right)\right]$$

Spherical coordinates:

$$\nabla^2\phi = \frac{1}{r^2\sin\theta}\left[\frac{\partial}{\partial r}\left(r^2\sin\theta\frac{\partial\Phi}{\partial r}\right) + \frac{\partial}{\partial \theta}\left(\sin\theta\frac{\partial\Phi}{\partial \theta}\right) + \frac{\partial}{\partial \phi}\left(\frac{1}{\sin\theta}\frac{\partial\Phi}{\partial \phi}\right)\right]$$

APPENDIX II

ATMOSPHERIC DATA

1. EXTRACTS FROM "U.S. STANDARD ATMOSPHERE SUPPLEMENTS, 1966"

Prepared under the sponsorship of the Environmental Science Services Administration, National Aeronautics and Space Administration, and the U.S. Air Force. Published by the U.S. Government Printing Office, Washington, D.C. 20402.

Note: The extract designated by the title "Mid-Latitude Spring/Fall" is widely known as the Standard Atmosphere.

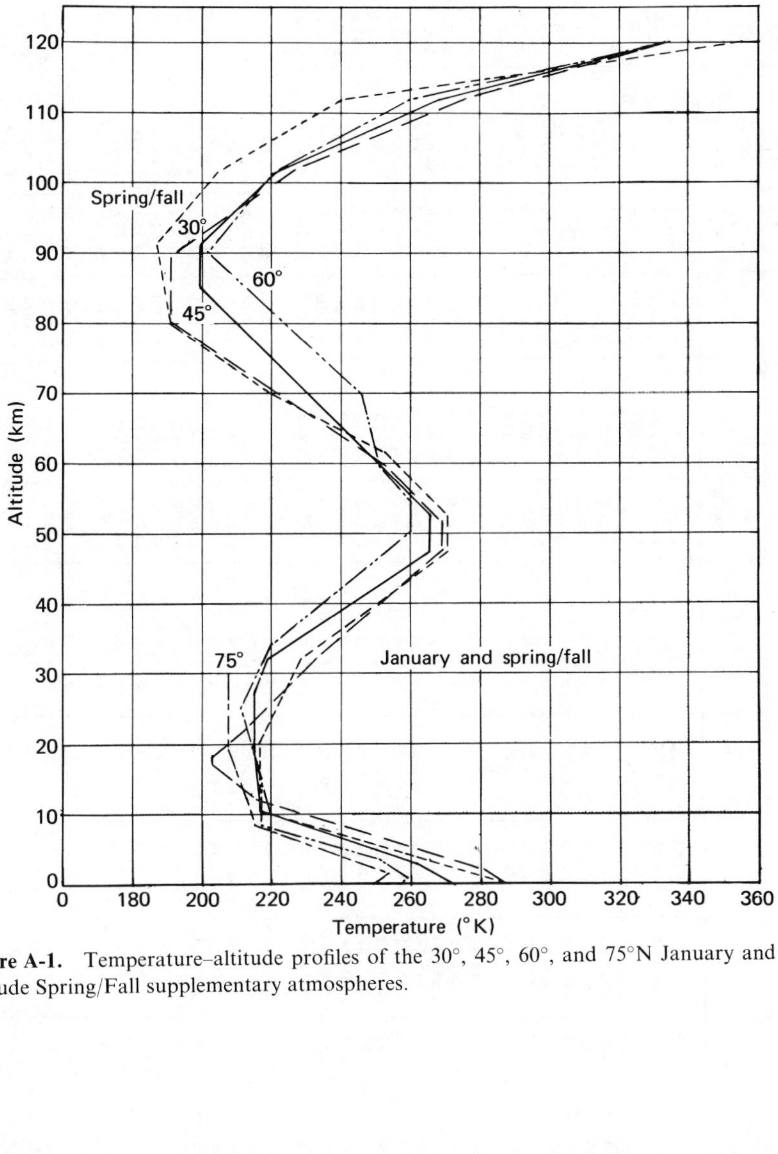

Figure A-1. Temperature–altitude profiles of the 30°, 45°, 60°, and 75°N January and mid-latitude Spring/Fall supplementary atmospheres.

Table A-1. The Standard (Mid-Latitude Spring/Fall) Atmosphere

Altitude		Temperature			Pressure		Density		Sound speed	Coefficient of viscosity	Thermal conductivity
Z, m	H, m′	T, °K	t, °C	$T-T_{std}$	P, mb	$\frac{P}{P_{std}}$	ρ, kg m^{-3}	$\frac{\rho}{\rho_{std}}$	C_s m sec^{-1}	μ kg m^{-1} sec^{-1}	k k-cal m^{-1} sec^{-1}(°K)$^{-1}$
0	0	288.15	15.00	0.00	1.01325 + 3	1.000	1.225 + 0	1.000	340.3	1.789 − 5	6.053 − 6
250	250	286.53	13.38	0.00	9.836 + 2	1.000	1.196	1.000	339.3	1.782	6.022
500	500	284.90	11.75	0.00	9.546	1.000	1.167	1.000	338.4	1.774	5.992
750	750	283.28	10.13	−0.00	9.263	1.000	1.139	1.000	337.4	1.766	5.961
1000	1000	281.65	8.50	0.00	8.988	1.000	1.112	1.000	336.4	1.758	5.931
1250	1250	280.03	6.88	−0.00	8.719	1.000	1.085	1.000	335.5	1.750	5.900
1500	1500	278.40	5.25	0.00	8.456	1.000	1.058	1.000	334.5	1.742	5.869
1750	1750	276.78	3.63	0.00	8.200	1.000	1.032	1.000	333.5	1.734	5.838
2000	1999	275.15	2.00	0.00	7.950	1.000	1.007	1.000	332.5	1.726	5.807
2250	2249	273.53	0.38	0.00	7.707	1.000	9.815 − 1	1.000	331.5	1.718	5.776
2500	2499	271.91	−1.24	0.00	7.469 + 2	1.000	9.570 − 1	1.000	330.6	1.710 − 5	5.745 − 6
2750	2749	270.28	−2.87	−0.00	7.238	1.000	9.329	1.000	329.6	1.702	5.714
3000	2999	268.66	−4.49	0.00	7.012	1.000	9.093	1.000	328.6	1.694	5.683
3250	3248	267.04	−6.11	−0.00	6.792	1.000	8.861	1.000	327.6	1.686	5.652
3500	3498	265.41	−7.74	−0.00	6.578	1.000	8.634	1.000	326.6	1.678	5.621
3750	3748	263.79	−9.36	0.00	6.369	1.000	8.412	1.000	325.6	1.669	5.590
4000	3997	262.17	−10.98	0.00	6.166	1.000	8.193	1.000	324.6	1.661	5.559
4250	4247	260.54	−12.61	0.00	5.968	1.000	7.980	1.000	323.6	1.653	5.527
4500	4497	258.92	−14.23	−0.00	5.775	1.000	7.770	1.000	322.6	1.645	5.496
4750	4746	257.30	−15.85	0.00	5.588	1.000	7.565	1.000	321.6	1.637	5.465
5000	4996	255.68	−17.47	−0.00	5.405 + 2	1.000	7.364 − 1	1.000	320.5	1.628 − 5	5.433 − 6
5250	5246	254.05	−19.10	0.00	5.227	1.000	7.167	1.000	319.5	1.620	5.402
5500	5495	252.43	−20.72	−0.00	5.054	1.000	6.975	1.000	318.5	1.612	5.370
5750	5745	250.81	−22.34	−0.00	4.886	1.000	6.786	1.000	317.5	1.603	5.338
6000	5994	249.19	−23.96	−0.00	4.722	1.000	6.601	1.000	316.5	1.595	5.307
6250	6244	247.56	−25.59	−0.00	4.562	1.000	6.420	1.000	315.4	1.587	5.275
6500	6493	245.94	−27.21	0.00	4.408	1.000	6.243	1.000	314.4	1.578	5.243
6750	6743	244.32	−28.83	−0.00	4.257	1.000	6.070	1.000	313.3	1.570	5.212
7000	6992	242.70	−30.45	0.00	4.111	1.000	5.900	1.000	312.3	1.561	5.180
7250	7242	241.08	−32.07	−0.00	3.968	1.000	5.734	1.000	311.3	1.553	5.148
7500	7491	239.46	−33.69	0.00	3.830 + 2	1.000	5.572 − 1	1.000	310.2	1.544 − 5	5.116 − 6
7750	7741	237.84	−35.31	0.00	3.696	1.000	5.413	1.000	309.2	1.536	5.084
8000	7990	236.22	−36.93	0.00	3.565	1.000	5.258	1.000	308.1	1.527	5.052
8250	8239	234.59	−38.56	−0.00	3.438	1.000	5.106	1.000	307.0	1.519	5.020
8500	8489	232.97	−40.18	−0.00	3.315	1.000	4.958	1.000	306.0	1.510	4.988
8750	8738	231.35	−41.80	0.00	3.196	1.000	4.812	1.000	304.9	1.501	4.956
9000	8987	229.73	−43.42	−0.00	3.080	1.000	4.671	1.000	303.8	1.493	4.923
9250	9237	228.11	−45.04	0.00	2.968	1.000	4.532	1.000	302.8	1.484	4.891
9500	9486	226.49	−46.66	0.00	2.858	1.000	4.397	1.000	301.7	1.475	4.859
9750	9735	224.87	−48.28	0.00	2.753	1.000	4.264	1.000	300.6	1.466	4.827

10000	9984	223.25	-49.90	0.00	2.650 + 2	1.000	4.135 - 1	1.000	299.5	1.458 - 5	4.794 - 6
10250	10233	221.63	-51.52	0.00	2.550	1.000	4.009	1.000	298.4	1.449	4.762
10500	10483	220.01	-53.14	-0.00	2.454	1.000	3.886	1.000	297.4	1.440	4.729
10750	10732	218.39	-54.76	-0.00	2.361	1.000	3.765	1.000	296.3	1.431	4.697
11000	10981	216.77	-56.38	-0.00	2.270	1.000	3.648	1.000	295.2	1.422	4.664
11500	11479	216.65	-56.50	-0.00	2.098	1.000	3.374	1.000	295.1	1.422	4.662
12000	11977	216.65	-56.50	-0.00	1.940	1.000	3.119	1.000	295.1	1.422	4.662
12500	12475	216.65	-56.50	-0.00	1.793	1.000	2.884	1.000	295.1	1.422	4.662
13000	12973	216.65	-56.50	-0.00	1.658	1.000	2.666	1.000	295.1	1.422	4.662
13500	13471	216.65	-56.50	-0.00	1.533	1.000	2.465	1.000	295.1	1.422	4.662
14000	13969	216.65	-56.50	-0.00	1.417 + 2	1.000	2.279 - 1	1.000	295.1	1.422 - 5	4.662 - 6
14500	14467	216.65	-56.50	-0.00	1.310	1.000	2.107	1.000	295.1	1.422	4.662
15000	14965	216.65	-56.50	-0.00	1.211	1.000	1.948	1.000	295.1	1.422	4.662
15500	15462	216.65	-56.50	-0.00	1.120	1.000	1.801	1.000	295.1	1.422	4.662
16000	15960	216.65	-56.50	-0.00	1.035	1.000	1.665	1.000	295.1	1.422	4.662
16500	16457	216.65	-56.50	-0.00	9.572 + 1	1.000	1.539	1.000	295.1	1.422	4.662
17000	16955	216.65	-56.50	-0.00	8.850	1.000	1.423	1.000	295.1	1.422	4.662
17500	17452	216.65	-56.50	-0.00	8.182	1.000	1.316	1.000	295.1	1.422	4.662
18000	17949	216.65	-56.50	-0.00	7.565	1.000	1.216	1.000	295.1	1.422	4.662
18500	18446	216.65	-56.50	-0.00	6.995	1.000	1.125	1.000	295.1	1.422	4.662
19000	18943	216.65	-56.50	-0.00	6.467 + 1	1.000	1.040 - 1	1.000	295.1	1.422 - 5	4.662 - 6
19500	19440	216.65	-56.50	-0.00	5.980	1.000	9.616 - 2	1.000	295.1	1.422	4.662
20000	19937	216.65	-56.50	-0.00	5.529	1.000	8.891	1.000	295.1	1.422	4.662
20500	20434	217.08	-56.07	0.00	5.113	1.000	8.205	1.000	295.4	1.424	4.670
21000	20931	217.58	-55.57	-0.00	4.729	1.000	7.571	1.000	295.7	1.427	4.680
21500	21428	218.08	-55.07	-0.00	4.375	1.000	6.988	1.000	296.0	1.429	4.690
22000	21924	218.57	-54.58	0.00	4.047	1.000	6.451	1.000	296.4	1.432	4.700
22500	22421	219.07	-54.08	-0.00	3.746	1.000	5.956	1.000	296.7	1.435	4.710
23000	22917	219.57	-53.58	0.00	3.467	1.000	5.501	1.000	297.0	1.438	4.720
23500	23413	220.06	-53.09	0.00	3.209	1.000	5.081	1.000	297.4	1.440	4.730
24000	23910	220.56	-52.59	-0.00	2.972 + 1	1.000	4.694 - 2	1.000	297.7	1.443 - 5	4.740 - 6
24500	24406	221.06	-52.09	-0.00	2.752	1.000	4.337	1.000	298.1	1.446	4.750
25000	24902	221.55	-51.60	-0.00	2.549	1.000	4.008	1.000	298.4	1.448	4.760
25500	25398	222.05	-51.10	0.00	2.362	1.000	3.705	1.000	298.7	1.451	4.770
26000	25894	222.54	-50.61	0.00	2.188	1.000	3.426	1.000	299.1	1.454	4.780
26500	26390	223.04	-50.11	-0.00	2.028	1.000	3.168	1.000	299.4	1.457	4.790
27000	26886	223.54	-49.61	-0.00	1.880	1.000	2.930	1.000	299.7	1.459	4.800
27500	27382	224.03	-49.12	-0.00	1.743	1.000	2.710	1.000	300.1	1.462	4.810
28000	27877	224.53	-48.62	0.00	1.616	1.000	2.508	1.000	300.4	1.465	4.820
28500	28373	225.02	-48.13	-0.00	1.499	1.000	2.321	1.000	300.7	1.467	4.830

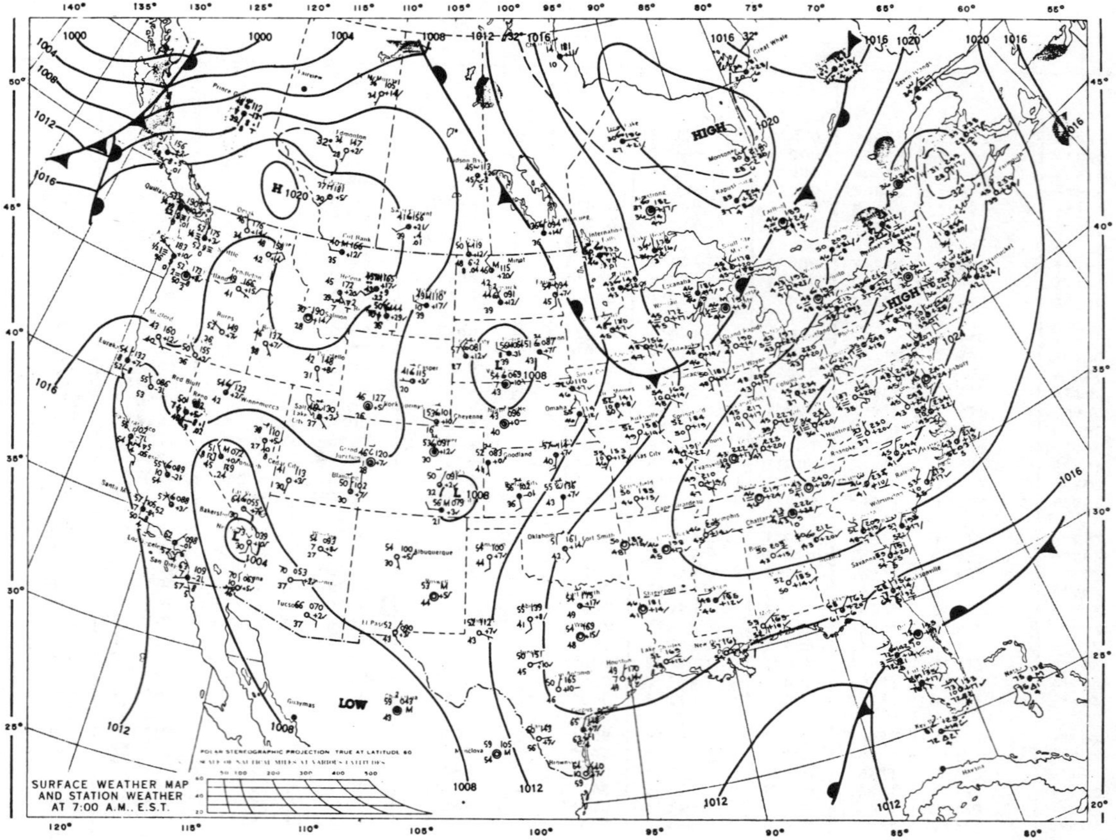

Figure A-2. The surface weather map end station weather at 0700 hr EST on October 2, 1972 is compiled and presented by the U.S. National Weather Service.

INDEX